Sneaking a Look at God's Cards

UNRAVELING THE MYSTERIES OF
QUANTUM MECHANICS

GianCarlo Ghirardi

Translated from the Italian by Gerald Malsbary

PRINCETON UNIVERSITY PRESS

PRINCETON AND OXFORD

Copyright © 2004 by Princeton University Press
Published by Princeton University Press, 41 William Street, Princeton, New Jersey 08540
In the United Kingdom: Princeton University Press, 3 Market Place,
Woodstock, Oxfordshire OX20 1SY

Originally published in Italian under the title *Un'occhiata alle carte de Dio*,
copyright © 1997 by il Saggiatore, Milano.

All Rights Reserved

Library of Congress Cataloging-in-Publication Data

Ghirardi, G. C.
[Occhiata alle carte di Dio. English]
Sneaking a look at God's cards : unraveling the mysteries of quantum mechanics /
GianCarlo Ghirardi ; translated from the Italian by Gerald Malsbary.
p. cm.
Includes bibliographical references and index.
ISBN 0-691-04934-3 (acid-free paper)
1. Science—Miscellanea. I. Title.

Q173.G4813 2004
539—dc21   2003043304

British Library Cataloging-in-Publication Data is available

Publication of this book has been aided by a grant from
the Italian Ministry of Foreign Affairs.

This book has been composed in Sabon

Printed on acid-free paper. ∞

www.pupress.princeton.edu

Printed in the United States of America

10 9 8 7 6 5 4 3 2 1

To my wife Laura and my daughters
Monica, Barbara, and Lucia

I WANT TO KNOW THE THOUGHTS OF GOD; EVERYTHING ELSE IS JUST DETAILS.

—*Albert Einstein*

PHILOSOPHY IS WRITTEN IN THAT HUGE BOOK WHICH FOREVER LIES OPEN TO OUR EYES (THE UNIVERSE, THAT IS), BUT CANNOT BE UNDERSTOOD UNLESS WE FIRST STUDY THE LANGUAGE AND BECOME FAMILIAR WITH THE CHARACTERS IN WHICH IT HAS BEEN WRITTEN. THE LANGUAGE OF THAT BOOK IS MATHEMATICS, AND THE CHARACTERS ARE TRIANGLES, CIRCLES, AND THE OTHER GEOMETRICAL FIGURES.

—*Galileo Galilei*

Contents

Preface xi

Acknowledgments xvii

CHAPTER ONE
The Collapse of the "Classical" World View 1

CHAPTER TWO
The Polarization of Light 25

CHAPTER THREE
Quanta, Chance Events, and Indeterminism 43

CHAPTER FOUR
The Superposition Principle and the Conceptual Structure
of the Theory 79

CHAPTER FIVE
Visualization and Scientific Progress 111

CHAPTER SIX
The Interpretation of the Theory 120

CHAPTER SEVEN
The Bohr-Einstein Dialogue 149

CHAPTER EIGHT
A Bolt from the Blue: The Einstein-Podolski-
Rosen Argument 165

CHAPTER NINE
Hidden Variables 195

CHAPTER TEN
Bell's Inequality and Nonlocality 226

CHAPTER ELEVEN
Nonlocality and Superluminal Signals 261

CHAPTER TWELVE
Quantum Cryptography 292

CONTENTS

CHAPTER THIRTEEN
Quantum Computers   313

CHAPTER FOURTEEN
Systems of Identical Particles   331

CHAPTER FIFTEEN
From Microscopic to Macroscopic   344

CHAPTER SIXTEEN
In Search of a Coherent Framework for All Physical Processes   377

CHAPTER SEVENTEEN
Spontaneous Localization, Properties, and Perceptions   416

CHAPTER EIGHTEEN
Macrorealism and Noninvasive Measurements   437

CHAPTER NINETEEN
Conclusions   448

Notes   455

Bibliography   473

Index   477

Preface

> The quantum question is so incredibly important and difficult that everyone should busy himself with it.
> —*Albert Einstein*, in a letter to J. Laub, 1908

AS THE GREAT PHYSICIST and Nobel Prize winner Isaac Rabi once said, "It's a great pity that the general public has very little inkling of the tremendous excitement—intellectual and emotional excitement—that goes on in the advanced fields of physics." Such a statement is particularly appropriate for quantum mechanics, the theory that underlies the modern scientific understanding of the universe. I say this for two reasons.

First of all, knowledge of quantum mechanics has not as yet managed to pass beyond the inner circle of the specialists who work with it, for reasons I will explain thoroughly in the present book. This very fact is astonishing when you consider that the overwhelming majority of the important technological innovations of recent times (for good or ill) have been based on conclusions specifically drawn from quantum mechanics, and incredible further developments are waiting in the wings. Even a partial list would seem to go on forever: it should suffice to mention nuclear energy, semiconductors, and consequently all the modern technology of computers, superconductors, and the rest. Digital clocks, the computer at which I am now writing, and so many other instruments we all use every day have been made possible only through the profound understanding of reality that sprang from this splendid theory. And (as we shall see) we are on the threshold of amazing new discoveries as well: for the first time the use of precise quantum effects provides us with an absolutely reliable method of sending encoded messages, and seems to open the way for realizing a dream that came years ago to another Nobel Prize winner, Richard Phillips Feynman: quantum mechanical computers. These will bring a "quantum" leap in the technology of such instruments. Although this advance cannot as yet be precisely assessed, it will certainly be revolutionary.

The second factor that makes desirable a more widely diffused knowledge of the elements of quantum formalism derives from the highly conceptual and philosophical relevance of this great conquest of modern thought. This is all the more surprising, for whereas various theoretical constructions have become an integral part of the common culture (such as Darwinian evolution, genetics and heredity, special and general relativity), nothing of the kind has as yet taken place in the case of this conceptual revolution, which has already dominated three-quarters of a century of modern science. The failure of the public to assimilate such an extraordinary cultural event is particularly serious, and aggravates the tragic rift, so typical of modern times, between the so-called two cultures of science and the humanities. In fact, one of the most characteristic aspects of the theory consists in the fact that, more than any other scientific scheme hitherto elaborated by man in the course of his long journey in the comprehension of reality, quantum mechanics poses philosophical problems that are at once very significant and very peculiar on the conceptual and epistemological level. Questions are raised in quantum mechanics that cannot help but interest anyone with any intellectual curiosity, and call for serious critical reflection on the part of humanists and philosophers.

Now it is all too easy to blame this unfortunate state of affairs on the peculiar nature of the theory itself, and say that the mathematical knowledge needed to understand it is so abstract and advanced as to doom every attempt to communicate its essentials to someone without the right mathematical preparation. No doubt, quantum formalism is particularly difficult, and the language of the theory appears to be a language only for initiates. But this language simply represents the most suitable and elegant device (or "formalism") for describing some very peculiar aspects of reality, and of course, these aspects, and these alone, are the elements that need to be made accessible to the public. This book intends to undertake the daunting but urgently needed task of making intelligible "the mysteries of the quantum world" to those who are not professionally engaged with it. Now, my American friend and colleague Tim Maudlin, in the introduction to his exciting book, *Quantum Non-locality and Relativity—Metaphysical Intimations of Modern Physics* (1994), maintains that any compromise between rigor and simplicity constitutes a pact with the devil. And thus I too am obliged to explain the peculiar pact I have chosen to sign. I have resolved not to give up anything on the level of conceptual rigor, if not formal rigor. I have put particular effort into not leaving any room in the text for hints and implications that can lead to those illegitimate evasions often made recently, even by authoritative figures, when it comes to quantum mechanics. Some individual cases are easy to identify: they have been listed in an important and interesting article by R. P. Crease and C. C. Mann, "Physics for Mystics" (1987). After cataloguing a whole series of rather successful books that claim to have discovered in quantum mechanics valid proofs for parapsychology, paranormal phenomena, oriental philosophies, and so forth, the authors of the article quite appropriately record the reactions of the scientific community to such mystifications. The second point on which I have committed myself to a position of extreme rigidity is that of guaranteeing the simplicity of the exposition from the technical point of view. I have chosen to carry out my task of didactic exposition in such a way that a reader without any special mathematical or scientific preparation should be able to follow all the arguments. The reader is asked only to comprehend the functioning of polarizing lenses, to have an idea of trigonometric functions, and to apply enough effort to follow the exposition of the many experiments (abundantly illustrated in the text by figures) that show the crucial points of the theory. Now, someone may ask, "So what is *your* pact with the devil? What compromise have *you* accepted?" Well, my own compromise consists in the fact that, even though the book is elementary from the formal point of view, it nevertheless requires serious study and a certain amount of application to be understood and digested. The text is certainly not an "easy read": it presupposes the reader's serious interest in acquiring an understanding of the scarcely imaginable behavior of microsystems, in watching the titanic confrontations of the brilliant minds that brought such peculiar features of nature to light, and in understanding the profound conceptual and epistemological implications of all this for reality. This should not be cause for surprise: the new and revolutionary vision of reality brought by quantum theory makes any attempt to understand its implications—not exhaustively, but correctly—a very challenging task. But a challenging book need not be a book "for experts only."

This book is also the fruit of my personal experience over recent years in presenting these themes to a large variety of audiences. These have included teacher conferences, high-school classes, panel discussions with philosophers and theologians, interviews on radio and television, and reports in various scientific journals. However, apart from some highly favorable recognition, these experiences have not, as a whole, been completely satisfying. I especially treasure the contact I have made with those who have read what I have written in newspaper articles, who have shown such lively interest by writing to me and starting an exchange of views, for these contacts have led me to gain an ever clearer understand-

PREFACE

ing of how very difficult it is to attain my goal. These very attentive and interested readers have forced me to return to the starting points that had seemed clear to me, and they convinced me that I need to do better. This is my attempt to do just that.

At this point I cannot avoid answering an obvious question that may be raised: why another book on this topic, when there are already so many? My answer is simple: there certainly do exist books of remarkable quality, especially by British and American authors, on the subjects I write of here. But some observations need to be made about this situation. In recent years, the debate on the most relevant issues of the theory has moved to a higher level. Our understanding of the implications of the formalism has changed, thanks in particular to some important technological innovations which have made possible many new experiments in our laboratories—experiments that only a few years back were described in textbooks as "thought experiments" helpful for illustrating some peculiarities of formalism. Secondly, in recent years there have been new and illuminating theoretical contributions. Even if some of these have already been analyzed in the most recent work, I believe it is difficult to find all the crucial conceptual points brought together into a single book, and that is my object. And this is the primary motive for reviewing this material: books that do not include an analysis of the new aspects do not permit an understanding of the truly innovative parts of the theory, so that the loss is not only in the quantity of information, but in the quality of conceptual understanding.

I would also like to mention a peculiar aspect of the fascinating theory I plan to analyze, something that concerns the possible modes of presenting it. During the more than seventy years that now intervene between its first definitive formulation in 1926 and today, the theory has been the focus of an unending debate. As I will explain in the course of the book, it is well known how some of the founding fathers of the theory, such as Albert Einstein, Max Planck, and Erwin Schrödinger, made clear their profound dissatisfaction with the scheme they helped bring to birth. Anybody with a passing interest in modern science has heard at least once of the titanic struggle between Niels Bohr and Albert Einstein. But not everyone is aware that this battle has not yet seen any victors or losers. And it is only in recent years that the debate has taken a new turn, with almost all researchers in the field agreeing that the so-called orthodox formulation of the School of Copenhagen (the one worked out by Niels Bohr, Werner Heisenberg, Max Born, Wolfgang Pauli, and Pascal Jordan) is fundamentally dissatisfying and requires reinterpretation if not substantial modification. Alongside this general agreement, however, there is a remarkable variety of positions on the best means to overcome the impasse. Naturally, each writer will explain the facts from a personal perspective. I cannot deny that I have my own position on these matters. However, just as I have firmly committed myself to writing a truly nontechnical work, so I have also made it my policy to set out the problems in the most objective manner possible, and present the reader with the best solutions that have been suggested. I leave it to the reader's own judgment to choose the one that seems most congenial.

Finally, allow me to emphasize once more the real motive behind this work. It is to enable a reader without special technical preparation, but with a certain amount of effort, to arrive at a conceptually correct and undistorted grasp of the essential elements of these fascinating mysteries—mysteries that the investigation of reality compels us to face. This is how this book is different from the rest. If it succeeds, then you and I together will be "sneaking a look at God's cards."

Acknowledgments

THE PERSONS AND EVENTS that have played an important role in my decision to write a book of this nature are so numerous that I certainly will not be able to list them all. But certain outstanding figures and facts I cannot fail to mention. Above all, there is my thesis director, Piero Caldirola, and his numerous colleagues and assistants. In the years when I was completing my studies at the University of Milan (the second half of the 1950s), in few Italian universities could be found a school so interested and engaged in the study of the foundational problems of science and of quantum mechanics in particular. It was from Caldirola himself and his collaborators that I learned the love of clarity, logical rigor, and interest in researches of this kind. Some of these collaborators, such as Pietro Bocchieri, Angelo Loinger, and Giovanni Maria Prosperi, in those years were making important contributions to the theme that forms the subject of this book. Afterward, the path of my scientific career for some years led me away from special research in the field of foundations, but never took away from me my taste for mathematical and logical rigor, nor my interest in themes of broad relevance. And it was at Milan in particular where my first encounter with Alberto Rimini took place, surely the most meaningful event of my scientific career. No other collaboration I have known has been so fruitful, lasting for so many years and bearing fruit in a long series of common projects.

A second event of great importance I must not fail to mention is my encounter with John Bell, the scientist who will frequently appear as a major protagonist in the book. For over ten years I was engaged along with my colleagues Alberto Rimini and Tullio Weber in researches on the foundations of quantum mechanics, and this led us to the elaboration in 1985 of the phenomenological model I discuss in the final chapters of the book. For anyone who was working in this field, Bell undoubtedly represented the world's greatest living authority, and his passionate interest in our researches certainly contributed greatly to the attention they received from the international scientific community. However, much more significantly on the existential level was the chance to develop a serious dialogue with this most lucid thinker, to share the passion that inspired him, and to treasure his exquisite sensibility. Even though my friendship with this great man of genius did not last long because of his premature passing, certainly his constant preoccupation with the foundational problems

ACKNOWLEDGMENTS

on which he had done such significant work and his indignation that the scientific community did not show a sufficient degree of interest in problems that he, like Einstein, considered fundamental and unavoidable—such things helped to create in me the desire to render the themes of the present book accessible to a larger public, precisely because they represent *par excellence* those aspects of science that make it a fascinating activity, of interest to anyone with a bit of intellectual curiosity. I am also grateful to Bell for inspiring Philip Pearle to spend a sabbatical year in Italy (at Trieste and Pavia), during which a fruitful interaction took place between us that would extend to years of publishing numerous scientific works in common.

The third event has been coming into direct contact with the international community of both physicists and philosophers of science. In the United States and England can be found groups and associations actively engaged in the study of the foundations of modern physics the like of which is difficult to find in Italy. Discussions, exchanges of viewpoints, and debates with such impressive figures as David Albert, James Cushing, Arthur Miller, Abner Shimony, and Bas van Fraassen (among epistemologists) and with Alain Aspect, Detlef Dürr, Bernard D'Espagnat, Shelly Goldstein, Anthony Leggett, David Mermin, Roger Penrose, and Euan Squires (among physicists) have represented valuable occasions of deepening my understanding of the themes that interest me. Participating as a speaker in almost all the most important conferences in the field from Columbia University to the University of Maryland, and (in Europe) from Erice to Bielefeld, from Cambridge to Copenhagen, at meetings of the Philosophy of Science Association (held every three years) and at meetings of the International Society of Logic, Methodology, and Philosophy of Science (held every four) has been an ideal opportunity to confront practically all the important positions in this discipline and has allowed me to enrich my knowledge to a remarkable degree.

With reference to the scientific situation in Italy, I must say that recent years have witnessed a renewed interest in these topics in my own country. Consequently, in numerous conventions, conferences, and meetings in the various Italian universities I have been able to interact with persons who have enriched my understanding. I would like to mention in particular Enrico Bellone, Enrico Beltrametti, Marcello Cini, Marisa Dalla Chiara, and Giuliano Toraldo di Francia.

At Trieste, the city where I live and work, important stimuli have come from members of the Dipartimento di Fisica Teorica (Department of Theoretical Physics) with whom I work directly, in particular (besides Tullio

Weber) from a brilliant student of mine, Renata Grassi, who with her profundity and lucidity and her total dedication to science, has been of great help to me in understanding topics I was aware of but had not fully mastered, and also collaborated with me and with Alberto Rimini in the research program that we began. I must also mention my student and collaborator Fabio Benatti, as well as various colleagues such as Giorgio Calucci, with whom I have had stimulating exchanges of ideas. Outside of the department, I have been able to maintain a fruitful interaction as well with Maria Carla Galavotti of the Department of Philosophy, which has led to stimulating international contacts; Galavotti's arrival in Trieste favored the birth of a constructive dialogue between worlds that have not always succeeded in communicating with each other. Other opportunities at Trieste have played no small role in encouraging me to write this book, especially the interdisciplinary laboratory of the SISSA which, in conjunction with the founding of the Master's Program in Scientific Journalism, has seen me engaged in the labor of communicating the themes I will be treating in the book. This undertaking has furnished the occasion for renewing contacts with many scientific journalists such as Gianfranco Bangone, Carla Ghelli, Pietro Greco, Fabio Pagan, and Franco Prattico, with most of whom I had established contact at earlier occasions as an author of newspaper articles or as a participant in broadcast radio debates (principally the one called Palomar).

I must also thank from the bottom of my heart Francesco de Stefano for his meticulous work in correcting the proofs, and I would like to mention the role played by Andrea Wedlin who took on (in the midst of his studies) the laborious task of being the guinea pig, and showed me in our conversations that a high-school student could truly understand what I had written.

Finally, there are other technical points to acknowledge. I have to thank the Corel Company, for my use of the Clipart Coreldraw for the images of Alice, Bob, and Charlie, and especially for "Bohm's witch" who appear in the illustrations of chapter 9 and following. Special gratitude is owing to the *American Journal of Physics* for granting me permission to use their images in Figure 3.3; to the Physical Society of Japan for their permission to use the logo (Figure 15.7) of the ISQM convention, the International Commission of Physics Education and World Scientific Publishers in Singapore for allowing me to use (respectively) the photo of Einstein and Bohr placed at the beginning of chapter 7, and the charming pictures drawn by John Bell that appear in chapter 15.

Sneaking a Look at God's Cards

CHAPTER ONE

The Collapse of the "Classical" World View

> I remember discussions with Bohr which went through many hours till very late at night and ended almost in despair; and when at the end of the discussion I went alone for a walk in the neighboring park I repeated to myself again and again the question: Can nature possibly be as absurd as it seemed to us in these atomic experiments?
> *-- Werner Heisenberg*

WE ARE GOING TO follow the fascinating trail that led to the scientific revolution of quantum mechanics in the first quarter of the twentieth century. Together with the theory of relativity, the conceptual structure of quantum mechanics is now the basis of the modern view of the physical world. Rarely in history has a new theory been so hotly contested, or called forth such great energies from such great minds. And even though its power to predict phenomena is unprecedented in the history of science, quantum mechanics has stirred up controversy—no less unprecedented—over its meaning. This will not seem surprising, once we penetrate the "secrets" of the new microcosm revealed by the theory. These secrets—the incredible feats we will discover microscopic systems can perform—are quite revolutionary when set beside the "classical" conceptions formerly elaborated to explain our macroscopic experience. It is only to be expected that the new theory would require so much effort and suffering to be worked out, or that such furious debate would still be raging today about its philosophical implications.

What the physicist Isidor Isaac Rabi said (quoted at the beginning of the preface) lamenting public ignorance of modern advances in physics, applies with particular force to quantum mechanics. In the words of the great Robert Oppenheimer, "As history, its re-creation would call for an art as high as the story of Oedipus or the story of Cromwell, yet in a realm of action so remote from our common experience that it is unlikely to be known to any poet or historian."

CHAPTER ONE

The prologue of such a drama would consist in the fundamental incapacity of "classical" conceptual schemes to explain certain fundamental physical phenomena. A complete list would be too lengthy, and an exhaustive analysis would require a discussion of sophisticated effects that would be inappropriate for the nontechnical spirit of the present work. Instead, I will limit myself to listing a few elementary processes that could not be explained within the conceptual system of the "classical" view of the physical world.

But before I begin, I should explain that such expressions as "the classical conception of the world" (or the equivalent) will be used in this work to designate the body of knowledge elaborated over the long course of development of scientific thought, from Galileo's revolution in the early 1600s to about 1800. This knowledge was synthesized into the two pillars of nineteenth-century physics, namely, mechanics and electromagnetism.

Classical mechanics, born in the early seventeenth century from the profound intuitions of Galileo, found its concrete realization in the inspired labors of Isaac Newton. Increasingly refined and generalized formulations were attained in the eighteenth century in the works of Joseph Louis de Lagrange, and in the nineteenth by William Rowan Hamilton. This superb theory, as is well known, treats the movement of material bodies as determined by forces acting upon them; such forces as the mutual attraction (or repulsion) between individual particles govern the motion of bodies down to the tiniest detail. Classical mechanics managed the unification of what appeared to be the most diverse phenomena: for example, the doctrine showed that the movement of the planets in the heavens was governed by exactly the same laws that regulate the movement of any physical object as it falls to the ground. And the theory would reach still farther, for, whereas scientists had formerly thought that the physical process involved in temperature exchange could not be explained in terms of mechanics, by the nineteenth century it was possible to show that even thermal processes had their origin in the disordered movements of material constituents. Classical mechanics saw one of its greatest triumphs in the nineteenth century, when Willard Gibbs, Ludwig Boltzmann, and James Clerk Maxwell realized a more profound unification of physical phenomena through the mechanical explanation of thermodynamic processes.

A parallel story can be told of the other great "classical" theory, electromagnetism. For a long time the phenomena of light seemed to have nothing to do with electrical or magnetic phenomena. The researches of Faraday, Maxwell, and others led to the recognition that these so disparate processes were nothing other than diverse manifestations of a single entity, known as the electromagnetic field: equations were worked out that governed precisely all the phenomena in question. In this way, the concept of the *field*—that is, as we shall soon see, of a physical entity continuously distributed in space and time—made its dramatic entry into science, requiring the recognition of an existence just as fundamental as one of the material particles. The electromagnetic field is capable of transporting energy through empty space in the form of light waves, radio waves, x rays, and so forth. This "wavelike" nature then became the fundamental characteristic of all processes governed by the laws of electromagnetism, as formulated by Maxwell.

This was the framework achieved by the end of the nineteenth century: a real existence had to be attributed both to discrete material particles ("corpuscles," from Latin *corpusculum*, "little body"), and to continuous fields. These physical entities were understood to evolve in a precise way in space, under the influence of their mutual interactions, and as codified in the equations of mechanics and electromagnetism. The equations, in turn, would permit the understanding of all other processes of the physical world.

Imagine, then, the crisis that occurred when some simple physical processes were discovered to be absolutely incomprehensible—to resist all attempts to reduce them to a classical understanding!

1.1. The Dependence of the Color of Objects on Temperature

It is common knowledge that a physical object, such as an iron bar, changes its color as its temperature changes. At low temperatures, the iron appears "natural" in color to us, but when its temperature is raised, it begins to give off heat (remember that thermal radiation is a form of electromagnetic radiation). Then the iron looks at first red, then yellow, and finally, white hot. This process involves thermodynamic effects, which bring an increase in thermal agitation of the constituents of the matter. These constituents, of course, are electrically charged particles, and the laws of electromagnetism teach us that charges in nonuniform motion release electromagnetic radiation. If the conditions are suitable, this radiation will appear as light. No matter how complex the process, in accordance with the previous remarks, it should eventually reenter the typical framework of "classical physics" and become perfectly intelligi-

CHAPTER ONE

ble. Unfortunately, such hope proved groundless. Despite the persistent efforts of the scientific community, even a process so common as this one did not admit of explanation in terms of classical laws of mechanics and thermodynamics. And it remained a mystery until Max Planck advanced an hypothesis of an absolutely revolutionary nature that would upset all classical ideas about light—or, to be more precise, about electromagnetic radiation.

1.2. Atoms and Their Properties

At the end of the nineteenth century, then, and at the beginning of the twentieth, a series of researches was undertaken (the most important being those conducted by Sir Ernest Rutherford) which would lead to a model of the atom very similar to the one used today.[1] An atom was conceived as a miniature planetary system, having a positively charged nucleus where almost the entire mass of the atom was concentrated. Around the nucleus, revolving like planets around a tiny sun, were electrons, negative in charge, and of such a number as to neutralize exactly the positive charge of the nucleus. The attraction between the opposing charges (according to Coulomb's law) functioned like solar-planetary gravitation,[2] and thus every electron was attracted by its nucleus (the law of attraction between two opposite charges has the same mathematical form as the attraction between two masses, that is, it decreases by the square of the distance that separates the two charges). But, once again, we will be surprised to learn that this analogy (which seems so natural to us) between atomic structure and a planetary system is, for various reasons, absolutely untenable within the classical scheme. Let us analyze a few of these reasons.

    1. *The constancy of atomic characteristics.* It is only natural to wonder how all the atoms of one element—oxygen, for example—exhibit absolutely the same physical properties, however these atoms may have been produced; or again, we may wonder how such properties can persist unchanged, while the systems in question are subjected to highly invasive procedures, such as fusion or evaporation, and then returned to their initial state. Something analogous would be impossible for a classical system, *qua* planetary system. In fact, the orbits of the planets, especially in a system with many bodies, depend in an absolutely critical way on the initial conditions. Different procedures of "preparation" of a planetary

system would inevitably lead to systems that are appreciably different. Interactions even of the most minute entity with other systems determine important changes in evolution and change the structure of any such system. Consequently, given a fixed nucleus of an atom and a fixed number of electrons to orbit it, the effective physical and chemical properties of the resultant system should show an enormous variety. There would have to be many atoms with a nucleus equal to that of oxygen (a mass sixteen times that of a proton with a positive charge equal to the eight electrons orbiting around it), and the variety of atoms would correspond to the different ways in which the nucleus has, so to speak, "captured" its electrons at the moment of its formation. But the opposite is true: all the physical and chemical phenomenology shows that the properties of an atomic element are absolutely identical, independently of preparatory conditions and any subsequent transformations.

2. These precise, persistent properties that characterize an atom also determine its behavior in physical and chemical processes. Nevertheless, as is well known, even in the case of an atom with many electrons, the characteristic properties change radically as we go from one atom to another with only one electron more or less. For example, we can consider an atom of xenon, a noble gas, which has a nucleus that contains (along with its neutrons) 54 protons, around which revolve an equal number of electrons. This atom is chemically inert, that is to say, it will be extremely difficult for it to form chemical compounds with other atoms. We only need to take away one electron and one proton (the relevant constituents for holding the system together) and a few neutrons to change it into an atom of the nonmetal solid iodine, a system with a very precise and conspicuous electronic affinity, and thus with radically different behavior and properties. This would certainly *not* happen in a planetary system with 54 planets: the passage from one system to another with one fewer planet (and with its sun correspondingly a little lighter in weight) would not, it would seem, bring with it such radical changes in the behavior of the system.

3. Finally, another fundamental fact generates an incurable conflict between the stability of atoms and the "planetary" model within the conceptual structure of classical physics. According to the equations of classical dynamics, a system of electric charges can remain in equilibrium only if the charges are in motion. The fact that the atom has a limited extension requires that the charges present in it should move in circular or elliptical orbits (similar to the planets), and thus have acceleration. Now, according to securely established laws of electromagnetism (Maxwell's

5

equations), an accelerated charge inevitably emits electromagnetic waves, or radiation. By radiating, the electron would lose energy and its orbit would decrease, causing it to fall into the nucleus within a very short period of time.[3] Exact calculation brings the conclusion that every atom ought to have a very ephemeral life, showing "constant" properties for extremely brief periods (fractions of a second). Of course, this would contradict all the familiar phenomena.

Many other facts too—in particular the specific modes of interaction between atoms and electromagnetic radiation—bring us into an irreconcilable conflict with classical conceptions. This is how a crisis began to prepare the way—as has often happened in the history of science—for a true scientific revolution. Thanks to the combined strength of a remarkable group of geniuses, that revolution arrived in the form of quantum theory, the theme of this book.

1.3. WAVE PHENOMENA

Once having derived the laws of electromagnetism, Maxwell came to the conclusion that an accelerated charge radiates electromagnetic energy. This energy propagates itself in space in waves characterized by a double field, electrical and magnetic. This was verified by Hertz in 1888, and in 1901 Marconi succeeded in transmitting electromagnetic waves across the Atlantic Ocean. This phenomenon—the transmission of radio waves—is probably the most well-known example of the propagation of electromagnetic waves. In this case, the variable current (that is, the accelerated movement of the electrical charges contained in it) which runs through a radio antenna produces the emission of waves (Figure 1.1), which are propagated in space. Afterward, the information that these waves transport can be, as it were, decoded by the receiving apparatus.

We now need to examine this phenomenon more profoundly, in order to grasp the essential points we need to know, for understanding what follows. In fact, this part of the book, including the rest of this chapter and the first sections of the next, is dedicated to the discussion of the phenomenon of *light polarization,* and will appear rather technical, even though the exposition of the important points will stay at an elementary level. I need to ask the reader for a little extra effort to master the few, simple concepts I am going to explain, since their correct understanding will pave the way for the rest.

COLLAPSE OF THE "CLASSICAL" WORLD VIEW

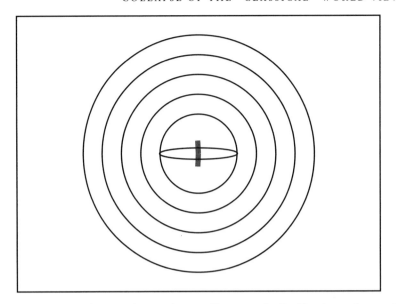

FIGURE 1.1. An electric charge that oscillates periodically along the vertical bar (antenna), being subject to an acceleration, radiates an electromagnetic wave in space that can be received by an appropriate receiver.

Suppose we have an electric charge, moving with a periodic motion along a certain segment, represented by the dark vertical line in Figure 1.1. The charge will radiate electromagnetic waves of the same frequency in the surrounding space. One electromagnetic wave consists of two field vectors, mutually perpendicular, and perpendicular to the direction of the propagation of the wave. These two vectors represent the electric field E and the magnetic field H, respectively (see Figure 1.2). The velocity of the propagation of the radio wave in a vacuum is the same as the velocity of light, which means that it is traveling at about 300,000 kilometers per second.

Let us study the electric field first. We can observe it from two perspectives: we can study how its magnitude and direction vary at a given point in space with the passage of time, or, alternatively, we can study what values it assumes at various points of space at one given instant of time. In the first case we have a sinusoid or "foldlike" shape like that in Figure 1.3, which tells us, for example, that at the considered point, and at time t_1 the field points upward, and has the value E_1. The time T that it takes the field to return to exactly the same value at the same point in space, is the period of oscillation, and coincides with the oscillation of the charge that generates it. In the second case (Figure 1.4) we have an analogous

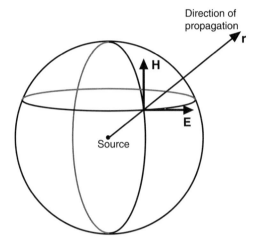

FIGURE 1.2. The electric and magnetic fields associated with the propagation of a spherical wave from a point source. The vectors are perpendicular to each other and perpendicular to the direction r of the propagation of the wave at the point under consideration.

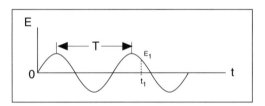

FIGURE 1.3. Variations of the electric field E in a fixed point of space, depending on change of time. The time interval that occurs, for example, between two maximal peaks represents the period of the radiation.

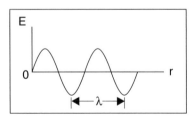

FIGURE 1.4. The variation of the electric field E at a fixed moment of time, depending on its spatial position. The spatial interval which intervenes, for example, between two troughs or between two crests represents the wavelength λ of the radiation.

representation, but giving us an image of the field at different points of space, along the direction of wave propagation, at one given instant. This also is sinusoidal in shape. We can define the wavelength λ of the electromagnetic field as the spatial distance between the two successive troughs, or crests, of the wave itself.

In place of the period T, it is also convenient to introduce its inverse, i.e., the frequency v of the wave:

$$v = 1/T \tag{1.1}$$

which represents the number of oscillations the field (or the charge which generates the field) performs in one second.

It is true in general of wave phenomena (whether electromagnetic, liquid, or sound waves) that the distance a wave is propagated in one second (its velocity) is equal to the distance λ covered by a single oscillation, times the number of oscillations per second (the frequency v). If the velocity of propagation is equal to c, then

$$\lambda v = c, \tag{1.2}$$

a formula that represents the relation between wavelength, frequency, and velocity of light.

In principle, electromagnetic waves can have any frequency between zero and infinity. The classification of radiations according to their frequency is known as the electromagnetic spectrum, as shown in Figure 1.5 below.

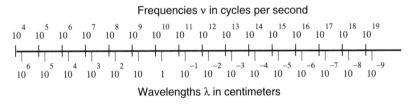

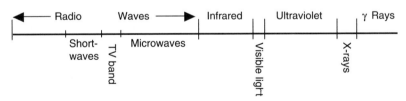

FIGURE 1.5. The electromagnetic spectrum. In the upper part of the graph are given the frequencies; below them, the corresponding wavelengths. Underneath both are shown the types of radiation that correspond to the various frequencies and wavelengths.

CHAPTER ONE

This is a good time to explain that the example considered in Figures 1.3 and 1.4 represents a very simple case, especially because a field with a definite frequency is assumed (so that, for example, if light waves were being considered, they would be the waves of a single color of light). Furthermore, the oscillation is taking place in a well-defined plane (that is, as we shall explain more fully in the next chapter, it exhibits plane polarization). In the most typical instance, however, the field can change in an arbitrary fashion: while still remaining perpendicular to the direction of the waves propagation, it can assume a variety of orientations and amplitudes from time to time, and from one point to the next. Without going into great detail, it is crucial that we understand at least one thing about wave processes, and that is the *linear* nature of the equations that govern them.

Suppose we have a field (let us say, an electrical field) $E_1(r,t)$, which represents a solution of the equations of wave propagation, under opportune conditions characterizing the production mode of the wave in question, for example, as emitted from a specific antenna (situated in New York City); then say we have a second field $E_2(r,t)$, which also represents a solution of the equations, relative to another production mode (as emitted from an antenna situated in Philadelphia). Now, the field which at every instant t of time and in every point r of space is the vector sum of the two fields (according to the familiar parallelogram rule: see Figure 1.6),

$$E(r,t) = E_1(r,t) + E_2(r,t), \qquad (1.3)$$

is the solution of the equations that define the real state of the electric field, when both sources are operative.

Before concluding this analysis we should recall that what interests us most in the physics of electromagnetic wave propagation is the fact that the wave carries energy. This energy is behind the various processes leading to the very interesting effects of, say, chlorophyllic functions, photographic phenomena, or the stimulation of our sense of sight. Now, in the case of an electromagnetic wave, the density of energy at a given point in space and in a given instant of time is proportional to the sum of the squares of the electric and magnetic fields at that point and in that instant. However, in vacuum, the electric and magnetic fields (with the appropriate units of measurement) are of equal intensity. This means that the electromagnetic energy in a certain volume of space at a certain time is proportional to the product of the square of the electric field and the volume—supposing, for the sake of simplicity, that the volume is small enough to eliminate any appreciable variations of the field within it

COLLAPSE OF THE "CLASSICAL" WORLD VIEW

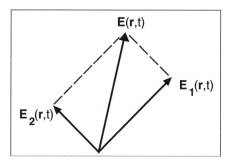

FIGURE 1.6. The sum of the values of the two electrical fields at a certain instant t of time and at the same point r of space. In the example, the fields indicated would both be solutions of the electromagnetic wave-propagation equation, with their characteristic dependence on r and t; their sum at each point of space and at each instant of time would also represent one solution. This fundamental fact derives from the linear nature of evolutionary equations.

(technically, in the case of fields that vary rapidly, the quantity that really matters is the integral of the square of the field, extended to the volume). Recalling Figure 1.4, then, where we represented the field at various points in space of a light ray, we now can graph (in Figure 1.7) the density of energy, simply taking its square (E^2). It follows that the area taken up under the curve at a certain portion of the wave Δr represents the energy concentrated at that interval. The figure shows how in those regions of space where the field is more intense, a significant concentration of energy will be found, while the weaker fields correspond to low energy densities. However, one of the cardinal points of Maxwell's theory of light, as already mentioned, is that the field, and consequently the energy trans-

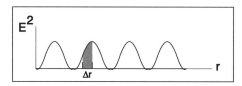

FIGURE 1.7. The square of the field at various points of space represents the volume density of energy transported by the field itself. From this it follows that the shadowed area of the curve represents the total energy of the field in the interval Δr. It should be noted that according to the graph, by varying from point to point, the energy is continuously distributed through the space.

CHAPTER ONE

ported by the field, is continuously distributed in space, as the field varies from point to point.

1.4. Diffraction and Interference

We began by saying that at the beginning of the twentieth century, the scientific community was in agreement in setting two kinds of process at the basis of the physical theory of the universe: waves and particles. In the foregoing section, much attention has been paid to the description of wave phenomena. Now, for the full comprehension of the argument to be developed, we will need to linger a little longer on this topic. And in order to underline the radical difference—really, the incompatibility—between the two aspects, it would seem particularly fitting to analyze in detail two phenomena (in reality, two facets of one and the same phenomenon) of great conceptual relevance, which are of major conceptual importance and are a direct consequence of the linear nature of wave equations: diffraction and interference.

The first is the phenomenon we can all observe, for example, on a surface of water in which a point reached by a wave disturbance becomes itself the origin of new waves. In this way, the wave is not compelled to propagate itself only in a rectilinear fashion,[4] but can, in a certain sense, "get around the obstacles." The second phenomenon can also be easily observed on a surface of water. If a certain point (or area) of the surface is reached by two waves (generated, for instance, by the passing of two motorboats or the tossing of two rocks into the water), the wave phenomena on every point of the surface will be governed by the "sum" of the two disturbances, so that it may happen that a certain point will remain still, if it is being lifted by one movement and lowered by the other. Thus the two waves at some points will get in the way of each other, in some places nullifying each other (destructive interference), in other places reinforcing each other (constructive interference).

It is important to realize that the possibility of demonstrating these characteristic aspects of wave phenomena depends, in an absolutely crucial way, on the relationship between the wavelength that characterizes the phenomenon and the dimensions of the obstacles encountered by the wave in its propagation. To illustrate the point it will be useful to move beyond the propagation of waves on a water surface, as analyzed above, and consider light phenomena and the familiar process of shadow formation. If a screen is illuminated, in which has been made a hole of macroscopic

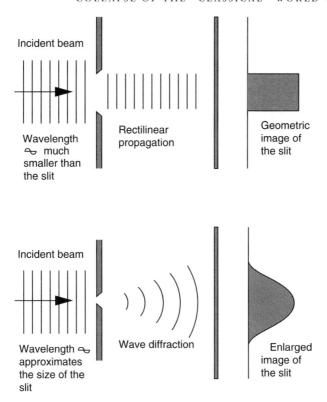

FIGURE 1.8. The phenomenon of the diffraction of light. While the length of the light wave is much smaller than the obstacles it encounters, the propagation is governed (in a very accurate way) by the laws of geometrical optics which predict rectilinear trajectories for luminous rays and distinct shadows of objects. When the objects encountered by the waves have dimensions approximating the wavelengths, diffraction phenomena occur, and the result is the enlargement of the image of the opening. In the illustration, the shaded areas to the right of the screen represent the square of the intensity of the field, and thereby the density of the electromagnetic energy at any single point.

size (one centimeter square, for example), behind the screen will be seen a geometrically exact image of the hole. The light does not diffract, but propagates in a rectilinear fashion.[5] However, if we make the hole so small that its dimensions become comparable to the wavelength, we then see the light "angle out," so that the shadow becomes larger. We all experience a similar diffraction when we look directly at a beam of light by squinting. The divergent behaviors of the two cases are pictured in Figure 1.8.

13

CHAPTER ONE

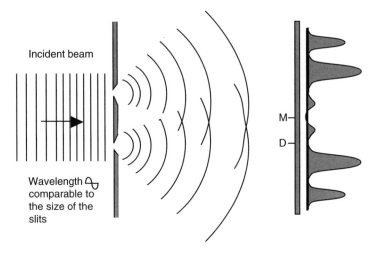

FIGURE 1.9. The phenomenon of interference. The two waves which are generated by the two holes interfere with each other in the region of the screen. At some places, as at M, the contributions of the two waves mutually reinforce each other, while in other places, as at D, they nullify each other. In this case, too, the amplitude of the shape on the right represents the density of energy at the various points of the screen, and therefore, if the screen were a photographic film, it would have an alternation of lighter and darker areas.

Analogously, Figure 1.9 shows the phenomenon of interference. The beam of light entering the two apertures gives rise to two waves, which propagate to the right of the screen. At every point in the space, the electrical field is thus the sum of the fields associated with the waves that emerge from the two holes.[6] It is intuitively apparent that at one point between the openings (such as at M) on the screen to the right, the two electrical fields, which have had equal paths to travel from their origins, arrive, so to speak, "in phase," and they add together, so that they reinforce each other. But at point D, for which the difference between the two paths from the openings is half a wavelength, the field coming from the first slit points upward, while the one coming from the second points downward, and thus their sum will be zero (it is assumed that since the screen is very far from the openings with respect to the extension of the wavelength, the two rays are in fact in line with each other). In both figures, beyond the screen are shown shadowed areas, representing the density of electromagnetic energy at the various points. Where these shapes have a larger amplitude, the energy is greater, so that if the screen were a photographic film those spaces would be more heavily exposed. With reference to Figure 1.9, we can say more exactly that it shows the interference fringes so characteristic of the process.

After these prefatory points, we can face the problem of the emission of electromagnetic energy on the part of a body when its temperature is varied, and thereby illustrate the revolutionary hypothesis that led Planck to its first theoretical explanation.

1.5. Planck's Hypothesis, and Its Later Elaboration by Einstein

As we saw in section 2 of this chapter, all the attempts to explain the radiance of a body as it changes temperature in terms of "classical" theories were miserable failures, even though this problem had engaged some of the most brilliant minds of the time. Planck showed that a solution could be obtained only by granting that exchanges of energy between radiation and matter did not occur continuously. For the classical model required continuity: the electric field associated with a wave acting on a charged particle makes it oscillate by giving it energy; but Planck assumed that a field with frequency v could exchange energy with matter only through discrete "quanta," that is to say, only in quantities that would be whole number multiples of the quantity hv. In this relationship, the constant h is the universal constant that now bears Planck's name, and whose value is extremely small, that is to say, equal to 6.55×10^{-27} erg-seconds. From the first moment of its appearance, this constant, also known as the "quantum of action," has played an increasingly important role in the interpretation of the most diverse physical phenomena, especially at the microscopic level. But Planck's hypothesis, which was surely a decisive step in the development of modern scientific thinking, did not immediately win the consensus it deserved, precisely because the quantum of action introduced an uncomfortable rupture into the continuity of natural phenomena universally accepted at the time. The hypothesis appeared to Planck himself as a mathematical artifice, an abstract construction that did not directly require a profound alteration in the understanding of reality. In a way, Planck saw himself constrained to introduce the quantum of action as soon as he realized that there was no other way to provide a theoretical basis for the phenomena of radiation.

But this state of affairs soon changed. In 1905 Einstein used Planck's discovery to explain the photoelectric effect (that is, the process by which it is possible, through the illumination of a metal surface under the right conditions, to pull electrons from the metal), at the same time extending the idea and remarkably increasing its significance: "the observations ... connected with the emission and transformation of light, are easily understood when we assume that the energy of the light is discontinuously distributed in space." It is interesting to note that Einstein did not receive the Nobel Prize for his monumental constructions (the theories of general and special relativity) but for his interpretation of the photoelectric effect.[7]

From that day forward, a beam of light was likened to a flow of quanta, of energy granules or *photons*, rather than to a continuous distribution of energy. This revolutionary hypothesis which, among other things, attributed corpuscular or particle properties to electromagnetic radiation, led eventually to quantum mechanics, the conceptual scheme to which this book is dedicated.

We can get an idea of the very small size of the quantum of action by trying to estimate, for example, how many photons pass over an opening of one square centimeter, situated one meter away from a 100 watt lamp. For the sake of simplicity we will disregard the electromagnetic emission produced by the lamp in the form of heat, and we suppose that the radiation has a frequency typical of the visible spectrum, such that $v = 5 \times 10^{14}$ cycles per second. We turn on the lamp for one second. This provides an energy of 100 joules, or 10^9 ergs. Since one photon carries an energy equal to

$$h v = (6.55 \times 10^{-27}) \times (5 \times 10^{14}) = 32.75 \times 10^{-13} \text{ ergs}, \quad (1.4)$$

the number of photons that, in one second, cross the sphere (with a diameter of one meter) surrounding the lamp will be equal to the ratio between the total emitted energy and the amount carried by one photon, that is, $[10^9 : (32.75 \times 10^{-13})] = 3 \times 10^{21}$. A sphere like this has a surface of $4\pi r^2$, or an area equal to 12.56×10^4 cm^2. From this it follows that the number of photons that passes in one second through an opening one centimeter square would be 2.4×10^{16}, which is to say 24 million billions (or 24 quadrillions) of photons. It should not be surprising that at this level the difference between a continuous distribution and a discrete, quantized distribution of energy will be difficult to discern, and this is why the experience with electromagnetic fields at the macroscopic level suggests a continuous distribution of energy in space. But, as we shall see, modern technology makes possible situations where, over a surface such as the one in question, in each second only a few photons, or just a single photon passes. In this case it will be possible to "see" them (that is, to detect them) one by one, and this should convince us that the hypothesis of Planck and Einstein is correct: the electromagnetic waves, right alongside the aspects that characterize them as undulatory or wavelike, possess other aspects as well, which in the classical understanding would be corpuscular or particlelike.

1.6. Bohr's Atom and Quantization

We have seen how Planck was driven to form a strange hypothesis in conflict with the conceptual scheme of classical physics in order to overcome the difficulties of the problem of radiation. We will now see how, in an analogous fashion, Bohr overcame a great part of the difficulties of the planetary model of the atom through a similarly "heretical" hypothesis: the quantization of material systems. The idea is rather simple: we assume, for whatever reason, that the classical vision, according to which all the "planetary" movements of the electrons are possible, is not in fact correct, but that there are some orbits that are "privileged" and that these and only these can be traveled by an electron. It may be remarked that this hypothesis no less than Planck's negates the fundamental continuity of physical processes, that is, the deeply rooted idea that "nature does not make leaps" (*natura non facit saltus*). While, according to the classical vision, and as the above discussion has maintained, continuous variations of the initial conditions of the "capture" of a planet by the sun would make for continuous variations in the orbit, according to Bohr's hypothesis this would not happen: an electron, by becoming connected to an atom, has to be "captured" in one of the peculiar conditions that are uniquely compatible with the formation of that system. Distinct values of energy correspond to these discrete orbits, and this energy will be quantized, and no longer continuously variable. The fact that only some orbits are possible explains above all the regularity of the atoms and their relative insensibility (at the appropriate scale) to external disturbances. For, however an electron connects itself to an atom, it must do so in such a way as to wind up on one of the orbits permitted to it, and not on just any arbitrary orbit. Furthermore, no small disturbance resulting from interaction with external systems will be able to change the atom, insofar as its own state can change only when the electron is refitted into another one of its permitted orbits. But the energy of the permitted orbit closest to the original orbit is separated from the latter by a finite amount. Unless this small amount of energy can be provided, the system remains unchanged.

Bohr found a precise mathematical formula for determining these privileged orbits, that is, the only ones that are possible for electrons circling the nucleus of an atom. For an understanding of the arguments from this point forward it will not be necessary to enter into the details of the formula, or make explicit the rules of quantization. Instead, it will be worthwhile to observe how, respecting the principle of the conservation of energy, and recalling the hypothesis of Planck, Bohr also succeeded in explaining another mystery of the behavior of atoms: the fact that they emit and absorb radiation at only very precise frequencies. The reason is fairly simple. If an atom can stand only in one of a series of states of precise energy, it will be able to emit or absorb electromagnetic radiation only by passing from one state to another. In a corresponding way, it will emit a quantum, or photon, whose energy will be such as to guarantee the conservation of total energy. Let us suppose therefore that an atom passes from an orbit with the energy E_2 to an orbit with less energy, E_1, and thereby emits a light quantum. The energy that the electron loses becomes transferred to the photon, and since the energy of this latter is proportional to its frequency $v_{(2 \to 1)}$ we will have

$$E_2 - E_1 = h v_{(2 \to 1)}. \qquad (1.5)$$

If we keep in mind that only certain specific energy states are permitted, it will follow that only precise frequencies can be emitted (or absorbed) by an atom; and this is just what explains the nature of atomic emission spectra (Figure 1.10).

Bohr's theory, based on a single, simple hypothesis involving a unique and precise rule for identifying permitted orbits, rendered intelligible an incredible quantity of data about the emission and absorption of the various elements studied by the spectroscopists of the time. Then it suddenly became evident to everyone that this peculiar hypothesis had a very profound significance. It has to be stressed that Bohr had the courage to present a model that was patently inconsistent. In fact, after using the classical laws to determine the orbits of the electrons, he introduced the contradictory hypothesis that not all orbits are possible! Not only that: he also violated the laws of classical electromagnetism, by assuming that the electrons, although moving in circular orbits (and thus possessing acceleration), did not radiate, so long as they stood in one of their permitted orbits.

FIGURE 1.10. Whereas the white light that comes from the sun contains, so to speak, all the visible colors and thus when analyzed by a crystal prism yields the familiar spectrum (i.e., the rainbow), the light emitted by an atom contains only precise wavelengths and thus, when analyzed by a spectroscope (an instrument like a prism except more sophisticated), gives us a series of "spectral lines." In the figure are represented the lines characteristic of the emissions of an atom of hydrogen heated to high temperature, in an interval of wavelength between 500 Å and 7000 Å. Reminder: an angstrom (Å) represents a unit of measurement used by spectroscopists and is equivalent to 10^{-8} cm.

1.7. DE BROGLIE'S HYPOTHESIS

For our purposes, it is not necessary to slow down our pace to describe the discomfort of the scientific community during those passionate years. We will summarize, mentioning only a few salient facts.

In 1924 Louis Victor de Broglie presented his doctoral thesis (which would earn him the Nobel Prize) in which he set forth an hypothesis that showed how we could advance our understanding of disconcerting aspects of reality through "increasing" the confusion. Not only do we have to recognize, with Planck and Einstein, that waves present some particle aspects; it is also useful to suppose that particles present some wave aspects! His idea could be expressed quite simply: a wave of a certain wavelength can be associated with any particle with a mass m and velocity v:

$$\lambda = h/mv, \qquad (1.6)$$

where h is the familiar and ever-present Planck's constant. Two comments are in order.

1. The reason why increasing confusion brings greater understanding is that once the de Broglie hypothesis is accepted, it will follow that the orbits permitted according to Bohr's rule of quantization are precisely and solely those orbits where the velocity of the electron is such that a *whole number* of wavelengths stay in exactly one orbit. If it is allowed that an electron can in a certain sense be effectively associated with a wave, this condition is in fact the only one which permits (see Figure

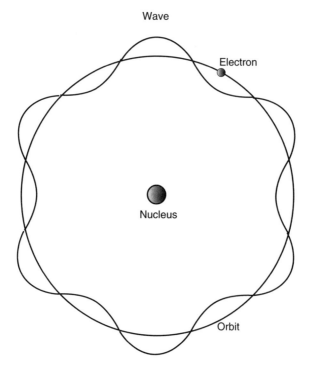

FIGURE 1.11. The orbit of an electron in an atom and the wavelength of the wave associated with it, according to the de Broglie hypothesis. The stationary condition of the process requires that an integral number of wavelengths be located perfectly within the circumference. This condition exactly identifies the same orbits as Bohr's theory.

1.11) a stationary situation on an orbit (something like a violin string, always vibrating at the same frequency)—that is, without variation over time. In turn, the stationary nature of the process is needed to explain how properties of atoms do not change.

2. The hypothesis now discussed immediately poses a problem: why the wave aspects postulated by de Broglie never emerge in experiments on material bodies. If de Broglie's hypothesis is true, why has nobody ever witnessed (for example) the diffraction of a particle? The answer is fairly simple, as long as we keep in mind that diffraction occurs only when the dimensions of the obstacles which the wave encounters are comparable to the wavelength in question. The solution to the problem is to evaluate the wavelengths of various particles at various velocities. I do not wish to slow us down with tedious calculations, even though, on

the basis of the preceding discussion, it would not be difficult to determine "de Broglie's wavelength" of a material particle moving at a certain velocity. I will confine myself to the consideration of a few simple examples. An atom of hydrogen that moves at the velocity characterizing its thermal agitation at normal temperature is associated with a wave whose wavelength is about 10^{-8} cm. As seen in Figure 1.5, this corresponds to the length of x rays. Since the length of the wave decreases with the increase of the mass, it will be apparent that all the particles of the "classical" world will be associated with waves whose wavelength is so small that, to reveal the effects of diffraction, they would have to encounter obstacles of dimensions well below the scale of the processes they involve. And this is why nobody has ever encountered the wave aspects of material bodies. But if the preceding calculation is carried out for an electron moving in an atom, it should be clear (as discussed before with reference to the de Broglie hypothesis, and from Figure 1.11) that the dimensions of the atom are comparable to the wavelengths that characterize the electrons themselves. From this it follows that, if the de Broglie hypothesis is true, it must necessarily be taken into account when describing such systems.

The line of thinking sparked off by the courageous idea of de Broglie would turn out to be extremely fertile, and, in less than a year, would bring Erwin Schrödinger to work out "wave mechanics" (one of the two equivalent formulations of quantum theory). The great importance of de Broglie's hypothesis is revealed in a statement by Albert Einstein in a letter he sent to Hendrick Antoon Lorentz in December, 1924: "A younger brother of ... de Broglie has undertaken a very interesting attempt to interpret the Bohr-Sommerfeld quantum rules [this was the business of fitting the electron waves around the Bohr orbits] ... I believe it is the first feeble ray of light on this worst of our physics enigmas. I, too, have found something which speaks for his construction. . . ."[8] Einstein played a key role in securing the cooperation of de Broglie's reluctant thesis supervisor Paul Langevin. Einstein made it clear, when Langevin asked about its merit, that he found it very interesting indeed. Abraham Pais aptly remarked that in this way Einstein was not only one of the fathers of quantum theory, he was also the "grandfather" of wave mechanics!

The important role of Einstein's thought for Schrödinger was fully recognized by the latter, who on various occasions said that "his theory meant nothing other than taking seriously the Einstein-de Broglie wave theory of particles in motion," and Schrödinger would always refer to the assumption I have been discussing in this section as "the de Broglie–Einstein hypothesis."

1.8. Overcoming the Crisis

The profound crisis over the vision of physical processes we have delineated in the preceding pages found its ultimate outcome a quarter century later. During 1925 and 1926 Heisenberg and Schrödinger, working independently and following two quite different lines of thought, succeeded in precisely formulating two theoretical models that signaled the birth of the new science of quantum mechanics.

Heisenberg, a disciple of Bohr and profound student of his work, was convinced that the master's insistence on visualizing the atom as a planetary system represented an obstacle to overcoming the contradictions of the theory. He therefore decided to focus his attention not on something that could not be experienced, such as the orbits of electrons, but instead on the radiation emitted by atoms, namely, the spectral lines of which we spoke in section 1.6. In this way, in 1925 Heisenberg arrived at the formulation of a mathematical scheme that would eventually be known as "matrix mechanics." The interpretation of matrix mechanics remained an open problem for a time, and its full understanding required the important contributions of Born, Jordan, and Pauli.

Like Einstein's 1905, Schrödinger's 1926 was a year of glory. By this time he was thirty-eight, with a brilliant although not really outstanding scientific career. He felt conscious of his worth, and was somewhat discouraged that he had not yet made any significant contribution to science. In November, 1925, he wrote to Einstein, speaking with enthusiasm about the idea of the wave nature of particles. Again, at the end of the same month he wrote a letter to Hans Thirring, in which, according to Thirring, one can find a distinct outline of the ideas that would constitute the basis of quantum mechanics. But it still lacked some decisive elements.

Schrödinger was at a curious period in his life: his marriage was foundering and he spoke often about divorce. At the end of the year he decided to spend Christmas vacation at Arosa, an Alpine spot of 1,700 meters altitude, not far from Davos. He wrote to "an old girl friend from Vienna," intending to meet her, while his wife stayed in Zürich. It is not possible to identify the girl friend: as Walter Moore commented in his biography of Schrödinger, this woman, like the "black lady" who inspired the sonnets of Shakespeare, will probably be forever unknown. But according to the explicit admission of the great scientist, she set in motion a period of creative activity unequaled in the history of science.[9] In fact, in those two weeks at Arosa, Schrödinger found his own path, and upon his return to Vienna, taking advantage of the subtle mathematical knowledge of his friend Hermann Weyl, in less than a year he brought to completion a series of six works containing the definitive formulation of wave mechanics.

We can conclude this biographical digression with a return to our theme: the birth of the new conception of physical processes. It is interesting to consider that the two formulations of the theory, that of Heisenberg and that of Schrödinger, do not appear to have any relationship to each other, and in their form they each use a very different language. The theory of Heisenberg, as I mentioned, was formulated in the language of matrices, and hence acquired the name matrix mechanics, while that of Schrödinger came to be formulated in the typical language of wave equations, and came to be known as wave mechanics. It would take some years for it to be rigorously proved, thanks to the further researches of Schrödinger himself and of Paul Adrien Maurice Dirac, that the two theories were really but two different mathematical modes of expressing the same laws.

1.9. THE WAVE/PARTICLE DUALISM

The new theory presents very peculiar aspects but, as the reader will probably have sensed, at the same time brings an important new unification of our conceptions of nature. Phenomena so diverse from the point of view of classical physics—such as waves and particles—become assimilated. Every physical process simultaneously involves these two faces of reality. The quanta of light, or photons, behave in many experiments just like particles (with peculiar properties, of course), and analogously, particles, under opportune conditions, behave like waves. The problem arises, then, of understanding how concepts that seem so irreconcilable and contradictory, can ever be integrated. Quantum formalism required an interpretation, and a new phase began, characterized by great interest and enthusiasm—a feverish time, which would bring the vision I shall illustrate in the following pages.

Of course, apart from the enormous successes that the theory met with quite soon in the explanation of a variety of physical processes, it imme-

CHAPTER ONE

diately stirred up an experimental problem of great interest: is it possible to get direct evidence of the wave aspects of particles in the laboratory? Various researchers cooperated in this endeavor and the response was in the affirmative.

I would now like to close this first chapter with a strange and delightful observation. In 1937, George Paget Thompson received the Nobel Prize for having shown that the electron was a wave. Exactly thirty-one years earlier, his father, Joseph John Thompson, had received the Nobel Prize for having shown, through experiments in radioactive processes, that the electron was a particle![10]

CHAPTER TWO

The Polarization of Light

> We begin with some simple experiments with a beam of light and three identical transparent disks of a rather special kind ... which can be acquired by purchasing two sets of polarizing sunglasses. ... The effect is particularly dramatic if, instead of shining light through the three disk sandwich, you look through it. After the middle disk is removed nothing at all can be seen. Conversely if the middle disk is reintroduced the sandwich again becomes transparent. By adding a disk to the pile you have succeeded in letting through more light ... this behavior, like a good magic trick, remains very striking to observe, no matter how long or how well you have understood the explanation.
> —David Mermin

WHY WOULD A BOOK on the history and philosophical implications of quantum mechanics need to devote an entire chapter to the single phenomenon of light polarization? For a very good reason: polarization phenomena—especially, the behavior of polarized states when combined—present a very close analogy to the combination of quantum states in general, giving us a clear and simple way to illustrate the principles of quantum formalism.

And this is not all: nearly all quantum experiments with major relevance for the debate directly involve either polarization states of photons or certain properties of material particles, such as *spin*, which present a formally perfect analogy with properties of the electromagnetic field. In the next chapter, we will supplement the classical analysis presented in this chapter, by adding Planck's hypothesis about quanta of light; after that, we will be ready to understand some important aspects of quantum formalism.

CHAPTER TWO

2.1. Polarization States of the Electromagnetic Field

In the preceding chapter we discussed states of the electric field that correspond to certain polarization states—that is, linear polarization. We saw how the electric field at one point of space varies with the progress of time (Figure 1.3), and how the same field varies at one given instant with the changes of position along the ray of the wave's propagation (Figure 1.4). It has been possible to use such simple graphs to represent the value and direction of the electric field because it has been tacitly assumed that we are dealing with linear polarization states. This means that the electric field (and the magnetic field perpendicular to it) oscillates in a plane. The same illustrations will also give us the opportunity of making more clear another topic we discussed in that chapter, namely, interference phenomena.

Let us now suppose that at a certain instant of time, a certain region of space is traversed by two different "electromagnetic waves," which for the sake of convenience we will now describe only in terms of the electric field. Let us then consider a light ray—a precise line along which the fields are being propagated—and suppose that the two waves are charac-

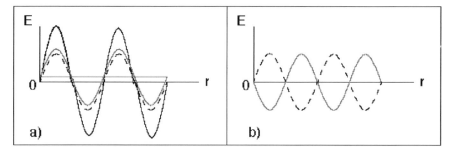

FIGURE 2.1. Constructive interference (a), and destructive interference (b), of two electrical fields with the same plane of polarization and of equal frequency and amplitude. The two fields are represented by the dotted and shaded lines, the resulting one by the solid dark line. As long as the fields are in phase (a), they give a field of double intensity (of quadruple energy density at every point). When their difference of phase is equal to half a wavelength (b), they will have (at every point and at every moment) equal and opposite values, so that their sum will be equal to zero, and there will be no field.

POLARIZATION OF LIGHT

terized by electric fields polarized in the same plane, with equal wavelength and amplitude.[1] We can imagine two different scenarios. First, we can imagine that the two fields are "in phase," which means that at a given moment and at a given point along the ray, they will both have the value zero, and increase when they move in the same direction along the ray. Alternatively, we can think of a case where they are "out of phase," which means that one has been moved with respect to the other by exactly one-half wavelength. As pointed out before, the "true" field at every point of the region in question is given by the sum of the two: in the first case we have the situation illustrated in Figure 2.1a, and the second case is shown by Figure 2.1b.

In the most general case, the fields we add to one another (according to the parallelogram rule) are diversely oriented in space. It is particularly interesting to analyze the possible combinations of two polarized plane waves of equal wavelength, propagated along the same direction but with polarization planes perpendicular to one another (and thus perpendicular to their supposed common direction of propagation). In Figure 2.2 two fields of this kind are represented, E_V and E_H, polarized in the planes xz (vertical) and yz (horizontal), respectively. We can now study the field resulting from the sum of the two fields. Since at every point the fields themselves are perpendicular and equal in intensity, their sum will

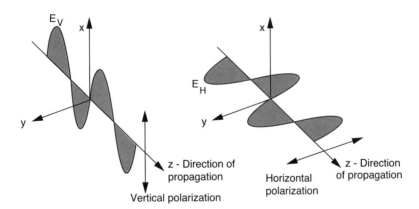

FIGURE 2.2. Propagation of a polarized wave. To the left, a wave is represented whose plane of polarization is vertical; to the right, one whose plane is horizontal. The origin of the system of reference is the same for the two fields, which for this reason, even though they are differently oriented, are in phase according to the terminology previously adopted.

27

CHAPTER TWO

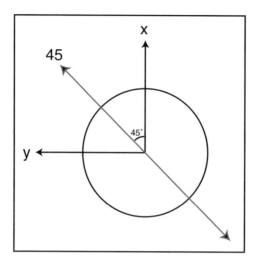

FIGURE 2.3. Summing two fields, polarized along the directions x and y, perpendicular to the direction of propagation, having equal intensity, and being in phase, we get a field (which itself has plane polarization) oscillating along the plane indicated by the shadowed line, at 45°.

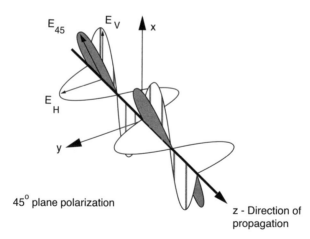

FIGURE 2.4. A field polarized at 45° is generated by the sum of two fields of equal amplitude and in phase, polarized along two perpendicular planes.

28

POLARIZATION OF LIGHT

always lie along the diagonal of the plane *xy*, and thus the resulting field $E = E_V + E_H$ will also oscillate in a plane and will be characterized by the same frequency and polarization.

The direction along which the field *E* oscillates is depicted in Figure 2.3, and for obvious reasons, we will call the resulting field E_{45}. The phenomenon can be shown more dramatically in Figure 2.4.

(The passage that now follows, and other smaller-type passages included as appendixes to the chapters, can be left aside in the first reading by anyone less interested in the technical details.)

. . . . .

We can now use a more precise mathematical language, to discuss more complex cases of combined polarization states along two perpendicular planes, where the two fields have different amplitudes and different phase relationships.

In what follows, it is convenient to choose a certain quantitative "standard" of reference for the amplitude of the electrical field (i.e., its maximum value), which we will indicate by *E*, and thus express amplitude *E#* of a (more or less intense) arbitrary field by multiplying E by a suitable factor of amplification or diminution. We need to posit this value because the density of energy transported by the field depends on the square of the field. The "standard" amplitude will correspond, therefore, to a certain chosen luminous intensity, and the intensity which characterizes a given field of amplitude $E# = aE$ will be, in every point of space and at every instant, a^2 times the intensity associated with an identical field, but with amplitude *E*. Such pedantic precision is essential for determining the precise mathematical relations between the component fields and the resulting field, so that we can assess immediately the amount of energy involved in the process.

With reference to the case at hand, we observe that, if it is assumed that the field E_{45} as well as the constituent fields E_V and E_H have a "standard" amplitude, the precise relationship that connects them can be expressed as

$$E_{45} = E_V^\# + E_V^\# = \frac{1}{\sqrt{2}} E_V + \frac{1}{\sqrt{2}} E_H. \tag{2.1}$$

In fact, if the amplitudes of E_v and E_H are both equal to *E*, then by the Pythagorean theorem we will have (see Figure 2.5)

$$E_{45}^2 - \frac{1}{2}E^2 + \frac{1}{2}E^2 - E^2, \tag{2.2}$$

from which it clearly follows that $E_{45} = E$, so that the resulting field has standard amplitude.

CHAPTER TWO

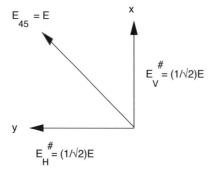

FIGURE 2.5. The analogy of the process of composition of Figures 2.3 and 2.4 is rendered mathematically more precise through standardizing the amplitudes. The figure refers to a point of space and an instant of time in which the three fields reach their maximum level. Moving along the ray, or allowing time to pass, the fields synchronize perfectly in their oscillations, whether decreasing, becoming zero, or negative, and then coming back to the situation represented here.

We can now go on to consider how the picture changes when we take advantage of a further degree of freedom by changing the phase relations between the two fields. As already specified and with reference to Figure 2.2, this involves moving one field along the ray in relation to the other. We begin with the simplest case: we move (see Figure 2.6) E_H one-half wavelength with respect to the position it had in Figure 2.2. The picture clearly shows that this is equivalent to

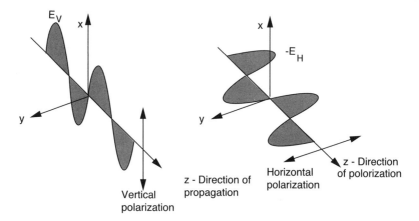

FIGURE 2.6. The analogue of Figure 2.2, in which, however, the relative phases of the two fields have been changed, moving field E_H by one-half a wavelength.

30

changing the sign of field E_H itself (of course, the ray is imagined as infinitely extended).

In this case, the resulting field will be indicated as E_{135}:

$$E_{135} = E_V^\# - E_H^\# = \frac{1}{\sqrt{2}} E_V - \frac{1}{\sqrt{2}} E_H, \qquad (2.3)$$

and becomes a field with a the polarization plane shown in Figure 2.7.

Let us combine now two fields of equal amplitude but out of phase by a quarter wavelength. When one of the two fields reaches it maximum level, the other will be zero, and as the first decreases the second increases, so that the length of the vector sum will remain constant. Nevertheless, it will change its angle and revolve in a circular motion to the right, with the same period as the two fields, along the direction of propagation. In the case when the two fields have different amplitudes, the resulting field, by twisting in a spiral movement in the direction of propagation, will change amplitude from point to point. When seen from the front, it will show an ellipse instead of a circle. Figure 2.8 shows precisely the frontal view of various types of polarization fields obtained by adjusting the amplitude or phase relations of the two fields E_H and E_V.

For our purposes in what follows, it is not necessary to consider the most general case, that of elliptical polarization. On the other hand, it is quite important to clarify that any field we imagine (with standard amplitude, and a precise wavelength and polarization) can be realized in an infinite number of ways as the sum of two fields of standard amplitude appropriately rescaled, characterized by the

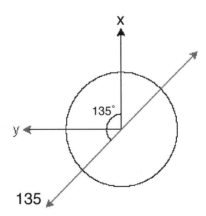

FIGURE 2.7. Subtracting field E_H from field E_V gives us plane-polarized light that oscillates on the plane indicated by 135° in the figure. It should be noted that this plane is perpendicular to the one previously indicated as 45°.

CHAPTER TWO

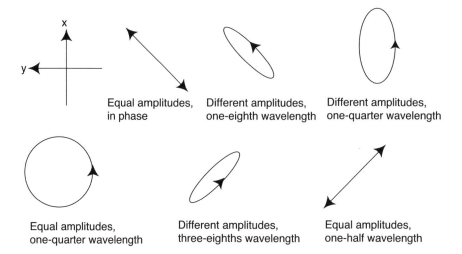

FIGURE 2.8. The various kinds of polarization generated by the sum of the two fields polarized along two perpendicular planes, with variations of their relative amplitude and phase displacement, measured in fractions of the length.

same frequency of resulting field, in phase with them and with a polarization plane along two arbitrary right-angled planes A and B, intersecting on the ray. This merely corresponds to the fact that any vector which lies in a plane can be written as the sum of two vectors on the same plane, perpendicular to each other, and with any arbitrary orientation. We will have, then, as shown in Figure 2.9,

$$E = aE_A + bE_B. \qquad (2.4)$$

Recalling our trigonometry, we can see that the reduction factors of the two standard fields a and b will be equal, respectively, to the cosine and sine of the angle formed between the sum field E and the axis A, an angle indicated by the symbol in the drawing. In other words, the preceding equation can be more appropriately written as

$$E = \cos\Theta\, E_A + \sin\Theta\, E_B. \qquad (2.5)$$

This important mathematical formula will suffice for understanding the majority of the arguments to follow.

(For those who may not be familiar with this argument, a basic review of trigonometric functions is provided in an appendix to this chapter.)

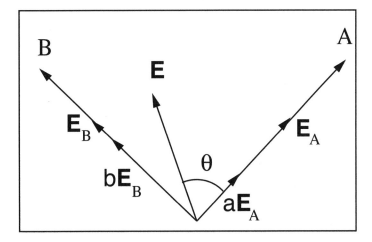

FIGURE 2.9. The decomposition of a plane-polarized field of standard amplitude, as the sum of two fields with the same frequency and with polarization planes along two arbitrary directions at right angles.

. . . . .

With these premises in hand, we are now properly prepared for exploring the actual phenomena of polarization, in terms of classical theory.

2.2. A Home Experiment with Polarized Filters

In this section, we will discuss a series of very simple but instructive experiments with beams of polarized light. The processes we are concerned with here are so simple they can be reproduced at home with a pair of polarized sunglasses, or in the more sophisticated case mentioned by Mermin, with two pairs, which anyone can readily obtain.

A natural light source (the sun, the filament of a lightbulb, a piece of glowing-hot metal) gives off luminous radiations that vary continuously in space and time. In general, a beam of light is neither monochromatic (a prism shows that it contains radiation of various wavelengths) nor polarized. The reason for this will be clear when we consider that a multitude of independent light sources (the atoms and molecules of the light source) work together to produce the light wave. These diverse sources emit radiation without precise relations of frequency, amplitude, or phase. The resulting field varies in a casual manner from one point to another and from one instant to another, and this means, especially, that light like this lacks a definite state of polarization. However, we can polarize the beam of light simply by letting it pass through a polarizing filter, which allows passage only to radiation with a precise polarization. The simplest instrument of this type is a polarizing filter, a sheet of special plastic used in the manufacture of sunglasses, that can quite noticeably reduce the glare from metal surfaces or snow, and prevent temporary blindness.[2]

With reference to Figure 2.10, we can now describe in a rather precise fashion how a beam of light becomes polarized. We can think of a light ray generated by a natural source, propagated in the direction of the axis z of the figure, and we can think of the wave surface at a certain instant, at a certain point of the ray. In this case, the vector E (which, in accordance with our convention, defines, together with the direction of the ray, its polarization plane) can possess any arbitrary orientation at its surface. When the light beam passes the polarized filter it becomes less intense (that is, only a fraction of the light that meets it passes beyond it), and it will now be polarized in the polarization plane of the filter, in whatever direction the filter may be aligned.

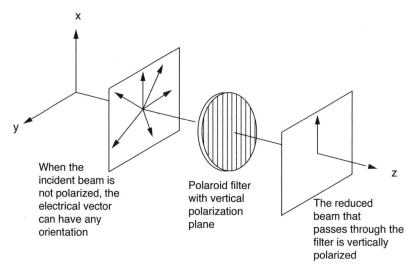

FIGURE 2.10. A beam of natural light (and therefore without any precise plane of polarization) passing through a filter with a vertical polarization plane, is decreased in intensity, and when it exits the lens, becomes vertically polarized.

POLARIZATION OF LIGHT

From now on, we will only consider experiments in which the light has already been polarized: let us say, for example, that it has passed through a polarizing filter with a vertical polarization plane. We then introduce this ray into a second polarizing filter. What happens to the light when, holding the first filter in a fixed position, we rotate the second one, looking through them at a light source? The intensity of light that meets our eyes changes from a maximum brightness to a practically total darkness.[3] The maximum of transmitted light (and hardly any absorption) occurs when the characteristic plane of the second filter coincides with that of the first. The minimum of transmitted light (total darkness), on the other hand, takes place when the two planes are crossed, or perpendicular to each other: that is, when the first filter is oriented to a vertical polarization and the second filter is oriented horizontally. But what about orientations other than these? We already know that the intensity of the light beam is lessened, but how, precisely? The various modalities are well known and can be expressed by the law of Malus, which connects the transmitted luminous intensity[4] I_T to the incident one I_0:

$$I_T = I_0 \cos^2\Theta, \tag{2.6}$$

where Θ represents the angle between the two polarization planes of the two filters.[5] This formula follows immediately from what was discussed in section 2.1 above. In fact, the decomposition of a field of plane polarization (as in Figure 2.9) along two perpendicular axes, combined with the observation that a polarized filter lets pass only what is parallel to the plane of polarization, and that the intensity of light depends on the square of the field, leads directly to the outcome we are now discussing.

Figure 2.11 provides an illustration: the effect obtained when vertically polarized light enters a filter with a plane of polarization that is (a) vertical, (b) horizontal, and (c) at 45°. As implied by the laws of Malus, and by the values of the trigonometric functions given in the appendix in Table 2A.1, we can see that in the first case the light passes without any change, in the second it is completely absorbed, and in the third, it is polarized at 45° and its intensity is reduced by one-half.

The reader should now be in a better position to understand the sense of Mermin's statement quoted at the beginning of the chapter. If the light emanating from a source is seen through two polarized filters, and if the first is held firmly in place, while the second is rotated, when the polarization plane of the latter becomes perpendicular to that of the first, complete darkness results (assuming, of course, that your filters are perfect). Now, take a third lens and hold it between the two (this is why you need

35

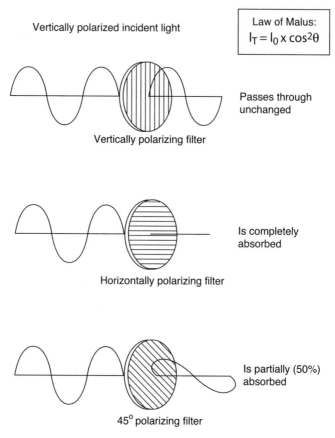

FIGURE 2.11. Behavior of a polarized beam which falls upon a polarizing filter with a polarization plane variously oriented with respect to that of the incident light.

two pair of sunglasses) at a polarization plane of 45° with respect to the first. What happens? As we explained, one-half of the light that emerges from the first lens passes through the second and becomes polarized at 45°. This light now falls onto the third lens. Since its plane of polarization forms an angle of 45° with the polarization plane of the light coming into it, once again, one-half of the light falling upon it will emerge from the third filter and will have a horizontal polarization. We see that there is a loss of one-half of the light when it passes through the second filter, and a similar loss at its passage through the third. With all three filters in play, then, one-fourth of the light emerging from the first will overcome the two successive filtrations, and we will be able to see the light that was invisible when the intermediate filter was out of the way. As Mermin observes, it is surprising to discover that adding an obstacle (something that by itself would absorb light) to the pathway of the light ray in fact favors the penetration of the third lens, which had previously completely blocked the light. It is worthwhile emphasizing that this phenomenon brings a very important characteristic into the open, and that is, that the polarizing filters cannot be considered as purely passive instruments (as ordinary dark-colored glasses are) in regard to the light rays that pass through them; in reality, they play an active role.

2.3. A More Versatile Instrument: Birefringent Crystals

We now consider a series of experiments involving an article not readily found at your local optician's: a birefringent crystal. Such an item is not difficult to produce, and some good optical firms can supply you with one. Like polarizing filters, these crystals owe their properties to the fact that they do not have an isotropic structure. This means that, thanks to their peculiar crystalline structure, they influence the propagation of the light that passes through them, in modalities that depend on the direction and polarization of the light beam. Previously, I did not explain why a polarizing filter only admits light with a certain polarization, nor will I now attempt to explain the specific mechanism which makes these crystals function the way they do. Instead, I will simply limit myself to the description of what happens when polarized light is sent into one of these crystals (properly cut, of course, with respect to its own crystalline structure).

If the incident ray of light is polarized vertically, it pursues its path undisturbed, and preserves its initial polarization. If, on the other hand, it is horizontally polarized, it becomes deflected upward within the crystal, and emerges, changing its direction once again, so that it is parallel to the incident ray, but displaced from that ray by a degree that depends on the length of the crystal. In this as well, the beam of light preserves its horizontal polarization. The situation is illustrated in Figure 2.12.

Now we can ask ourselves, what happens when a light ray strikes a crystal like this with a polarization plane neither vertical nor horizontal but of, say, 45°? The answer is simple and predictable: the ray divides into two rays (in the present case, the two new rays will be equal in intensity to each other and each exactly one-half of the intensity of the incident ray).

CHAPTER TWO

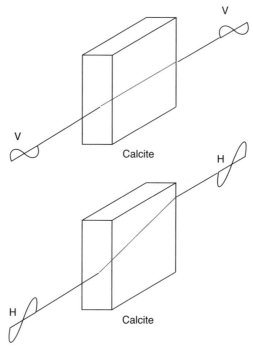

FIGURE 2.12. A birefringent crystal, such as calcite, differently transmits a light ray that falls on it, according to the polarization of the incident light. In particular, if the crystal has been properly cut and is held in the right position, it transmits a vertically polarized light beam without change, but a horizontally polarized ray will be displaced by an amount that depends on the length of the crystal.

Furthermore, one of them will be polarized vertically (the one that has not been deflected, known as the ordinary ray), and the other horizontally (the deflected one, called the extraordinary ray). It is clear that here too, the crystal is acting on the light actively, and not passively. It should also be clear that this experiment has an advantage over the simpler one with the polarized filters: instead of permitting only one "component" of the light ray, with the proper polarization, to pass through, while absorbing the rest, these crystals preserve all the light and produce two beams with two polarizations, reduced proportionally to the vertical and horizontal components of the incident field. This is why such crystals are useful tools for showing the composition of the incident ray, with regard to its polarization. In the typical case, we will have a situation like that illustrated in Figure 2.13.

POLARIZATION OF LIGHT

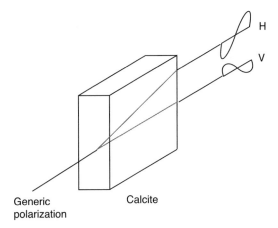

FIGURE 2.13. In the general case, when the incident ray presents neither vertical nor horizontal polarization, by passing through a birefringent crystal, it becomes divided into two distinct rays characterized by precise polarizations.

The discussion in this section refers to cases when the light beams are intense enough to justify using the classical point of view, which conceives radiation as distributed continuously in space. The modifications one needs to bring to this picture, when one takes into account the quantum aspects of radiation sketched out in chapter one, will be analyzed in the following chapter, and will allow us to understand certain conceptually crucial points of the new theoretical scheme. Before moving on to that analysis, I would like to recapitulate the essential formal elements of this chapter, to provide a point of reference for subsequent discussions.

. . . . .

2.4. THE BEHAVIOR OF FILTERS AND CRYSTALS: A SYNTHESIS

In this last section I bring together the principal aspects of the experiments in which a beam of light polarized at a certain plane enters a polarized filter L or a birefringent crystal B. The resulting processes are absolutely clear, as long as we take account of the modalities of decomposition of plane-polarized beams, discussed in Section 2.1. In fact, it should suffice to recall that a polarizing filter only lets pass the light polarized along its own characteristic plane and that a birefringent crystal transmits light unchanged when it is polarized along the plane corre-

CHAPTER TWO

sponding to the ordinary ray, and transmits it laterally displaced, when the light is polarized at a right angle to that plane. Thus it is enough to decompose the incident beam into the sum of the two beams, with the specified planes of polarization, to grasp the process clearly. In Figure 2.14, we have indicated (with P) the polarization plane of the incident beam of light, and have assumed that this forms an angle with the plane Q, characteristic of the polarizing filter L. Analogously, we have indicated with O and E (for "ordinary" and "extraordinary") the two characteristic planes, at right angles to each other, of the birefringent crystal B, and have used ϕ to designate the angle between the plane P of polarization of the incident light and the plane O of the crystal.

The classical phenomenology of the processes at work can be described as follows:

- When the beam of light passes through the filter L, its intensity is reduced; only the fraction $\cos^2\Theta$ of the incident light passes the filter, and at its exit, is polarized in the same plane as the filter.
- When the beam passes the birefringent crystal, it is subdivided into two beams of reduced, and, generally, different intensities. From the crystal emerges a so-called ordinary ray, which is not displaced with respect to the incident ray, and which bears the fraction $\cos^2\phi$ of the incident light, and is polarized in the plane O. A second ray emerges from the crystal, the so-

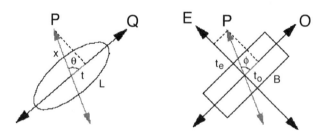

FIGURE 2.14. A beam of polarized light in plane P (figure to the left) enters a filter polarized at plane Q. The fraction of light passing through it will then be equal to the square of the relationship between the segment t and length x ($=\cos^2\Theta$). The same beam enters a birefringent crystal (figure to the right), with "ordinary" and "extraordinary" axes O and E, respectively. A fraction of light, equal to the square of the relationship between length t_O and length x ($=\cos^2\phi$), follows a path in the direction of incidence and becomes polarized in plane O, while a fraction equal to the square of the relationship between length t_E and length x ($=\sin^2\phi$) is laterally displaced with respect to the incident direction and becomes polarized in plane E.

40

POLARIZATION OF LIGHT

called extraordinary ray, parallel to but displaced with respect to the incident beam. Its intensity is equal to the fraction $1 - \cos^2\phi = \sin^2\phi$ of the intensity of the incident beam,[6] and will be polarized in the plane E.

Before concluding this chapter it will be a good idea to display the equations of section 2.1, which associate the fields E_{45} and E_{135} with the fields E_V and E_H:

$$E_{45} = \frac{1}{\sqrt{2}}E_V + \frac{1}{\sqrt{2}}E_H,$$
$$E_{135} = \frac{1}{\sqrt{2}}E_V - \frac{1}{\sqrt{2}}E_H,$$
(2.7)

and their inverse equations:

$$E_V = \frac{1}{\sqrt{2}}E_{45} + \frac{1}{\sqrt{2}}E_{135},$$
$$E_H = \frac{1}{\sqrt{2}}E_{45} - \frac{1}{\sqrt{2}}E_{135}.$$
(2.8)

. . . . .

Although this chapter has probably been tedious, the above treatment contains practically all the formal elements necessary for understanding all that follows.

APPENDIX 2A: TRIGONOMETRIC FUNCTIONS

In this appendix I would like to complete the review of essentials for readers who may have forgotten some of their mathematical training, by recalling the simple definition of the trigonometric functions sine and cosine, indicated by the symbols "sin" and "cos" respectively. In a circle (see Fig. 2A.1) a radius is inscribed which forms an angle Θ with the horizontal, and we consider the two segments OA and AB. The ratios between the measures of these segments and the radius OB (that is, OA/OB and AB/OB) represent the sine and cosine, respectively, of the angle in question. In the discussion to follow, we will be making reference almost exclusively to a case in which the angle between the two planes of polarization assumes one of the following values: 0°, 45°, 60°, 90°, 120°, 135°. The corresponding values of the sine and cosine are indicated in Table 2A.1, and can be easily derived with reference to the geometrical analysis of simple triangles.

CHAPTER TWO

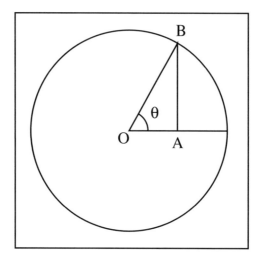

FIGURE 2A.1. The values of the trigonometric functions sinΘ and cosΘ relative to the angle Θ indicated in the figure, are given simply as the relations between the lengths of the segments OA/OB and AB/OB, respectively.

TABLE 2A.1.
Values of the trigonometric sine and cosine functions for certain selected values of the angle that will be useful in what follows.

Angle θ (deg)	Sine	Cosine
0	0	1
45	$1/\sqrt{2}$ (or 0.7071)	$1/\sqrt{2}$ (or 0.7071)
60	$(\sqrt{3})/2$ (or 0.8660)	1/2 (or 0.5)
90	1	0
120	$(\sqrt{3})/2$ (or 0.8660)	−1/2 (or −0.5)
135	$1/\sqrt{2}$ (or 0.7071)	$-1/\sqrt{2}$ (or −0.7071)

CHAPTER THREE

Quanta, Chance Events, and Indeterminism

> This discovery [i.e., the quantum theory]
> set science a new task: that of finding a
> new conceptual basis for all of physics.
> —*Albert Einstein*

IN THE PRECEDING CHAPTER, wave processes were analyzed from a classical perspective. We now need to take account of the revolutionary conclusions achieved by the "heroic age" of quantum theory. The analysis of light polarization phenomena just concluded will now be reconsidered from the perspective of the new understanding of electromagnetic phenomena; in particular, we need to take account of the particle nature of radiation, since, as explained in the first chapter, radiation is now conceived as a stream of particles, known as photons. This new way of looking at the problem will bring us to conclusions of major conceptual significance. Furthermore, if we recall that the new conception of physical processes puts waves and particles at the same level, the conclusions we obtain from the analysis of the first sections of this chapter can be applied immediately to processes involving *material* particles. Once these points have been treated, we will move to a brief description of some of these processes and show how the phenomenology of photons finds its exact parallel in particles. Of course, this will also require introducing the procedures and apparatus in the realm of material particles that correspond to polarization processes and birefringent crystals in the realm of light phenomena. We will then be in a position to consider the most significant conceptual implications of the theory: indeterminism and the irreducibly probabilistic prediction of future events.

3.1. THE PHENOMENOLOGY OF LIGHT QUANTA

The first question to ask concerns the implications of the particle character of light for the processes discussed in the foregoing chapter. We can begin with the experiments involving polarized light. Let us imagine a monochromatic beam of light (that is, light characterized by a precise frequency ν) with a vertical polarization plane. In section 2.1 we saw how if a beam of light like this impinges onto a filter with a different polarization plane, the intensity of light (the energy, that is) that passes through the second filter will be reduced when compared with the original, incident light. Now let us suppose, for simplicity's sake, that the electromagnetic field has an extremely weak intensity, so that the energy it carries into the filter during one second is equivalent to a single quantum, that is, hν. In other words (taking the next logical step and moving into the language of quanta) we suppose that a single polarized photon enters the polarizing filter each second.[1] Since we know that the energy emerging from the filter is reduced (according to the law of Malus) by the factor $\cos^2\Theta$ with respect to the incident light, we must ask ourselves, "What happens to a single photon entering the filter?" Should we think, perhaps, that the reduction of the beam's intensity is due to the fact that every photon has been somehow "divided" into two parts, and that only one of these "parts" has overcome the barrier? The precise answer to the question is "No," and it follows directly from the analyses of Planck and Einstein: electromagnetic radiation is quantized, and for a given frequency ν, it cannot help but be transported in granules, each of the value hν. Since passage through the filter does not change the frequency of the radiation, the sole possible way to account for the reduction in the light beam's intensity is to assume that not all the photons manage to pass through the filter, and that the number of those that do emerge are reduced exactly by the proportion required by the law of Malus.

For illustration, we consider a case where the vertically polarized light beam passes through a polarized filter whose polarization plane is, by turns, vertical, horizontal, and 45°. In Figure 3.1 we will now have the analogue of Figure 2.11, but this time, by depicting the single photons, we show the particle structure. When the filter is oriented vertically, all the photons pass through it; when oriented horizontally, none pass through it; at 45° (which, we know, reduces the intensity by one-half) every other photon passes through, on the average.

Two observations need to be made at this point. The first refers to the precise significance of the assertion, "a photon passes through the filter." What exactly do we mean by this expression? What specific, physical event does it describe? The question touches one of the crucial points of the theory to which we will return so often in the course of the following discussion: namely, the way in which a genuinely microscopic process has to be amplified to the macroscopic level to become perceptible. For the moment, not wishing to get involved in a problem that later chapters will

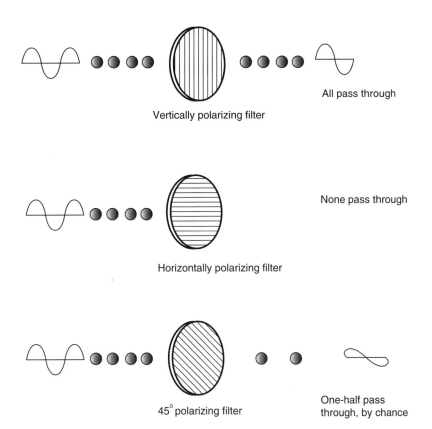

FIGURE 3.1. Passage through a polarizing filter by the quanta of the electromagnetic field, or photons. The figure represents the correct way to visualize what was illustrated in Figure 2.11, once we understand that the energy of the electromagnetic wave is not distributed continuously in space but is concentrated into quanta. It should be noted that the lessening of the beam's intensity is not to be understood as a reduction of the energy transported by each single photon, but as a diminution of the number of quanta (each one of which carries the same amount of energy, hν). The only characteristic of the photon modified by the filter is its plane of polarization, which, as it exits, is rotated so as to coincide with the polarization plane of the filter.

address, we can limit ourselves to saying that the technology does exist that can register the arrival of single photons. An apparatus can make a single photon set off an avalanche of photons, which causes the emission of a sound (as in a Geiger counter), or the movement of a needle on a gauge. Alternatively, we can suppose that highly sensitive photographic film has been placed behind the second filter, which can register the arrival of a single photon with a tiny dot. At any rate, we can imagine, as in the figure, that behind the polarizing lens exists some suitable means of detection (not pictured here, but shown explicitly in later chapters). The statement, "a photon has passed through the filter" really means something like "the apparatus has registered an event" or "the detector just clicked."

The second refinement we need to make is relevant not just technically but also conceptually. It will be observed in the illustration that, when there is a case of some photons passing through and others not passing through, the wording is "by chance, one-half pass through the filter." What is the precise significance of this expression, "by chance"? "By chance" is not just there "by chance"! I say it to underline one of the most important aspects of the new theory: the fundamentally *random* nature of microscopic processes. In our example we have a sequence of photons traveling toward the polarized filter. We know that some of these will meet a "happy fate," that is, that they will overcome the obstacle placed in their path, and that others will be absorbed, and disappear. The question arises, whether the passing-through or not-passing-through of a photon depends on some special characteristic of that photon? Are the photons that pass already somehow different from the ones that do not? Well, a cardinal point that the new formalism compels us to accept is that there is no, absolutely no, difference between one photon and the next. In other words, not only is it practically impossible to distinguish a priori the photons that will pass from those that will not, but, if it is accepted that the formal quantum description is correct and complete (this is a point to which we will return more than once) it is quite incorrect to think that the photons possess—even in some way unknown to us—some special characteristic before they enter the filter that determines whether or not they will pass through it. This point, as already hinted, will prove to be of great conceptual importance and will be discussed in detail in section 3.6 below.

A similar analysis can be repeated for the case when polarized light is sent with components on both the characteristic planes, on a birefringent crystal (compare Figures 2.12 and 2.13). In this case, too (Figure 3.2), the

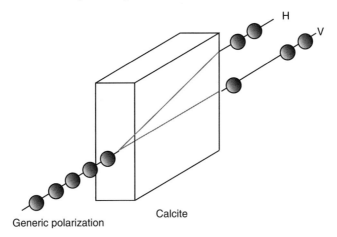

FIGURE 3.2. When a polarized beam of photons falls upon a birefringent crystal, every photon, behaving as an indivisible unity, must wind up activating a register, either on the path of the ordinary ray or on the path of the extraordinary. The choice will take place according to the laws of probability, which guarantee the appropriate intensity for each ray.

beam of light is constituted of individual photons, and, once again, a photon cannot be divided into two or three parts. The experiment at the level of classical physics shows two separate and differently polarized beams emerging from the crystal. This would now be reformulated to say that every photon will activate one of the two detectors, placed on the paths of the ordinary and extraordinary rays, respectively. Furthermore, since the photons at their exit must reproduce the intensity of the two beams, these will activate the sensing device placed at the ordinary ray $\cos^2\phi$ of the times, and will activate the other one (at the extraordinary ray), the remainder of the times (see Figure 2.14). Once again, if the theory is true, and its description of the process exhaustive, we are not permitted to think that the photons already have some distinguishing characteristic before they enter the crystal, that determines their fate once they meet it.

From these preliminary considerations we can already discern the fundamentally random nature of the processes according to the new theory. As a natural corollary of this inescapable randomness of events, it follows that the theory will be instrinsically probabilistic. And that also means that, unlike the situation in classical physics, we will not in general be able to predict the outcome of a process: we have to be content with knowing the probabilities of the various alternative outcomes.

CHAPTER THREE

3.2. Photons in Processes of Diffraction and Interference

Clearly, the analysis just now sketched can be repeated, step by step, for the processes of interference and diffraction, while keeping the perspective of the corpuscular conception of light. For these processes, too, we can consider the case when the intensity of the light beam is so low that only one photon per second passes through the opening (as in the second diffraction experiment, shown in Figure 1.8). Once again we can ask ourselves, "What happens to the screen when a single photon passes through?" Is it perhaps that a bell-shaped figure like the one represented is formed: very faint (because of the low energy involved), but gradually becoming more pronounced as more photons are added? Again, the reply given by the theory (which perfectly confirms our experience in other respects as well)—is in the negative. A single photon could not be dispersed over an extended region of the screen but must end up in a precise point. The bell shape to the right side of the figure must then become constituted over time, as more and more photons are added. Each one gets "assigned" to a precise place within the system, and the photons "select" the point where they will end up, in such a way that more photons go where the density of energy on the screen is greater. Once again, initially there is nothing, no conceivable characteristic, existing previous to the experiment that would determine which photons proceed unimpeded, and which end up on the margin of the diffraction image.

Our final step is to use the same perspective to analyze the interference experiment of Figure 1.9. In this case as well, the photons do not produce, so to speak, a "pale reflection" of the interference pattern. Every photon goes to a precise point of the screen. However, the probabilistic laws which guide the photons make them end up predominantly in those places where the intensity of light is high, and never end up in the places where the interference is destructive, because the electrical field is null. We need not linger on a discussion of further analogies to clarify the behavior of single photons, since by now, every reader can see the rules of the game. In the next sections we will move on to consider material particles instead of photons, especially in order to illustrate the analogy between the experiment just now described (the progressive formation of a figure of interference through the accumulation of individuals), and another experiment with electron interference. The development of the process will confirm the hypothesis of de Broglie (that particles present wave aspects), but will also reveal that this wavelike nature does not require them to be diffused in space the way waves normally are, since in every experiment undertaken to identify their positions (such as photographic detection on a screen), they stand revealed as fundamentally pointlike objects, each with a precise localization.

3.3. From Light Quanta to Material Particles

We now consider processes which, instead of photons, involve elementary particles such as electrons, protons, neutrons, atoms, and so forth. And now I will follow a reverse order from the one I followed before, beginning with experiments of diffraction and interference, and afterward moving to experiments comparable with polarization phenomena. The reason for this is simple. The hypothesis of de Broglie, discussed in chapter 1, was specifically verified for electrons by G. P. Thompson and others, for neutrons by Enrico Fermi and Leona Marshall, and later for a variety of other microscopic systems by many other researchers. According to this hypothesis, elementary particles are associated with waves, and when given the appropriate experimental conditions, can produce diffraction and interference phenomena. The experimental requirements for demonstrating these "wave" aspects of "particle" phenomena do not differ at all from those discussed with reference to beams of photons. It is simply a matter of preparing streams of particles with precise velocities (according to de Broglie, one particle of a given mass has a precise wavelength if and only if it possesses a precise velocity), then passing them through suitable openings or slits of a size comparable to the wavelengths. We then study their final positions, replacing the screen of the earlier figures with mechanisms made to reveal the particles as they arrive at various places. It is not necessary to repeat the analyses as before. Everything happens exactly as with the photons.

The reason for commenting explicitly on these experiments is that they offer an opportunity—thanks especially to the splendid illustrations of experiments made by A. Tonomura and others—to show by a time lapse series of photographs, how single electrons end up in places apparently selected at random, and how it is only the accumulation of the processes caused by the successive impacts of many electrons that finally causes the interference pattern to emerge (see Figure 3.3).

A few observations are in order. Above all, it is important to stress that the incident beam is weak enough for every single electron to have ample

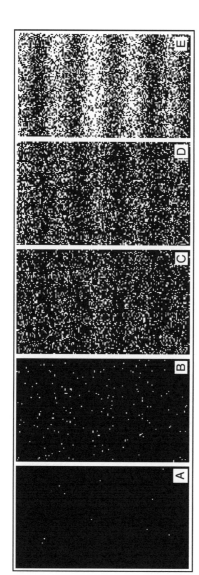

FIGURE 3.3. The emergence of a pattern of interference on a photographic film placed behind a screen with two slits, and bombarded with a stream of electrons. Picture (a) shows the appearance of 10 electrons; pictures (b) to (e) show 100, 3,000, 20,000, and 70,000, respectively.

time to pass through the apparatus and come to rest on the film, before the next electron enters. This shows clearly how embarrassing the double, wave-particle aspect of physical processes can be. For the electron, which is difficult for us to imagine *not* passing by one of the two apertures in Figure 1.9, seems always to "know" whether or not the other passage is available. If only one of them were open, we would not see interference but only diffraction, and consequently a slightly enlarged image of the opening. Similarly, if the other passage were the only one open, there would be another, single bell shape of diffraction. If the electrons did not behave like waves (and observe carefully that every single electron should behave so, since there is only one electron in the apparatus at a time), there would be two separate bell shapes, and not the series of interference bands so typical of the process. This capacity for self-interference on the part of an object, so that every time we try to reveal its presence it appears to us as a particle, constitutes the first clear indication of the revolutionary character of the theory. As Einstein aptly emphasized, *it requires finding a new conceptual basis for physics as a whole*. However, as we shall shortly see, the surprises of the microcosm do not end here. Among other things, quantum systems will show themselves capable of further, yet more disturbing processes.

3.4. Spin: The Analogue of Polarization for Material Particles

As anticipated, some material particles possess properties that are analogous to the polarized states of photons. These properties are connected with a characteristic called spin, which has no classical analogue. To understand what spin involves, we need to begin by realizing that atomic systems in general possess what technically can be defined as an "angular momentum." This expression is used to assert the existence of a privileged axis around which the system rotates, something like a spinning top. A characteristic of the momentum of quantities of motion (following Anglo-Saxon usage, we may call it angular momentum) consists in its ability to be quantized, both with respect to its magnitude,[2] and with respect to its possible spatial orientations. Taking up an analogy with the classical system, this would be equivalent to asserting, for example, that a top could not turn on its axis with just any arbitrary angular velocity, but only with a velocity corresponding to typical quantized values. In fact, spatial rotation can give rise, in quantum terms, only to angular mo-

CHAPTER THREE

menta whose absolute value L will be proportional to the quantity h/2π (with h as the ever-present Planck's constant), according to the formula

$$L = \sqrt{\ell(\ell+1)}\,\frac{h}{2\pi}, \text{with } \ell = 0, 1, 2, \ldots \qquad (3.1)$$

But this is not all. It is not only a question of the spin's magnitude, since the orientation of the angular momentum in a certain direction also becomes quantized.[3] In classical terms it would mean saying that a top could not rotate so that its axis formed just any arbitrary angle with the vertical, but that this angle could assume only certain fixed values. To be precise, when L assumes the value which corresponds, in the equation just given, to a specifically chosen positive integer ℓ, the spinning top would be found in only one of the $(2\ell + 1)$ allowed orientations with respect to the vertical. This is because the component of L in that direction has the value m(h/2π), where m can assume any integer value (positive, negative, or null) from $-\ell$ to ℓ. Once again, we cannot avoid being surprised at this: the familiar movement of a spinning top (gradually augmenting the angle taken by its axis with respect to the vertical, until it falls to the ground), when described in quantum terms, behaves in such a way that its axis makes discontinuous leaps, for example from the value m = ℓ (in units of h/2π), to the values $\ell - 1$, $\ell - 2$, and so on, until the value of m = 0 is attained. In Figure 3.4 are indicated the five possible orientations with respect to the vertical axis of an angular momentum for which $\ell = 2$.

We are now in a position to discuss "spin." This term refers to the fact that some elementary particles in spite of the fact they are punctiform, or pointlike, are comparable to microscopic spinning tops. The hypothesis that electrons have spin was originally formulated in 1925 by Samuel Goudsmit and George Uhlenbeck, when they were trying to account for some features of the atomic spectral lines left unexplained by Bohr's model. Their hypothesis was later proven by many experiments. Spin is analogous to angular momentum, with a few significant differences. Most importantly, it can assume more values than angular momentum. The formula that expresses the magnitude of the spin is formally the same as before. Nevertheless, in the case of spin, the analogue of the quantity *l* (which we will now indicate by s) can assume, in addition to the values 0, 1, 2, . . . also the positive half-integers, so that s = 1/2, 3/2, 5/2, . . . and so on. Another important difference derives from the fact that the absolute value of the spin is an individual characteristic of any particle, and as invariable as its mass or charge. In other words, while an

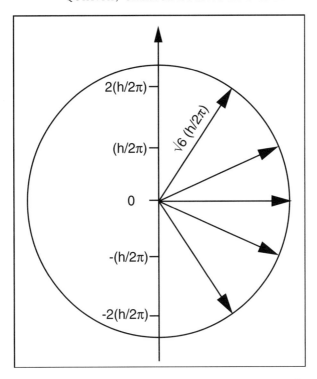

FIGURE 3.4. The five (and only five) possible orientations with respect to the vertical axis of an angular momentum, that is, of the axis of a spinning top, when $\ell = 2$.

electron can revolve around the nucleus with one of the values permitted by the formula above (3.1) for its angular momentum (with various angular frequencies of arbitrarily large value), it "rotates on itself" in only a precise way—a way that corresponds to a specific property, invariable for the type of particle it is. In fact, all electrons, protons, and neutrons invariably have s = 1/2, which means that the magnitude of their spin momentum is $S = \sqrt{s(s+1)}\,(h/2\pi) = \sqrt{3/4}\,(h/2\pi)$ and implies that they can be oriented in a certain direction in only two modes, corresponding to the projections of $+(h/2\pi)/2$ and $-(h/2\pi)/2$, respectively. Therefore we would say that one particle of this type can only have spin "upward" or "downward" in the given direction. In fact, because the magnitude of the spin (in the customary units) is equal to $\sqrt{3/4}$, while its projections turn out to have only the values +1/2 and −1/2, the spin is never perfectly aligned in any direction. In Figure 3.5, which is the exact analogue, in the case of spin 1/2, of Figure 3.4, we have more graphically depicted the

53

CHAPTER THREE

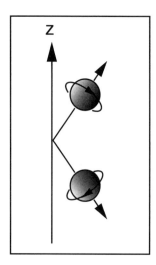

FIGURE 3.5. The two possible modes in which the spin of an electron can be oriented with respect to a certain physical direction, which in this case has been conventionally identified with the vertical, and with the axis z of some system of reference. These correspond to projections on the axes equal, respectively, to + 1/2 and –1/2, in units of h/2π.

possible orientations of the tiny top, represented by a spinning electron, with respect to a precise, physically specified direction.[4]

For our analysis something else becomes relevant, too: any angular momentum is associated with a certain *magnetic* momentum. This means that the tiny spinning tops that interest us at the moment also behave like tiny compass needles as they interact with a magnetic field. Furthermore, in each atom there are diverse angular momenta in consequence of the orbital movements of the electrons, and their states of spin combine with each other into a vector sum, according to the parallelogram rule. Now the case that interests us is that of an atom, let us say, of silver, in which all the angular momenta, except for the momentum of the spin of a single electron, combine in such a way that their sum is null. Thus the atom possesses an angular momentum (and the related magnetic momentum) equal to the spin of the one electron that remains by itself on the outermost shell. This system is particularly interesting in that it is electrically neutral, but behaves like the tiny needle of a compass, which, thanks to quantization, can be oriented only upward or downward along some physically identified direction.

The analysis I have just concluded will probably seem obscure. In any event, for future reference, the really important point is that there are microscopic systems (particles or atoms) which behave like miniature spinning tops and like miniature compass needles, but present quantum phenomena. This means that, with respect to a precise direction, they can only be oriented up or down. Formally, the states "up" and "down" with respect to a certain axis (such, for instance, as the vertical axis) are equivalent to the vertical and horizontal states of polarization of a photon. In a similar way, the states in which spin is aligned upward or downward with respect to some other direction are the analogues of other states of plane polarization of the quanta of the electromagnetic field.

3.5. The Stern-Gerlach Apparatus: The Analogue of Birefringent Crystals

Taking the example now of an electrically neutral material body (such as a neutron, or whole atom, of silver) with spin 1/2, we know, as explained above, that it could have spin only "up" or "down" along a given (oriented) direction. We also need to keep in mind that in close association with this spin is a tiny magnetic needle, lined up with the axis of rotation. Let us now introduce this particle into a region with a vertically oriented magnetic field with an intensity that varies (say, by increasing) the more it moves upward. A field of this kind can be created by using a specially shaped magnet, as shown in Figure 3.6, named after the physicists Otto Stern and Walter Gerlach, the first to suggest experiments of the type we are about to describe.

Suppose we introduce a neutral particle, with a spin of 1/2, into a field of this kind. Since the vertical direction is privileged, we already know that the particle will be able to orient itself in only one of the two modes possible for it, that is to say, upward or downward. If the former occurs, we will denote its state as |z up>, if the latter, as |z down>.

But first, a brief digression. In our discussion of spin, we have started using the notation invented by Dirac. This notation is particularly convenient for showing the possible states of a physically quantized system. When the state corresponds to a definite property, the symbol |...> will be used, which means "the state vector of the system," with a letter or brief phrase (where the dots are) to signify the property possessed by the system. Accordingly, with this notation, which we will use systematically from now on, |V> or |H> will denote the states corresponding to the

CHAPTER THREE

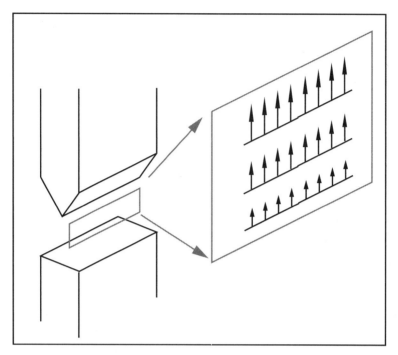

FIGURE 3.6. By carefully adjusting the position of a magnet (left) we can create a magnetic field which increases by moving upward from below. To the right are represented the values of the magnetic field at three different levels, aligned on the vertical plane that passes through the angular edge of the upper magnet.

property that characterizes photons as capable of "passing a test" of vertical or horizontal polarization, respectively.

We already know that the state |z up> corresponds to a particle with an upward spin, with the north pole of its associated magnetic needle pointing up and its south pole pointing down.[5] The Stern-Gerlach magnetic field acts on both poles, but since the field is more intense in its upper region, the force that the magnet exercises on the north pole of the particle is greater than the force exercised on the south pole. Consequently, the net force on the particle is directed upward. If we introduce a particle of the specified state into the magnetic field, the particle will be deflected upward along a fairly precise trajectory (see the discussion of the uncertainty principle, below in Section 3.7, to understand why we need this qualification). In the case of the state |z down>, exactly the opposite will happen. The situation is illustrated in Figure 3.7. The parallel with the

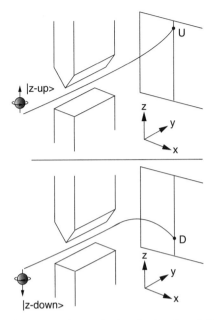

FIGURE 3.7. As a photon with vertical polarization follows the ordinary ray when it encounters a birefringent crystal, while one with horizontal polarization follows the extraordinary ray (displaced with respect to the ordinary), in the same way an antineutron with upward spin along axis z, when encountering a Stern-Gerlach magnet with vertical orientation, undergoes an upward deflection, while one with a downward spin along the same axis is deflected downward.

case of the birefringent crystal (Fig. 2.12) will come immediately to mind, when light was introduced into the crystal that was polarized along one of its two privileged planes.

Like the polarization states of the photons, the states of a particle's spin can be combined by multiplying, adding, and subtracting (the following chapter will be entirely dedicated to illustrating the implications of this facet of the theory). For, just as the linear combination of states |V> and |H> (see the concluding formulas of the last chapter) that we have labeled |45>:

$$|45> = \frac{1}{\sqrt{2}}|V> + \frac{1}{\sqrt{2}}|H> \qquad (3.2)$$

corresponds to a simple incident state, i.e., that of a photon which has overcome a polarization test at 45°, just so, the analogous combination

CHAPTER THREE

of states |z up> and |z down> corresponds to a particle with a precise spin property; the corresponding state will be indicated as |x up>, to express the fact that its axis of rotation points in direction x:

$$|x\ up\rangle = \frac{1}{\sqrt{2}}|z\ up\rangle + \frac{1}{\sqrt{2}}|z\ down\rangle. \quad (3.3)$$

In the same way,

$$|x\ down\rangle = \frac{1}{\sqrt{2}}|z\ up\rangle - \frac{1}{\sqrt{2}}|z\ down\rangle, \quad (3.4)$$

which is to say that the analogue of the state |135>, of the preceding chapter, describes a particle with spin oriented toward the negative of the x axis.

With reference to these states, then, we can pose the same questions we posed at the beginning of the chapter, about the photons polarized at 45° or at 135°, before they entered the birefringent crystal. What happens when a single particle, without spin upward or downward with respect to the z axis, enters a Stern-Gerlach apparatus with a vertical orientation? And how, in particular, will a particle in the state |x up> or |x down> behave, when it undergoes this test? Once again, since we are considering microscopic systems, this question might be more properly formulated if we had in mind an apparatus predisposed to reveal the position at which the particle strikes a screen placed behind the magnet. We already know from the previous experiments that when the state is |z up> (or |z down>) the trajectory intersects the screen at point U (for "up") or D (for "down") in Figure 3.7. As a consequence of this phenomenon—the quantization of the spatial orientation of spin momentum—the (predictable) answer to the question regarding the particle in the state of |x up> is absolutely identical to the answer given for the experiment with photons: one-half of the incident particles will end up at point U which corresponds to the "upward trajectory," and one-half will finish at point D, in a completely random manner (see Figure 3.8). To understand the peculiarity of quantization, it is illuminating to contrast the classical situation. In classical terms, a magnetic needle can have any orientation whatever with the vertical. Consequently, the difference between the forces exercised on its south pole and on its north pole varies in a *continuous* relationship with the variation of the needle's orientation.[6] This force is at its maximum, and is directed upward when the needle is oriented upward, then it decreases to zero when the needle is horizontal (the two forces being equal) and is at maximum and directed downward

QUANTA, CHANCE EVENTS AND INDETERMINISM

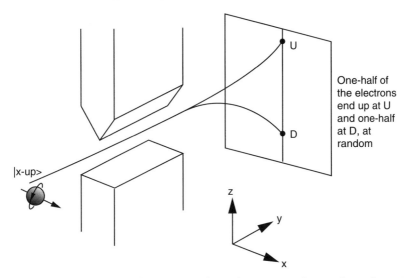

FIGURE 3.8. In the case where a particle with an upward spin along the axis x is sent through a magnet oriented toward z, the particle will terminate by chance (but with equal probability) in point U or point D. The perfect analogy with the case of Figure 3.2 should be noted, which describes the passage of a photon with polarization at 45° entering a birefringent crystal.

when the needle is oriented downward. Therefore, the "classical" outcome of an experiment of the type now considered would lead us to a continuous distribution (in the vertical direction) of the trajectories of the particle with respect to its original direction of propagation, and the screen would register particles at all the points between U and D. The fact that all the particles end up at the two points only is incomprehensible at the classical level and contradicts all the principles of dynamics that govern such a system.

It is also worthwhile to observe that, according to quantum mechanics, if a particle in the same state |x up> were to be introduced into a Stern-Gerlach magnet with an orientation along axis x, it would certainly be deflected into the positive direction of this axis (see Figure 3.9).

At this point, the formal part of the text is completed. All the structure of the theory, its profound conceptual and epistemological implications, can, it seems to me, be fully and correctly understood by anyone who has been able to follow the behavior of the photons and particles discussed in this chapter. We can therefore pass immediately to an analysis of the first relevant conceptual implications of quantum formalism.

59

CHAPTER THREE

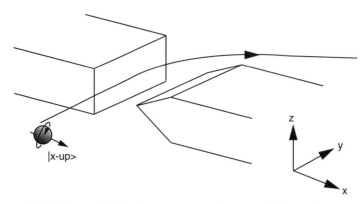

FIGURE 3.9. The particle in the same state |x up> of that of Figure 3.8, if it is brought into the gap of a magnet oriented toward the positive of axis x, is certainly deflected upward with respect to this axis.

3.6. The Fundamentally Random Character of Physical Processes

As we have emphasized more than once, the outcomes of measurements on quantum systems are, in general, genuinely random. This important aspect of quantum formalism needs to be brought out more clearly at this point. Clearly, an assertion concerning the randomness or stochasticity of a series of events has a rigorous meaning only if it has been formulated in a precisely defined context.[7] We can return to our example of the photon polarized to 45° and entering a filter with a vertical polarization plane: that the event "photon overcoming the obstacle" is genuinely random, that it has the same probability of occurring as of not occurring, and that these two alternatives present themselves in a chance distribution, are assertions that require a very large number of tests before they can be made. It would have to happen that in one-half of the cases, the photon effectively overcame the filter, and in the other one-half of the cases did not. No regularity could be identified in the succession of the events. Finally, there would have to be no known way to prepare a photon to be in such a state as to ensure that it would be able to pass with certainty both a test of vertical polarization and one at 45° polarization. But even if these conditions could be verified, the genuinely "random" nature of the process is still not proved. It is only in the precise conceptual context of the theory that the question can be correctly formulated, and a nonambiguous answer be given to the question.[8]

As a matter of fact, the theory asserts that the state which ought to be used to describe a photon of the type in question turns out to be precisely that which we indicated as |45>, and that for the given system, no further specification is logically possible. The genuinely random nature of quantum processes is therefore implied—even apart from experimental verifications—by the assumption that, as said in technical jargon, the theory is *complete*, an expression which asserts that the theoretical description is exhaustive, meaning that the specification of the state vector represents the most complete information that can be obtained (in principle, that is, and not only in practice) of a physical system. The rules of the theory tell us that once this state has been specified, it is possible to evaluate the probability that the photon will overcome *any test whatsoever* of polarization, but also that *nothing more than this* can ever be known to us. In particular, the relationship analogous to those of section 2.4 above, which expresses the state of polarization in plane P that forms an angle Θ with plane Q and the complementary angle $(90° - \Theta)$ with plane R at right angles to Q, in Dirac's notation for state vectors, would be

$$|P> = \cos\Theta \, |Q> + \sin\Theta \, |R>, \qquad (3.5)$$

and gives us precise, probabilistic information about the outcomes of a process studied to reveal if the photon would pass a test of polarization along Q—that is, that it has a probability $\cos^2\Theta$ of overcoming it and $\sin^2\Theta$ of not overcoming it. On the other hand, the theory also asserts that the knowledge of this probability of the outcomes constitutes the only thing we are able to know about the process.

Two observations, then, are in order. First, as observed in the previous chapter (section 2.4), since any state whatsoever of plane polarization can be decomposed into the sum of two states of plane polarization in two arbitrary right-angle planes (of course, once the state of the photon has been assigned—that is to say, its polarization plane P—the angle Θ of the above equation varies with the variation of Q), the theory furnishes us with the probability of the outcomes of any measure of plane polarization.[9] Secondly, in some cases the probability of an event can assume the value 1 or the value 0, which is to say the event itself can become "certain" or "impossible," even within the genuinely probabilistic context of formalism. As we have already pointed out, these two situations present themselves, respectively, when the plane Q coincides with, or be-

CHAPTER THREE

comes perpendicular to, the polarization plane of the incident photon (see Figure 3.1).

Concluding this brief analysis, and before we confront other types of chance events, we can assert that wherever the quantum description of physical systems is assumed to be valid and complete, the quantum probabilities, in the language of the philosophy of science, are *nonepistemic,* which means that they cannot be attributed to ignorance, or to a certain gap in information about the system which, if we could only fill that gap, would transform probabilistic assertions into certain ones.

In fact, the question whether microscopic processes ought to be considered fundamentally stochastic, or whether it is possible (with the addition of further specifications in the description of the system) to complete the theory in a deterministic sense, was a question raised at the very beginning of the lively debate on the implications of quantum mechanics, and has had a very peculiar fate, as will be fully illustrated in later pages. Here I would like to attempt to make clear the significance of this irreducible randomness of microscopic processes, and bring in a comparison with other possibilities, that is, with probabilistic processes in which, nevertheless, the probabilities are epistemic, which means that they can be shown to be caused by our ignorance of the real state of affairs of the physical system in question. This analysis becomes particularly relevant at a time like the present, when the immense novelties represented by the discovery of deterministic chaos and all the discussion about it (for and against) even at the level of popular publications make it particularly difficult for the nonexpert to grasp the sense and significance of probabilistic conceptions, which are essentially altogether different.

We can begin with a very simple example. Everyone knows it is impossible to predict the toss of a coin, provided the coin is a normal, balanced coin, and not improperly weighted. Now to describe this process we have recourse to the theory of probability that tells us that the two possible outcomes, "heads" or "tails," present themselves by chance, and have an equal probability of happening. Let us now investigate the type of randomness implied in the process, situating ourselves within the world view of classical mechanics. It should be obvious to all that if we assume this perspective, the probabilities involved in the process are epistemic, which is to say, are due to our ignorance of the precise background conditions and of all the various conditions involved in the process of flipping a coin. In other words, if we assume that the fall of the coin is governed by the classical laws, then we can assert that if we knew with absolute precision the rotation that was first given to the coin, the precise distribution of the molecules of the air the coin collides with in its trajectory, the structural details of the surface on which it falls, and so on, we would be able, in principle, to foresee with certainty if the result is going to be "heads" or "tails." Now, here is an example of a process which, in the light of the theory which (it is supposed) describes it correctly, requires a probabilistic description; it turns out, however, that this description is imported for accidental reasons only—namely, merely because in practice it is impossible to take account of *all* the elements that determine the outcome (which, in reality is perfectly determined). The situation now being illustrated—and the consequent position vis-à-vis the probabilistic structure of the process—exactly recalls the mechanistic philosophy of the great eighteenth-century mathematician Pierre-Simon de Laplace. Laplace maintained that once the position and velocity of all the particles of the universe were known, it would be possible to foresee its future evolution to eternity. In 1776 he wrote,

> The present state of the system of nature evidently follows from what it was the moment before, and if we were to imagine an intelligence that at any given instant could comprehend all the relations within the entity of this universe, this same intelligence could know the respective positions, movements, and general dispositions of all entities, in whatsoever instant of the past or future. . . . But the ignorance [Note: here is where the epistemic argument begins] of the various causes which concur in the formation of the events, their very complexity, together with the imperfection of our analysis, prevent us from attaining the same certainty when it comes to the great majority of phenomena. There are some things that are uncertain to us, things more or less probable, and we seek to counteract the impossibility of knowing them by distinguishing various degrees of likeliness. Thus it comes about that the very weakness of the human mind gives rise to one of the most refined and ingenious of mathematical theories, the science of chance or probability.

It would be hard to find a more lucid statement of the view that takes probability in the description of physical processes as something accidental, and not something fundamental and intrinsically inevitable.

But mechanics too has undergone a profound revolution in recent times, making necessary certain important modifications of Laplace's position. Indeed, it has become possible to identify many processes where we have what is called "extreme sensitivity to the initial conditions," which in turn implies so-called deterministic chaos or complexity.[10] Technically speaking, movements are called chaotic which are extremely complicated, and manifest an incredibly rapid growth of errors, and which,

CHAPTER THREE

despite the perfect determinism which is supposed to govern them, make long-term predictions impossible. This contemporary understanding was prefigured by the great mathematician Jules-Henri Poincaré at the beginning of the century, when he quite properly introduced a theoretical distinction between (1) the unpredictability that arises from the extreme complication of various factors that come into play, and (2) the extreme sensitivity of initial conditions, even for relatively simple systems. In 1903 he wrote,

> A very tiny cause which escapes our attention, determines a rather considerable effect which we cannot help but see, and so we say that the effect is brought about by chance. If we could know the laws of nature exactly and the situation of the universe at the instant when it began, we could exactly predict what the situation of the same universe would be a moment afterwards. But even if the laws of nature held no secret from us, even in that case we would only have an approximate knowledge of the initial state of affairs. If we were permitted to foresee the succeeding state of affairs with the same approximation, we would need nothing more, and would have to say that the phenomenon was foreseen, and governed by laws. But it is not always this way: it can happen that tiny discrepancies in the initial conditions produce great discrepancies in the ultimate phenomena. A small error at first produces an enormous error later. Prediction becomes impossible, and we have a chance phenomenon.

I have two reasons for spending a little extra time on this point: above all, in order to underline how much the modern view concerning the predictability of natural phenomena differs from the understanding of former centuries. Only recently have we learned how to evaluate appropriately the difference between two things: between (1) processes which, as Poincaré said, permit us to make predictions with the same margin of error as we had concerning the initial conditions, and (2) cases where the initial discrepancy gives rise to an exponential increase. It is very interesting to observe that this latter case can present itself even in extremely simple systems, such as, for example, in a billiard table, to which extra cylindrical bumpers have been added. Unlike the normal billiard table, these bumpers amplify the differences between the various trajectories of a ball that strikes them, causing the balls to go in completely different directions, even in dependence on extremely similar initial conditions. In the same way, only recently has due attention been given to the fact that the "chaotic" character of the deterministic Newtonian dynamics can render illusory and unjustified the common expectation that one can study a system as if it were isolated, by ignoring little disturbances of the environment. This fact can be seen clearly with regard to a billiard table, in this case, even a traditional one.

To make the point even more surprising, let us pretend it is possible to make a perfect billiard table, one on which the balls roll without any friction and bank against the walls without any distortion. We can also suppose that the edges of the table are perfect, and perfectly meet the balls, making the angles of reflection exactly equivalent to the angles of incidence. We consider a case where some balls are on the table, let's say about ten, and that the billiard player who wants to know precisely the effect of his strokes has absolute control over them. To evaluate the predictability of the process, we consider first a case where nothing else exists in the universe except the billiard table. Now, without changing the perfection of the billiard table in any way, or the absolute precision of the billiard player, we try to change the universe just a little. It may surprise many readers to learn that if the change in the universe consisted only in adding a single electron as far away from the billiard table as the moon is from the earth, within a minute, the trajectories of the balls would be noticeably—i.e., macroscopically—different from what they were to begin with!

What lesson do we gain from this? A very important one. The examples illustrate what Poincaré said: that *prediction becomes impossible and we have a chance phenomenon*. In fact, it is relatively easy to demonstrate that deterministic systems exist with such sensitivity for their initial conditions that the prediction of their behavior after even short periods of time would require such a mass of information (the initial imprecisions increasing exponentially) that it would not be possible to enter it into a computer which used all the particles of the universe for chips and if you could load one bit into every chip. The conclusion is that we have to take account of the fact (and this is certainly an important conceptual conquest in itself) that situations are not unusual where in fact it is impossible to predict the behavior of a system even for a relatively brief period of time.

This little digression is meant to clarify the subtle differences between today's understanding of unpredictable processes and that of bygone centuries. The modern understanding requires the use of (as Laplace said) *one of the most refined and ingenious of mathematical theories, the science of chance or probability*. I wanted to enter into this fascinating theme, mostly to allow the reader to grasp the difference between epistemic and nonepistemic probability. The difference is not just a difference "in practice," but a real conceptual difference. The fact that even if the entire universe were turned into a computer, such a computer would still not be sufficiently powerful to allow us to enter enough information to be able to predict the evolution of a simple system for more than a minute or so — all this does not affect a more serious issue: the fact that, according to the theory underlying the dynamics of the process, the need for probabilistic description derives from our ignorance concerning the precise initial conditions. On the contrary, in quantum mechanical theory, it is simply not the case that, for example, the state vector is never determined with an absolute precision, nor that the dynamics is of the type that exponentially increase the errors introduced,[11] making necessary a probabilistic prediction of the results of measurement. The randomness of the outcomes is incorporated into the very structure of the formalism, which, if assumed complete, would simply not allow us to think that, in general, the results are predetermined, even in a way as yet unknown to us.

I have attempted to underline a very important aspect of the quantum mechanical formalism, an aspect to which we shall soon return. It constitutes the central theme of the interesting research now being done under the name of "hidden variable theories."

We can now move to a detailed discussion of another aspect of the formalism closely connected with the foregoing: quantum indeterminism.

3.7. The Uncertainty Principle

We can now resume our analysis of the diffraction experiment, analyzed in chapter 1 (see Figure 1.8), and discussed further in section 2 of the present chapter. We begin by observing, before anything else, that it is irrelevant to specify whether the experiment involves photons or electrons. In both cases, the image on the screen does not simply match the slit, but becomes enlarged, and in both cases, the process of exposure on the film to the right of the opening makes apparent the particle nature of the process: every single particle (whether photon or electron) comes to a finish at a precise point on the screen. The other extremely important point to keep in mind is that, once again, while the knowledge of the state of the system perfectly defines the probability that every single constituent lands in one place rather than in another, the single processes are genuinely random. If the theory is true and complete, there is nothing, nor can anything be thought of, that determines the point of impact of the single particles on the screen.

QUANTA, CHANCE EVENTS AND INDETERMINISM

But now look at the process from a different perspective. Before the beam of light enters the opening, it already consists of particles (see Figure 3.10), possibly very distant from one another, propagated with equal speed in a direction perpendicular to the screen.[12] The light beam itself, which can be thought of as originating from a relatively remote source, has a breadth of extension D, much greater than the size of the slit. It is thus natural to interpret the process as a *measurement* of the position of the incident particle in the vertical direction, indicated by x.

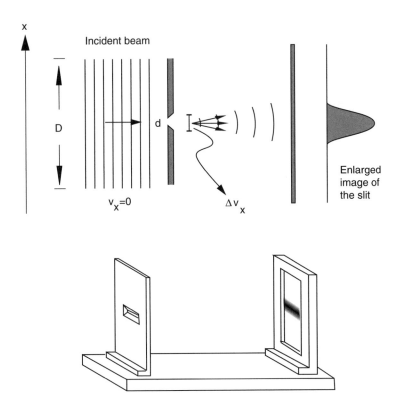

FIGURE 3.10. An illustration of the uncertainty principle. If we attempt to define the position of the particle with greater precision (by reducing the frontal surface of the particle-associated wave from D to d) we lose our knowledge of its velocity. This is an inevitable result of the wave aspects of the process, and its diffraction phenomena. A sketch of the apparatus appears below, which will be useful for appreciating some of the points of the debate between Einstein and Bohr.

CHAPTER THREE

We now can say that, before the wave associated with the particle strikes the screen that has been placed in its path, the wave (or better, to use the appropriate language, the state vector) describes a particle for which, in a measurement of position in direction x, all the positions corresponding to the interval D of the figure are equally probable (in fact, if the screen had not been placed there, on their arrival, the particles would be uniformly distributed—the single events being completely random—on the region of amplitude D on the final screen). The introduction of the slit implies that, at the moment of measurement, that is, in the act of passage itself, the particles that pass the test, and pass through, are being confined to a very tiny space. Our knowledge of their position in the given direction is remarkably increased. All the same, as we know very well, to the right of the slit, the particles propagate themselves in such a way as to form a figure of diffraction much larger than the opening itself, according to individual processes governed by genuinely random laws.

We can attempt to interpret the facts just now described: Before the particle arrives at the screen its position is known only with the precision of D. On the other hand, it is known that the particle is propagated in a direction perpendicular to the plane of the opening, or, to say the same thing, that the component of its velocity in the direction x is definite, and equal to zero. The position of a single particle that passes the opening is known at the moment of passage with a precision much greater than before. Nevertheless, this particle can now, with an appreciable probability and in an entirely unpredictable way, end up at one of the points of the bell-shaped figure to the right. It is clear, for example, that in order to reach a point along the upper edge of the diffraction figure, the particle has to be propagated in such a way as to have some component of its velocity in direction x. An analogous argument applies (apart from the change of sign of the velocity) for a particle that comes to a stop along the lower margin of the figure. However (and I apologize for repeating it so much that I seem pedantic) the theory implies that nothing, absolutely nothing, differentiates the ones that end up in one place on the screen from the ones that will end up at another among the places for which there is an appreciable probability of the particle ending there.

Now, the laws of diffraction precisely identify the size of the figure of diffraction, as a function of the wavelength and the size of the opening. In particular, these laws imply that the narrower the opening, the greater the figure of diffraction. What do we conclude from that? That our action taken to know more accurately the *position* of the particle in direction *x* (by narrowing the opening) has made us lose the accurate knowledge we had of its *velocity* in that direction. I shall not go into the technical details that permit us to recover in quantitative fashion the indeterminacy relations which I am atttempting to describe. At the passage of the opening, the position of the particle is indeterminate of the quantity $\Delta x = d$. The fact I do not know, and cannot know, is in what possible direction the particle is being propagated, and that makes it impossible for me to know in a precise way the component of its velocity in the same direction. Using the laws that govern diffraction, it can be deduced that the (nonepistemic) imprecision v_x which characterizes the component x of the velocity is of the order of the ratio between Planck's constant and the product of the particle's mass with d:

$$\Delta v_x \geq h/md. \qquad (3.6)$$

If we correctly signify by Δx the imprecision of our knowledge about position, we have the fundamental uncertainty principle of Heisenberg:

$$\Delta x \Delta v_x \geq h/m. \qquad (3.7)$$

The analysis we have just made tells us that the attempt to improve our knowledge about the position of the particle has made us lose information about its velocity. Some comments are in order:

- The uncertainty principle does not put limits on the precision with which we determine the position, because the opening can be made as narrow as we want it to be.
- The analysis just now completed shows that it is precisely the double particle and wave nature of physical processes that forces us to conclude, on the basis of a correct critical analysis of the physical process of measuring a position, that every attempt to determine this variable more accurately makes us lose knowledge of the variable velocity that corresponds to it.
- In a similar way, it can be demonstrated that any experimental procedure studied for the purpose of furnishing us with more precise knowledge of its velocity inevitably makes us lose our knowledge of the position, and the equation above will always be verified.
- The conclusion is not in any way linked to a specific mode we may have chosen for carrying out a measurement, but has an absolutely general va-

CHAPTER THREE

lidity. In fact, it constitutes one of the most immediate consequences of quantum formalism.

. . . . .

While not wanting to slow down the discussion at this point, I would nevertheless think it appropriate to illustrate in a little more detail the process I have just discussed. First of all, the state of the particle at time t is characterized by a mathematical entity, the state vector, which perfectly specifies it. This can be conveniently represented by referring to the notation of Dirac, as $|\Psi, t\rangle$, where the symbol Ψ specifies the property of the system. (Dirac's language, since it is very abstract and general, is particularly convenient for speaking of properties like spin or polarization.) When we are interested in the properties of a particle's position, it is convenient to give an explicit representation of the abstract entity $|\Psi, t\rangle$, a representation that becomes in each and every respect equivalent to it (i.e., to $|\Psi, t\rangle$), but puts into relief the spatial variables. In this way there is a description of the state of the system by means of a wave function which depends on the spatial variables, in addition, of course, to the time variable. For the sake of simplicity, let us suppose that the particle is bound to movement along a certain line, say the axis x. In this case, the wave function will now assume the form $\Psi(x,t)$. This representation is particularly significant for the magnitude that interests us (that is, the position where the particle can be found when a measurement of its position is performed). The wave function in fact plays a role perfectly analogous to that of the electric field for photons; that is, it provides[13] (with its square, $|\Psi(x,t)|^2$) the probability density of finding the particle in various places. In other words, if, at a certain predetermined instant, the wave function is represented by the function $\Psi(x)$, then the area subtended by $|\Psi(x)|^2$ in an interval (a,b) represents the probability that in a measurement of position, the particle will be found to be within the interval in question (of course, the area under the whole curve would have to be equal to 1, because the particle will certainly be found somewhere).

The wave function, like the state vector in its abstract form, gives us probabilistic information about all the possible measurements of the system we can imagine being taken. In particular, the function tells us about the probability of finding a certain value of the speed of the particle along the x axis. The mathematical prescription requires some technical points, such as the passage to a new function, called the "Fourier transform" of $\Psi(x)$. We are not interested now in specifying how effectively we move from $\Psi(x)$ to this new function, which we will name as $\Phi(v)$. We will limit ourselves to two important items. Above all, $\Phi(v)$ plays exactly the same role for velocity that $\Psi(x)$ plays for position, which means that $|\Phi(v)|^2$ gives us the probability density for finding the value v in a measure-

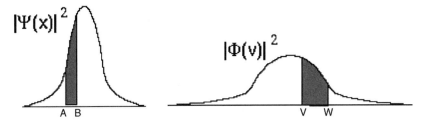

FIGURE 3.11. The probability of finding a particle in the spatial interval (a,b) in a measurement of position is governed by the square of the modulus of the wave function Ψ(x); the probability of finding a particle's speed, confined between the values v and w, is governed by the square of the modulus of its Fourier transform Φ(v).

ment of velocity, and thus, in an analogous fashion (see Figure 3.11) the area subtended by this function within an interval (v,w) gives us the probability of finding a velocity belonging to the interval indicated, if we measured for it. Secondly, the mathematical transformation which brings us from Ψ(x) to Φ(v) is such that the more concentrated one of the functions, the more expanded the other. This means that if there is some probability of finding the particle in only a small interval (that is, if its position is well defined), then many velocity values are probable as results of measuring this variable, and vice versa (see Figure 3.12). This argument

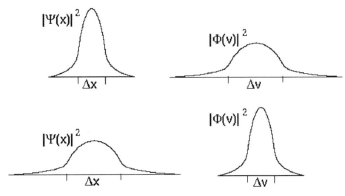

FIGURE 3.12. It follows from the mathematical relations that connect the two functions Ψ(x) and Φ(v) that while one of them is concentrated (thus restricting the relative variable to a narrow range of values), the other is enlarged, with the relative variable becoming noticeably undetermined. The argument is a precise mathematical method of showing that the incompatible variables, position and velocity, must satisfy Heisenberg's uncertainty principle.

CHAPTER THREE

represents another (but equivalent) formal mode of looking at the structure of the theory and has exactly the same implications as discussed before. In particular, it makes absolutely precise the argument for the uncertainty principle.

. . . . .

Thanks to his profound analysis, Heisenberg was able to conclude that there exists an uncrossable limit of precision with which a pair of variables can be measured, such as the position and velocity which constitute the prototypes of so-called incompatible variables. The existence of this limit (if the theory is accepted as true) is not due to practical difficulties, but has a fundamental character: it is a direct and inevitable consequence of the strange double nature, both wave and particle, of all physical processes.[14]

It is not surprising that Heisenberg's conclusions threw the scientific community into yet more confusion, and immediately aroused the lively interest of all the brilliant thinkers of the time who were actively engaged in clarifying the meaning of the formalisms then undergoing elaboration, which, due to their revolutionary implications, were still not fully understood at that time.

Before closing this section, it seems a good idea to avoid misunderstandings by emphasizing that not all physical variables of a system are incompatible, and thus subject to the uncertainty principle. To understand what I mean by this, consider a process which involves two successive measurements, meant to improve our knowledge of the position of a particle in two directions at right angles to each other, x and y. We suppose that we have an incident particle associated with a wave (which can be an electromagnetic wave, in the case of a photon, or a wave function, in a more general case). This wave is going to be appreciably different from zero only on the square of side D, at right angles to the direction of propagation (see Figure 3.13). Since, as observed before, the square of the wave function governs the probability by which a particle can be found at a certain point in a measurement of position, and it is supposed that the wave function of the wave itself is different from zero (having almost the same value on the square in question), we can assert that, before every measurement, the positions both in direction x and in direction y have values that are indeterminate for an amount D. In other words, the particles associated with the wave in question have equal and nonepistemic probabilities of hitting a screen perpendicular to the wave, at any one of the points of the square. Suppose, now, that we want to improve our knowledge of the position of the particle in direction x. The most

QUANTA, CHANCE EVENTS AND INDETERMINISM

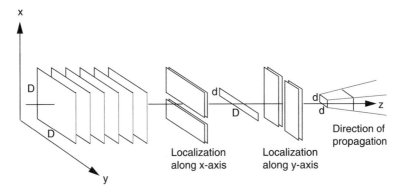

FIGURE 3.13. Illustration that measurements of x and y are not incompatible.

natural procedure for obtaining such knowledge would consist, as discussed earlier, in placing on the path of the light beam a screen with a straight slit, oriented in the direction of the y axis, and of opening d, much smaller than D, along the x axis. The particles that pass the test will now have a position confined within[15] the rectangle of sides d and D, one oriented like the axis x, the other oriented like the axis y. We now consider a second process of measurement, intended to reduce the indeterminism of position in direction y. This, clearly, will mean introducing a slit going in direction x but with a width d in direction y. For the particles that pass the test, this process has the effect of cutting down their wave function, by confining it to the strip of width d in direction y, but this does not alter the fact that, before the second test, the wave function of the particles has already been reduced to extension d in direction x. Thus the second process does not disturb in any way the wave function in direction x. It is understood that the two measurements happen immediately in succession (or, equivalently, they can happen simultaneously if we place a screen with a square aperture in the path of the beam). The particles that pass both measurements will have, at the end, position indeterminacies of Δx and Δy, both of the order of d, and d can be made as small as we please. In other words, the two observable positions "x" and "y" are compatible, the relative measurements do not disturb one another, and the theory does not put any conceptual limit upon a determinacy that is both simultaneous and as accurate as desired, concerning the given quantities.

Obviously, to place two slits in front of the wave rather than one produces diffraction in both directions (as shown in the right side of the fig-

73

ure) and thus a loss of knowledge, along both the x axis and the y axis, of the velocity components. The corresponding indeterminacies will satisfy Heisenberg's relations.

I hope this discussion has better clarified the meaning of quantum-mechanical indeterminism. The formalism implies that there exist pairs of variables in such a way that it is impossible to reduce the uncertainty of their values simultaneously, at a level that would violate Heisenberg's relations. Other examples of incompatible observables can be immediately identified in the numerous experiments already discussed. We have seen what happens when a beam of light with a vertical polarization plane is passed through a filter with a polarization of 45°: the photons that pass the test are polarized precisely at 45°. The process can be likened to a process of measurement that identifies the systems that have polarization at 45°. Our ignorance beforehand about the properties of the incident photons—whether they were, or were not, able to pass a test of polarization of this kind—has disappeared, for now we know that we have a beam of light that will undoubtedly pass a test of 45° polarization. But we also know that a beam like this has a one-half probability of passing or not passing a test of vertical polarization. Similarly, our initial, certain knowledge about the behavior of systems when confronted with a test like this has been lost. In fact, we now have a total uncertainty as far as regards this property, in so far as it is just as probable that the photon passes the test as fails it. The properties of polarization which belong to the vertical direction and to the 45° direction are incompatible: by improving our knowledge relative to one of them, we lose it relative to the other. The polarized filter which, as I pointed out, does not act simply in a passive way on the photon, but changes its state, allows us to obtain information about one variable (polarization at 45°), but makes us lose the precise information we had before (vertical polarization), incompatible with the former.

Immediately after the publication of Heisenberg's work, questions about its meaning set off a lively debate. We shall have to treat this debate in what follows, but it is now time to illustrate some important elements of the uncertainty principle and to explain the position assumed in response to it by Bohr, with his "principle of complementarity."

3.8. Bohr's Complementarity

We can now understand the discomfort produced by an analysis that constrains us to accept the existence of conceptual (and not practical) limitations to the simultaneous knowability of elementary physical magnitudes relative to a system, such, for example, as the position and velocity of a particle. But, even more astoundingly, the new formalism requires that we include in the same image mutually irreconcilable physical aspects, such as wave and particle. Reflecting on these obscure aspects of the theoretical schemes then being advanced, Bohr made an observation of great significance. He emphasized that (on the one hand) the experimental procedures necessary for determining magnitudes with more precision than the uncertainty relations could allow them, and (on the other hand) the experimental procedures necessary for making clear the particle and wave aspects of physical processes, were, in fact, impossible to realize simultaneously. Every possible conceptual apparatus for getting information on the position (to any arbitrary degree of precision) necessarily fails to inform us with the same precision about the velocity and, in a perfectly analogous way, every experimental situation in which particle aspects are being revealed at the same time fails to show wave aspects.

To appreciate this fact fully, it is necessary to go back to the experiment on interference, with the two slits, repeatedly analyzed in the pages above. In an experiment of this kind, according to Bohr, the formation of interference patterns on the screen shows that wave aspects play an important role in what happens between the slits and the screen. On the other hand, as I have repeatedly emphasized, the detection of the points of arrival on the screen shows the particle aspects of the process. The two contradictory aspects never emerge, so to speak, together. And Bohr also correctly pointed out that any attempt to show particle aspects along with wave aspects in the process of passing through the slits, for example, to ask oneself a typically "particle" question (such as, "By which opening did the particle pass?"), inevitably destroys the interference pattern (see Figure 3.14, which is an elaboration of a figure repeatedly used by Bohr in his debate with Einstein), and thus it is impossible to have access to both "complementary" aspects in the same experiment. Bohr's observation had a remarkable importance, and he was so enthusiastic about this idea that he proposed it as a paradigm of absolute generality, valid even outside the bounds of microscopic processes. The idea is that nature is extremely rich, multifaceted, and mysterious. We humans have been privileged with understanding various aspects of this complex reality, but it has not been given to us to know them simultaneously. Therefore, the procedures needed to gain access to one of the many faces of the real will be incompatible with those needed to gain access to other aspects complementary to the first. Bohr went very far in stressing the validity of this

CHAPTER THREE

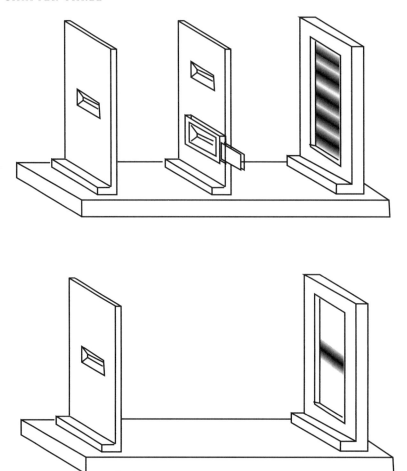

FIGURE 3.14. Any attempt whatsoever to put into evidence particle aspects in a process of interference, usually to identify by which of the two slits the particle has passed, suppresses the wave aspects of the process. On the screen, instead of bands of interference, an image of the slit is formed, even if slightly enlarged.

idea in reference to profoundly different phenomena, such as procedures for ascertaining if microorganisms or single-celled organisms were alive or dead, saying that every procedure intended to show that a cell is alive actually kills it.

A few comments are in order. No doubt, there is profound insight behind the observation that it would be impossible even to imagine ideal experiments where the uncertainty principle could be violated, or to bring aspects of reality into the open that seem so contradictory. But I also must say, along with widely accepted opinion, that Bohr was never able to provide a clear and convincing formulation of his principle of complementarity. Schrödinger's comment was pithy: "When something is not understood, we invent a new term and we think we understand it." The most insightful observations, in my view, about the philosophy that stands behind the idea of complementarity were expressed recently by John Stewart Bell. He wrote as follows:

> Bohr went further than pragmatism, and put forward a philosophy of what lies behind the recipes. Rather than being disturbed by the ambiguity in principle, by the shiftiness of the division between "quantum system" and "classical apparatus," he seemed to take satisfaction in it. He seemed to revel in the contradictions, for example, between wave and particle, that seem to appear in any attempt to go beyond the pragmatic level. Not to resolve these contradictions and ambiguities, but rather to reconcile us to them, he put forward a philosophy which he called "complementarity." He thought that complementarity was important not only for physics, but for the whole of human knowledge. The justly immense prestige of Bohr has led to the mention of complementarity in most textbooks of quantum theory, but usually only in a few lines. One is tempted to suspect that the authors do not understand the Bohr philosophy sufficiently to find it helpful. Einstein himself had great difficulty in reaching a sharp formulation of Bohr's meaning. What hope then for the rest of us? There is very little I can say about complementarity, but I wish to say one thing. It seems to me that Bohr used this word with the reverse of its usual meaning. Consider for example the elephant. From the front she is head, trunk, and two legs. From the back she is bottom, tail, and two legs. From the sides she is otherwise, and from top and bottom different again. These various views are complementary in the usual sense of the word. They supplement one another, they are consistent with one another, and they are all entailed by the unifying concept "elephant." It is my impression that to suppose Bohr used the word "complementary" in this ordinary way would have been regarded by him as missing the point and trivializing his thought. He seems to insist rather that we must use in our analysis elements which *contradict* one another, which do not add up to, or derive from, a whole. By "complementarity" he meant, it seems to me, the reverse: contradictoriness. Bohr seemed to like aphorisms such as "the opposite of a deep truth is also a deep truth"; "truth and clarity are complementary." Perhaps he took a subtle satisfaction in the use of a familiar word with the reverse of its familiar meaning.

"Complementarity" is one of what might be called the "romantic" world views inspired by quantum theory.

Bell's illuminating observations will conclude this chapter. Many of the points touched upon may now have led the reader to begin to ask him- or herself questions about the surprising physical phenomena the scientific research of this century forces us to consider. The randomness of physical processes, uncertainty, the impossibility of simultaneously executing procedures of measurement that are perfectly legitimate in a classical context—all these represent the fascinating challenges that now face us. A way out is needed. But, as Bell wrote in the same article I have quoted, this way out must be "nonromantic" in the sense that it will require mathematical labor by theoretical physicists rather than philosophical interpretations. We shall see a little later, in fact, how a succession of precise investigations—in particular, those developed by Bell himself—have contributed to a radical transformation of the theory's basic conceptual framework.

CHAPTER FOUR

The Superposition Principle and the Conceptual Structure of the Theory

> The assumption of superposition relationships between the states leads to a mathematical theory in which the equations that define a state are linear in the unknowns. In consequence of this, people have tried to establish analogies with systems in classical mechanics, such as vibrating strings or membranes, which are governed by linear equations and for which, therefore, a superposition principle holds. Such analogies have led to the name "Wave Mechanics" being sometimes given to quantum mechanics. It is important to remember, however, that the superposition that occurs in quantum mechanics is of an essentially different nature from any occurring in the classical theory, as is shown by the fact that the quantum superposition principle demands indeterminacy in the results of observations in order to be capable of a sensible physical interpretation. The analogies are thus liable to be misleading.
> —*Paul Adrien Maurice Dirac*

WE CAN NOW DEEPEN our analysis of the most innovative point of the new theory: the superposition principle. In the preceding chapters we have shown that the formal structure of the theory is such as to permit the "summation" of quantum states. In particular, our attention has been drawn to the fact that, for instance, the polarization state |45°> of a photon is the "sum" of the states |V> and |H> (for "vertical" and "horizontal," respec-

CHAPTER FOUR

tively). Analogously, it was asserted that the state of a particle with upward spin along the direction of the x axis is the "sum" of the states corresponding to upward and downward spin along the z axis. But the analyses provided have not been detailed enough to permit us to understand readily how, according to Dirac, "the superposition that occurs in quantum mechanics is of an essentially different nature from any occurring in the classical theory." Indeed up until now we have limited ourselves to considering superposition of states with reference to properties such as polarization or spin, which are not immediately related to our normal visualization of physical processes. It should prove much more illuminating to see the principle of superposition at work with reference to properties that are familiar to us, and about which we have deeply rooted convictions, thanks to our everyday experience—especially concerning the spatial position of things.

Accordingly, in the present chapter we are going to undertake the illustration of a fundamental aspect of the quantum mechanical formalism, presenting what David Albert of Columbia University has described (1992) as "the most unsettling story, perhaps, to have emerged from any of the physical sciences since the seventeenth century." We shall tell the story with reference to processes involving photons and birefringent crystals, and the attentive reader will understand that the same holds for systems of spin passed through Stern-Gerlach magnets. The story we are about to tell is true (as so many other, even more striking and surprising analogues): the actual experiments have been carried out and have confirmed the results here presented to an extraordinary level of certitude.

4.1. ANALYSIS OF A SERIES OF EXPERIMENTS

We begin again with the process discussed before, whereby a photon of a precise polarization impinges onto a birefringent crystal that has been specially cut and positioned. We can review some of the experiments already described in order to lay out the whole sequence of analysis and help the reader become more familiar with the phenomenology of the process. For simplicity, the pictures of crystals and polarization states will now be more schematic.

a. Description of the Setup for the First Group of Experiments

We analyze what happens when a photon strikes the crystal (i.e., a stream of photons so weak that practically every photon has enough

SUPERPOSITION THEORY

time to pass through the whole apparatus before the next photon can enter). Behind the crystal are placed two photon detectors, precisely situated to meet the rays that emerge—the ordinary (O) and extraordinary (E).

A1. THE INCIDENT PHOTONS ARE VERTICALLY POLARIZED

This situation is shown in Figure 4.1 (a1), where the detector behind the ordinary ray registers all the incident photons, while the other registers nothing at all. We can check that *all* the emergent photons have vertical

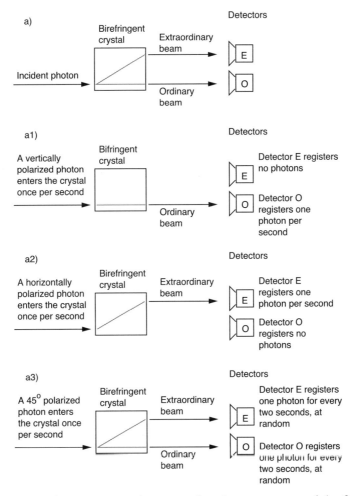

FIGURE 4.1. The experimental setup and various outcomes of the first series of experiments.

CHAPTER FOUR

polarization: if a filter with a vertical polarization plane is held in front of the counter, there will be no decrease of light hitting it.

A2. THE INCIDENT PHOTONS ARE HORIZONTALLY POLARIZED

The detector placed behind the extraordinary ray detects all the incident photons and the other one detects nothing. Once again, the emergent photons have the same polarization state as the incident photons, and this is easily verified by holding a polarized lens at the proper angle between the crystal and the counter.

A3. THE INCIDENT PHOTONS ARE POLARIZED AT 45°

Each photon detector registers, on the average, one photon for every two incident photons. No photon is lost, but the order in which the photons succeed one another when detected, on one ray or the other, is completely fortuitous.

b. Description of the Setup for a Second Group of Experiments

In this set, we use only photons polarized at 45°. However, to eliminate any confusions that might arise with reference to experiments we will consider in what follows, and to attempt to make clear the true physical meaning of the processes under examination, it is interesting to consider experiments of the type presented in Section 4.1.a3, but now adding an obstacle: a light-absorbing screen, placed directly in one or both the paths of the photons.

B1. AN ABSORBING SCREEN ON THE ORDINARY RAY

We have the same apparatus, with the same initial conditions as in the experiments of Section 4.1.a3, but now we place an absorbent screen behind the crystal [Figure 4.2 (b1)] in the path of the ordinary ray. What will happen? Certainly, everyone knows the answer: Detector O will detect nothing, since the photons that were supposed to set it off have all been absorbed by the screen. Detector E (on the extraordinary ray) will register, on the average, one photon for every two, just as in the preceding experiment. The photons enter in a random succession, and all are horizontally polarized. This can be verified by checking to see that nothing changes when a filter with horizontal polarization is held in front of detector E (on the extraordinary ray). Another way to verify the same thing would be to observe a reduction of photons in accordance with the law of Malus with any adjustment in the filter's orientation.

SUPERPOSITION THEORY

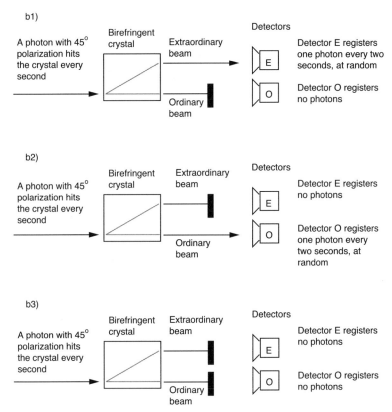

FIGURE 4.2. Variations on the theme of experiment a3. A beam of light with 45° polarization falls upon a birefringent crystal. Absorbing screens are placed, in turn, on the ordinary ray (b1), the extraordinary ray (b2), and both rays (b3), and the results at the detectors are studied. They are as predicted.

B2. AN ABSORBING SCREEN ON THE EXTRAORDINARY RAY

We need not waste any time describing this case. It is analogous to what has been said about the previous one.

B3. TWO SCREENS, ON BOTH RAYS

When the time is right, I will comment on this experiment, designed to eliminate any possible escape-routes from an embarrassing situation we will soon have to face. The result of the experiment should not be surprising: neither detector detects a photon.

83

CHAPTER FOUR

c. A New Tool of Investigation: The Birefringent Crystal, Reversed

Before we take up a third series of experiments, we must add that there exists a way to undo what a birefringent crystal does—the method is simple but will prove very interesting. All we have to do is reverse a crystal identical to the ones already used. As a matter of fact, it is the very simplicity of this maneuver that led me to depart from the usual procedure followed in textbooks on quantum mechanics, and to use birefringent crystals and polarized photons (rather than Stern-Gerlach magnets and spin analysis) to explain this "most upsetting" development to have emerged in the scientific world since the time of Galileo. Even though the incomparable success of the theory in the description of microscopic processes assures us that the behavior of particles with spin is identical to the behavior of photons, the experiments we will discuss are not easy to perform in practice since it is extremely difficult to make a magnet function exactly the reverse of another. Therefore, when it comes to spin, the experiments I describe still belong to the family of "thought experiments" (*Gedankenexperimente*), but there is not the slightest doubt that if the experiments could be carried out, they would show the very same characteristics that we will see in the crystal experiments I am now going to describe.

C1. EXPERIMENTS WITH VERTICALLY POLARIZED PHOTONS AND TWO BIREFRINGENT CRYSTALS SET IN OPPOSITE DIRECTIONS

Here a beam of vertically polarized photons enters a birefringent crystal and immediately afterward enters another crystal, oriented in the opposite direction [Figure 4.3 (c1)]. Behind the second crystal has been placed a detecting device on the line of the incident ray. The first crystal does not deflect the photons that pass through it, and they follow the line of the ordinary ray, emerging with the same polarization they had to begin with. These photons now enter a second crystal. This one undoes what the first one did, but just as the first crystal did not in fact change either the direction of propagation or the state of polarization, the second one likewise has no influence. At the exit of the second crystal is placed a detector that registers exactly as many photons as were incident. Furthermore, a polarizing filter, placed in front of the detector, verifies that the photons are vertically polarized on leaving the crystal.

SUPERPOSITION THEORY

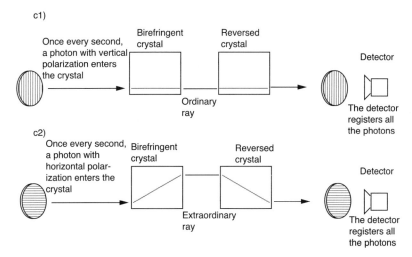

FIGURE 4.3. A combination of two birefringent crystals, with the second reversed. The second crystal undoes what the first did. Unlike the preceding figures, the images of the polarizing filters are explicitly represented at the entrance and exit, in order to show that the polarization properties are the same for the incident rays as for the exiting rays.

C2. EXPERIMENTS WITH TWO CRYSTALS WITH HORIZONTALLY POLARIZED PHOTONS

This is the exact analogue of the previous one except that the photons have a horizontal polarization. As we know and as we have ascertained in a rather painstaking way in the experiments of Sections 4.1.a and 4.1.b, the photons will issue from the first crystal along the extraordinary ray and will have the same polarization—that is, horizontal. The second crystal then undoes what the first one did, and then brings the photons back to their original path, preserving their horizontal polarization. The detector, placed behind the second crystal and lined up with the incident ray, will register all the photons. Furthermore, even in this case, the presence of a filter with a horizontal polarization guarantees that all the photons are horizontally polarized [Figure 4.3 (c2)], since there is no decrease in light.

Taking account of all the experiments of this section, we can now face the central problem of this chapter.

CHAPTER FOUR

4.2. A New Experiment—and a Logical Problem

With the premises laid down in the foregoing section, we now reach the heart of the matter, and pose the million dollar question. Let us consider the last experiment, but with two important changes (Figure 4.4). The incident photons will now be polarized at 45°, and at the right, in front of the detector, has been placed a polarizing filter, whose plane is at 45°. The crucial question now is the following: what is the probability that the detector will register a photon? In the same way, we can raise the question: if the experiment is repeated many times with a beam of light so weak that only one photon at a time could complete the entire journey from beginning to end (at the detector), how many photons would be registered, after a sufficiently long interval of time?

We can trace the logical thread connecting all the experiments as follows:

1. *Phase one.* As shown in experiment a3, and confirmed beyond all possible doubt in experiments b1 and b2, we can legitimately assert that in passing through the first crystal, each photon "decides," by chance, whether to go on the ordinary path or the extraordinary. We also know, having explicitly verified it, that the photons of the ordinary ray are vertically polarized, while those of the extraordinary are polarized horizontally.
2. *Phase two.* In consequence of the above, we further conclude that, on the average, just as many photons enter the second crystal on the ordinary ray (with vertical polarization) as enter it on the extraordinary ray (horizontal polarization). Experiments c1 and c2 demonstrate that the photons of the first type will continue their journey by preserving their polarization, and that those of the second type will at first be deviated and then restored to

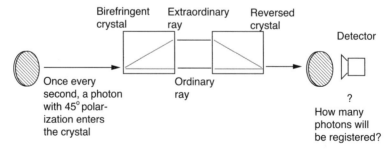

FIGURE 4.4. Experimental setup that leads to an uncomfortable outcome, but will also allow us to grasp the real meaning of the superposition principle.

their path along the original line of incidence by keeping their horizontal polarization.
3. *Phase three.* We must therefore acknowledge that, at their exit from the second crystal, there will be one photon for every incident photon, but each photon will have an equal probability of having a vertical or horizontal polarization: the distribution of states of polarization as the photons succeed one another is completely random.
4. *Phase four.* In order to reply to the question posed at the beginning of the section, all we need do is take account of the proven fact (see the discussion in chapter 2) that a photon with vertical polarization has a 50% probability—exactly as much probability as a photon with horizontal polarization—of passing a polarization test at 45°. Since one-half of the incident photons are vertically polarized, and one-half are horizontally polarized, only one-half of these will pass the test of 45° polarization. Our conclusion? The intensity of the beam will be reduced by 50% from its original level—that is, in particle terms, the detector at the right will reveal, on the average, one photon for every two incident photons.

But the outcome of the experiment contradicts our expectations: *all the incident photons pass the test and the detector will count just as many photons as entered the crystals.* Other than being unexpected, this fact is completely incomprehensible and logically contradictory if it is assumed that the preceding arguments (which represent a certain interpretation of what is really going on) are all correct. To convince ourselves of this, we can ask the following question, "What path do the single photons take in their journey from point of entry to point of exit and ultimate arrival at the detector?" Here are the possibilities.

1. They follow *either* the ordinary *or* the extraordinary ray. This alternative is not possible, since we have experimental proof that in both these cases, the photon emerges with vertical or horizontal polarization, respectively, and we also have experimental proof that photons in either state of polarization cannot always pass a test of 45° polarization (in fact, they pass it only 50% of the time). If this possibility were valid, the beam at its exit would in fact be reduced by 50%.
2. They follow *both* paths. Again, this is not possible, because on every test designed to identify the path followed by a photon, we discover either one photon or none. Half-photons or partial photons on one path with their "other half" or "remaining part" on the other path are never encountered.
3. They follow *neither* path. With this assertion we contemplate the logical possibility that photons can reach their point of impact (that is, the detec-

tor) by traveling paths other than the ones we have considered. But experiment b3 guarantees that this is not possible: if both pathways are blocked, not a single photon reaches the detector.

I must emphasize that alternatives 1, 2, and 3 do in fact exhaust the logical possibilities. How do we get over this impasse? What lesson can we draw from this startling truth? The answer is simple, if revolutionary: it shows unequivocally that microsystems (we are speaking here of photons but the same argument holds for electrons, neutrons, etc.) have modes of being and behaving that cannot be grasped through the conceptual schemes we have elaborated on the basis of our experience with macroscopic objects. This new mode of being is formally expressed in the language of the theory when we say that the photons are in the *superposition* of being along the ordinary path and of being along the extraordinary path:

$$|superposition\rangle = |O\rangle + |E\rangle = |photon\ on\ the\ ordinary\ path\rangle + |photon\ on\ the\ extraordinary\ path\rangle.$$

It becomes extremely important, conceptually speaking, to emphasize that the preceding analysis shows that the assertion "the photon is in the superposition $|O\rangle + |E\rangle$" is logically different from all the following statements: "it propagates itself along path O or along path E" or "it follows both O and E" or "it follows other paths."

4.3. Superposition of States with Definite Positions

In the course of the above analysis, a contradiction has been explicitly brought to light between the assumed logical line of thought and the actual experimental outcome of Figure 4.4 (this requires that in place of the question mark and the question under it, we should now put the unambiguous reply nature gives us, namely, that all the photons reach the detector). Clearly, the inconsistency is in keeping with a certain attitude we have adopted for interpreting the results of our experiments. This attitude is, first, to make tacit assumptions that appear obvious and inevitable in a classical framework, but which then turn out incorrect, in the light of the understanding of physical reality implied by the new formalism. In fact, in all these experiments we have tacitly (and without warrant, when we consider how nature herself instructs us) been treating as logical equivalents two assertions: (1) "The O (or E) detector registers

SUPERPOSITION THEORY

a photon the moment when the photon strikes it" and (2) "the photon, before it arrived at the detector, was in a state that corresponded, in all certainty, to its propagation along the O path (E path)."

When we assume a critical attitude toward this issue, we can better understand the meaning of the statement that microsystems have modes of being and behaving that cannot be grasped in the conceptual schemes we have elaborated on the basis of our experience with macroscopic objects. Therefore, "to be in superposition |O> + |E>" is logically incompatible with any of the following: "to be on path O or E," "to be on both O and E," and "to be elsewhere than O and E."

To reinforce this last point, we can discuss the situation at the exit point of the first crystal from the perspective of the wave function. This involves leaving behind the language of polarization or particle spin states along two paths (such language is irrelevant for the kind of analysis we are doing presently, even though it has an important role when wave functions corresponding to propagations along different paths are joined, and thus interfere with one another). The situation at the exit of the first filter can be represented in Figure 4.5. The illustration is meant to show that the paths are not merely geometrical lines, but have a certain "thickness," or transverse extension. At the exit of the crystal, the two regions are indicated (by shading) where the wave function differs from zero, and a + sign is placed between the regions to indicate precisely that we are dealing with the superposition of two wave functions.

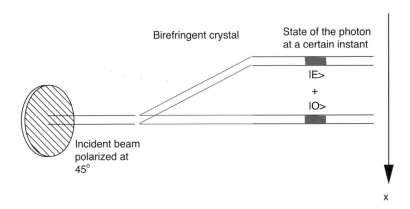

FIGURE 4.5. The photon (or, in an experiment with matter, the particle) can be, at a certain instant, in the superposition |E> + |O>, of the two states which correspond individually to two different spatial positions.

CHAPTER FOUR

We will use the language of wave functions, then, and adopt its corresponding graphic representation (which tells us the probability of finding a particle in a certain spatial region when its position is measured). Figure 4.6 shows the square of the wave function in the situations corresponding to experiments a1, a2, and a3, respectively. The diagram takes account of the thickness of the ray and the values of the square of the wave function along the x axis. In a1, the wave function is different from zero only in the interval O, which corresponds to the position of the ordinary ray with respect to the x axis. The wave function is standardized

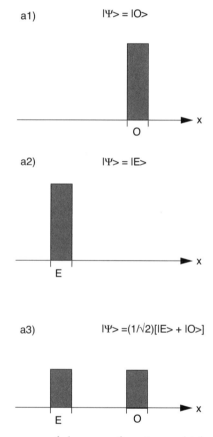

FIGURE 4.6. The squares of the wave functions which describe the particle at the exit of the birefringent crystal, in situations a1, a2, and a3 of Figure 4.1. The x axis has been chosen, as in Figure 4.5, perpendicular to the direction of the rays and oriented downward.

so that the area of the square (that is, the gray rectangle in the picture) is equal to 1, meaning that if its position is measured, there is a probability of 1, that is, a 100% probability (i.e., we are *certain*) of finding the particle in interval O. An analogous statement can be made in the case of a2 where the particle has the probability of 1 (again, 100% certainty) of being found in area E. As we know (see chapter two, especially the concluding formulas), the wave function of a3 will be the sum of the wave functions of the two preceding cases, but reduced by a factor of $1\sqrt{2}$. Consequently, the resulting square assumes for both E and O intervals a value equal to one-half the functions entering into the superposition.

This corresponds to the fact that the probability of finding the particle in E or O is therefore equal to 1: the particle can never be found outside one of the trajectories. But because recombining the two beams gives us interference (as implied by the behavior we discussed beforehand), it would not be logically correct, even before the measurement, to say that the particle is really in O or E. To affirm that the system is in the state |O> + |E> is conceptually different, and, if I may be excused for repeating myself one more time, is logically incompatible with the assertions that it is "in the state |O> or is in the state |E>," "sometimes in |O> and sometimes in |E>."

I trust that the last refinement has made clearer the distinctive meaning of the principle of superposition, and that this quantum principle is fundamentally nonclassical in nature. The argument just developed can be applied by the reader to the phenomenon of interference with two slits, discussed in the previous chapter. The only difference between the case of the two slits and the case at hand is that, whereas the waves that emerged from the slits were superposed spontaneously, in the present instance the two light rays were not sufficiently dispersed to cause a superposition, and we had to produce the effect ourselves by means of the reverse crystal. But even in the case of interference, the situation as regards the wave function when it leaves the slits is exactly the same as in Figure 4.6 a3. One could argue in the same way with reference to that experiment as I do here, showing how the emergence of the figure of interference implies a logical contradiction to the idea that the particle passes by one or the other slit. The plus sign that appears between the two states that correspond to a definite passage from one of the slits makes it illegitimate (since interference occurs) even to think that at its passage through the slits the particle is definitely in one or the other slit. It is in the state $(1\sqrt{2})$ [|in one > + |in the other>].

CHAPTER FOUR

4.4. Variations on a Theme: Other Simple Experimental Procedures for Superpositions of Spatially Separate States

There are other methods and other simple experimental techniques that illustrate the principle of superposition more fully. Spending some time considering these should enable us to understand the peculiar experiment of "delayed choice" suggested by the great physicist John Archibald Wheeler.

To begin with, we will continue to work with photons but will not concern ourselves with their polarization behavior. Instead, we will turn to another extremely simple instrument, a half-silvered mirror. We do not need to enter into the details of precisely how an apparatus of this kind acts the way it does; it should be enough to understand *what* it does: a photon is propagated in direction x, in whose path a half-silvered mirror has been set up, at an angle of 45°. Behind this mirror are placed two photon registers, the first one lined up along the incident ray, the second on a path at 90° with respect to the other, on the vertical at the point where the incident ray meets the mirror. The phenomena of the experiment can be interpreted quite simply: once again, the two detectors have an equal probability of registering the arrival of a photon, and they go off half the time, in a random order. The similarity with the crystal experiment should be noticeable, but two important differences call for mention. Above all, the half-silvered mirror behaves as it does for all photons independently of their polarization, which does not enter into consideration here. Secondly, the undeflected ray and the deflected ray differ in an important respect which is more easily expressed in "electromagnetic" language: namely, the two fields are no longer "in phase" but are displaced from each other by a certain definite quantity[1]; to be precise, the deflected light beam undergoes a slowdown of one-fourth a wavelength compared to the undeflected beam. In Figure 4.7 we can see the process in question, along with a graph of the electric fields along the x axis (the axis of incidence) and along the y axis (at right angles to the former), beginning at their common point of origin, O.

The fields propagated along the two lines act as the wave function of the photon, and their squares represent the probability (equal probability) with which they activate the detector placed on axis x or axis y. Exactly as with the preceding experiments, as long as the intensity of the field is such that, for instance, one photon per second strikes the half-

SUPERPOSITION THEORY

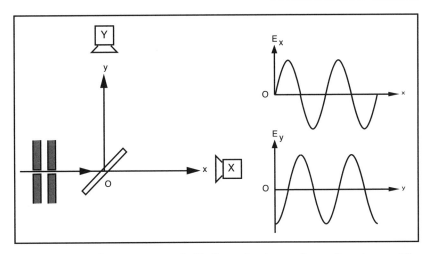

FIGURE 4.7. A beam meets a half-silvered mirror along direction x. The electric field is divided into two fields which are propagated along the directions x and y, respectively. As shown in the right side of the picture, the field that propagates vertically is out of phase by a quarter wavelength with respect to the field that pursues its own path.

silvered mirror, at its exit the photon is in a quantum superposition of state |X>, which describes the photon as being along the x axis, and state |Y>, which describes it as being along the y axis. As before, the actual state—that is, $(1\sqrt{2})$ [|X> + |Y>]—makes it illegitimate to say that the photon "is on x or y" or that "it is going in both directions" or "in some other direction altogether." In other words, we are again concerned with the superposition of two states, corresponding to spatially diverse localizations.

With these premises in mind, we can briefly study the so-called Mach-Zender interferometer (see Figure 4.8). The underlying idea should be fairly clear by now. The first stage of the experiment anticipates a situation analogous to that of Figure 4.7. The two fields, at their departure from a first half-silvered mirror (H-S1) are propagated in different directions and thus cannot give rise to the phenomena of interference. To put into evidence processes of this kind, we have to bring the rays together, as described in section 4.2 above. Now this can be done rather simply: by introducing, on both beams and at an equal distance from O, two *fully silvered* mirrors (S1 and S2) which not only reflect the whole beams but are also positioned to rotate them by 90°.

CHAPTER FOUR

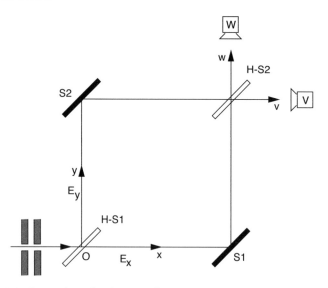

FIGURE 4.8. A Mach-Zehnder interferometer.

At this stage of the experiment we are not concerned with the problem of a possible change of phase under reflection, since both fields undergo the same process and their half-wavelength delay remains unchanged. But now we insert another half-silvered mirror (H-S2) at the point where the two fields cross again. Obviously, both fields, as before, are able to pass through the mirror as they pursue their course, or they can be reflected, and in this way undergo yet another adjustment of one-fourth a wavelength. We can now study in detail the two fields in two directions perpendicular to each other, designated as v and w.

The field E_x, which has passed the first half-silvered mirror (H-S1) unchanged, pursues its path without any retardation of phase, is then reflected by the fully silvered mirror (S1), and after that meets the second half-silvered mirror (H-S2), at which point it crosses the field that followed the other path. Let us suppose that the light is reflected. After the mirror, it will be propagated along axis v, but will have undergone a retardation of one-quarter wavelength. Alternatively, we can suppose that field E_x has passed through H-S2 directly, and then will be propagated along direction w without any delay of phase. Field E_y, we note, will first have undergone a deflection and delay of one-quarter wavelength at its impact with H-S1, then was reflected by S2, and finally, if it continues along direction v will not undergo any further retardation, but will re-

SUPERPOSITION THEORY

ceive a second retardation if it is deflected at H-S2 into direction w. The analysis shows that the two fields get into perfect phase along v (insofar as they *both* have suffered an equal change of phase), whereas along direction w, field E_y will have undergone *two* retardations, each a quarter wavelength in size. The two fields will thus be out of phase by one-half wavelength along direction w, and will destructively interfere with each other. At the same time, along direction v we have two equal fields in phase, which add together (constructive interference: it may help to review Figures 2.1a and 2.1b). Recalling that the square of the field governs the probability of finding a particle along a path when a measurement for position is carried out, we can affirm that, of the two detectors placed at the ends of the two paths v and w, respectively, the latter (w) will not register any photons, while the former (v) will register all of them.

It is also of interest to observe that if the second half-silvered mirror (H-S2) is removed, field E_x, which is propagated vertically after its reflection, continues undisturbed along direction w, while field E_y continues along its path in direction v. The two fields would then have equal intensity, and the detectors would show on average an equal number of photons, equal to one half of all the incident photons. This point will play a crucial role in the experiments on delayed choice, as we shall see.

4.5. FURTHER VARIATIONS ON A THEME: THE TUNNEL EFFECT

In this section we briefly consider a simple experiment that yields a superposition of different position states. A particle meets a potential barrier. To understand the process, we consider the classical situation first. We have a little ball that rolls (to the left in Figure 4.9) down an inclined plane of height H. It then strikes a mound of height 2H. Now, according to classical physics, the ball will climb up the mound as far as H, gradually losing its velocity until it stops and rolls backward. After a certain amount of time we can be sure that the ball will be to the left of the mound and that it will be found in region L.

In quantum terms, things are quite otherwise. In fact, it can happen that the ball will mysteriously manage to activate a counter to the *right* of the mound! This is known as the "tunnel effect" and it plays an extremely important role in some fundamental nuclear processes, such as the radioactive disintegration that accompanies the emission of an alpha particle. The choice of a certain height and breadth of the mound can cre-

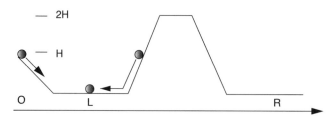

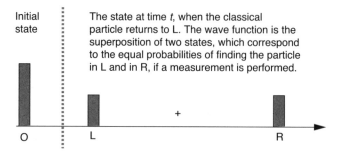

FIGURE 4.9. The tunnel effect. Classically, a ball with energy equal to one-half of that necessary to overcome a barrier (or a little hill or mound) will not be able to go over it, and turns back. In quantum terms, there exists, in general, a nonzero probability that an electron can be found to go beyond a barrier of this kind. For the values of the parameters considered in the text, this probability is equal to 1/2 (50%), and the electron ends in the superposition (√2) [|L> + |R>] of two states, spatially far removed from each other.

ate a situation wherein the little ball, in this case, an electron, has an equal probability of climbing over or turning back. For example, if there is an electron with an energy of 10 eV (where the expression eV is the abbreviation of "electron volt," the energy acquired by an electron when it falls through a potential difference of one volt) that encounters a hill of 20 V and one-half angstrom in length, there is exactly a 50% probability that the electron will be found to the right or left of the hill, once a measurement of its position is taken. Once more, the process has brought about the superposition (1√2 [|electron on the right> + |electron on the left>]. Now this does not permit us to say that before the measurement the electron was "on the left" or "on the right," or "in both places at once," or "in neither place." I need not dwell on this example but instead shall move on to describe how analogous situations with neutrons have recently led to the development of some extremely refined instruments that make use of the interference between the different (and distant) parts of a wave function that afterwards become superposed.

4.6. NEUTRON INTERFEROMETRY

Thanks to recent developments in the technology of crystal growing, crystals can now be made of such high quality that they permit the construction of an instrument analogous to a Mach-Zender interferometer, but using a beam of neutrons. This requires a few moments to be described, not so much to deepen our understanding of quantum particles (a topic we are already treating at length) as to show that recent technology has made feasible in practice what a few years before were only considered thought experiments. Beyond that, we will be able to see how the interference capacity of these microscopic constituents of nature makes possible the creation of very sophisticated and ever more accurate measuring tools.

We start with a virtually perfect crystal. To give an idea of the achievements of solid state physics, I will only say that by the expression "virtually perfect" I mean that the careful elimination of impurities and defects of a solid makes possible an alignment of the reticular planes along which the atoms are arranged to the accuracy of a billionth of a centimeter over a distance of ten centimeters. These atoms, when hit by a wave of incident neutrons, cause the wave to refract, bringing constructive interference in appropriate directions. It is not necessary to enter in the details of how a crystal like this can generate privileged directions of wave propagation for the wave function: what it does is very similar to the effect of the half-silvered mirror (see above, section 4.4) upon the electromagnetic field. Instead, we can study its manner of being cut and positioned.

We begin with a crystal of parallelipiped form, with a length of 20 cm and a section of 10 cm². This is cut as shown in Figure 4.10, in such a way that on its upper side it consists of separate slabs joined at a common base. The base serves to ensure that the reticular plane of the three walls retains the very precise alignment described above. Into this crystal is then introduced a beam of neutrons (or better, a single neutron at a time) along a direction that forms the proper angle with the first plane (Figure 4.10b represents a view from above, looking down on the crystal and the beam). The wave function will then be other than zero only along the two paths shown. When these pathways meet the second wall, they again bifurcate, and thus two rays which were separated toward the inside of the

CHAPTER FOUR

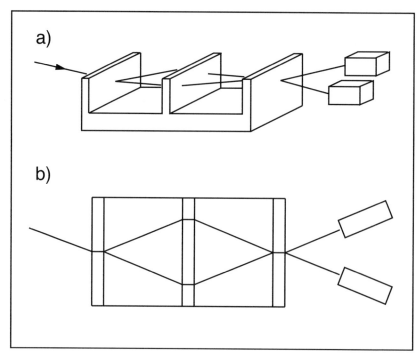

FIGURE 4.10. Neutron interferometry. Above, a view of the whole apparatus. Below, a view from above which shows the possible trajectories and the point of interference. The two rectangles represent, as always, detectors that can put into evidence the phenomena of interference.

crystal now intersect one another at the last wall, where they interfere with each other. It is not important to specify exactly how the interference is shown: the only relevant fact is that there is constructive interference at one extremely well-defined point.

It is important to note carefully the scale of the process: the dimensions of the neutron, the transverse extension of the trajectories (that is, the thickness of the rays to which the wave function is confined), the dimensions of the crystal, and the distance that separates the neutrons that enter the crystal one after the other. The neutron has dimensions on the order of 10^{-13} cm. The so-called wave packet, which represents the width of the region in which the wave function along a ray is different from zero (and also the dimensions of the transverse extension of the ray), is about 2×10^{-7} cm, or one ten-millionth of a centimeter. On the other hand, the crystal has a length of about 20 cm. The neutrons that enter the crystal are monochromatized (that means, are all of the same frequency, with the same de Broglie wavelength), and arrive at the crystal separated from each other by 300 meters, on average. To imagine the scale at work, we can multiply all the dimensions by ten million. Then the neutron would be a millionth of a centimeter; its wave function would differ from zero only in the two privileged pathways (or permitted trajectories), each of which would have a breadth of 2 cm. These trajectories would be apart from each other by about 500 to 1,000 km, and the successive neutrons would be about 3 million kilometers apart. Despite these enormous distances (compared with the 2 cm breadth of the two branches of the wave function), the wave function itself maintains such strict relationships of coherence that it brings about patterns of interference so exactly situated as to permit the marvelous accomplishments I am about to describe.

In the experiment of Figure 4.10, the two trajectories define a plane parallel to what we can call the "ground level," on which the interferometer lies. Suppose we rotate the apparatus (as in Figure 4.11) in such a way that the neutron's direction of incidence, and consequently the two branches of propagation "Low" and High" are parallel to the ground level. The hypothetical pathways L and H of our neutrons will now be found at two different distances from ground level (a difference, we recall, amounting to about ten centimeters). But at these different distances from ground level the gravitational field of the earth will be different, and

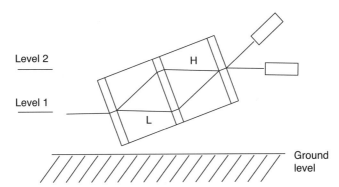

FIGURE 4.11. Positioning the interferometer as shown in the picture makes it possible to observe the difference of gravitational potential, that is, the attraction of the earth upon a neutron, along the two pathways which differ in height by about 10 cm. In this way the interferometer becomes a very refined instrument for measuring gravitational fields.

the neutron, because it possesses a tiny bit of mass, will be sensitive to this difference.[2] We now want to show that this infinitesimal difference of gravitational field is sufficiently powerful to change the propagation of the wave and thereby alter the phenomenon of interference. Still, it will be necessary to take account of an important effect. The crystal is now immersed in a nonhomogeneous gravitational field (its left superior angle is higher with respect to the ground than its right inferior angle) and is consequently deformed. Of course, such deformation would be completely irrelevant for us, but we must not forget that the interference occurs at a precise point because the crystal is virtually perfect and its crystalline planes are perfectly and regularly arranged. Therefore the deformation by itself brings a certain modification of the interference pattern. How do we identify it? For this purpose we can make use of photons, of the same wavelength as the neutrons, and introduce them into the crystal. Having no mass, the photons would experience practically no difference of gravitational potential.[3] We then proceed to introduce photons into the apparatus and identify the change in the pattern of interference that this causes. It will of course be slightly changed with respect to the previous instance, but the modification will be owing entirely to distortion of the crystal, and not gravity. We will then be able to take account of this distortion effect and adjust the outcome of the experiment accordingly. The experiment is repeated, but with a stream of neutrons. The pattern of interference is thereby displaced with respect to that of the photons, precisely because the two branches of propagation of the neutron wave function pass through regions of differing gravitational field (to see the highly refined nature of the effect, we should recall that the difference in elevation is 10 centimeters and the mass of the neutron is on the order of 10^{-24} grams). From this last modification of the interference pattern it becomes possible to measure the difference between the force of attraction of the entire earth upon a particle such as a neutron, when the neutron is displaced 10 centimeters.

But the story is not all told. We now refine the experiment still further, as illustrated in Figure 4.12. As widely known, the earth with its movement of rotation causes a centrifugal force (the Coriolis force), which varies according to latitude. We now contemplate placing the interferometer as pictured. Apart from the fact that the two pathways are found at different distances from the ground, they are also at different distances (about 10 cm) from the axis of terrestrial rotation and are thus subjected to different Coriolis forces. Of course, once we make corrections for preceding effects (distortion of the crystal and difference of gravitation), a

SUPERPOSITION THEORY

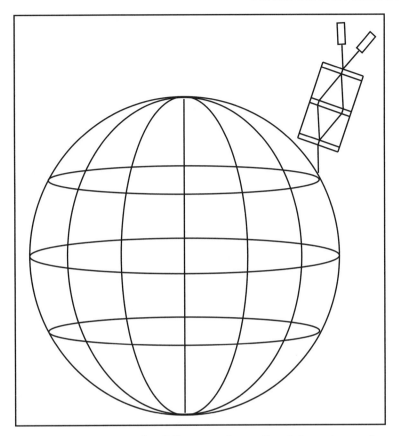

FIGURE 4. 12. Measuring the difference of centrifugal force due to the rotation of the earth at a distance of ten centimeters, through a refined experiment involving neutron interferometry.

further displacement emerges, in the position of the point of interference, which varies with the longitude. In brief, the interferometer with neutrons can reveal and measure the changing effect of the Coriolis force upon two possible neutron pathways, which are only 10 cm apart (it is not really necessary to recall here that the axis of the earth's rotation for latitudes near the equator, is off by about 5,000 km).

This concludes the analysis of interferometry with neutrons. I wanted to take the time with this matter to show the reader the level of accuracy that has been attained in our understanding of wavelike aspects exhibited by particles (here, neutrons) and how it is quite possible to take advantage of these aspects, combining them with important technological in-

CHAPTER FOUR

novations (here, the creation of virtually perfect crystals) to make powerful new tools and carry out astounding new experiments.

4.7. THE GENERAL FRAMEWORK, AND—ESPECIALLY—TEMPORAL EVOLUTION

It is now time for us to present the conceptual framework of the new theory with greater precision. Like all scientific theories, quantum formalism must yield a description of the three essential stages of every physical process: the preparation of the system, the evolution of the system, and the anticipation of the outcomes of any future measurements on the system. How are these three stages articulated in the new theory?

Preparatory Stage

This stage is realized by way of certain procedures of measurement, the results of which tell us what we need to know about the state of the system. This information should enable us to determine the quantum "state" associated with the system at an initial instant of time, conventionally assumed as t = 0. For simplicity we can indicate this state in Dirac's language as $|\Psi, 0\rangle$. For an example of a preparatory process we can think of the filtering of a photon through a polarized lens. If the photon passes the test it is polarized in the plane of the filter.

Evolution

Next, the theory should contain rules (dynamic or evolutionary equations) which permit us to use appropriate mathematical procedures for calculating what state must be associated with the system at any particular instant of time, let us say *t*, after the preparatory state. This task is accomplished by Schrödinger's equation (which I will not reproduce here). The equation has two very important features. First and foremost, the quantum evolution is *rigidly deterministic* at the level of the state vector. In other words, the knowledge of $|\Psi, 0\rangle$ determines the state $|\Psi, t\rangle$ at time *t* in an unambiguous fashion. Clearly, as with all theories, the explicit determination of the evolution of the initial state can be very difficult, if not almost impossible to determine in practice. But this is not what really matters.[4] What really matters from a conceptual point of view is that the solution of Schrödinger's equation, given the state at time

$t = 0$, is perfectly and unambiguously determined at every successive instant.

The second essential characteristic of quantum evolution is its *linearity*: i.e., it preserves the superpositions. To be more explicit, this means that if two initial states are considered, let us say $|\Psi,0\rangle$ and $|\Phi, 0\rangle$, and we know the evolution of these states at time t:

$$|\Psi, 0\rangle \to |\Psi,t\rangle,$$
$$|\Phi, 0\rangle \to |\Phi, t\rangle \tag{4.1}$$

(here we use an arrow to indicate the Schrödinger evolution from time $t = 0$ to time t), then the evolution equation implies that the initial state $|\Omega, 0\rangle = a|\Psi, 0\rangle + b|\Phi, 0\rangle$, corresponding to the superposition of the two states under consideration,[5] evolves into the same superposition of the evolved states at time t:

$$\{|\Omega, 0\rangle = [a|\Psi,0\rangle + b|\Phi,0\rangle]\} \to \{[|\Omega,t\rangle = a|\Psi,t\rangle + b|\Phi,t\rangle]\}. \tag{4.2}$$

In other words, the superpositions are preserved by the evolution.

We have seen this characteristic at work in the previous chapters. It is precisely through the linear nature of the evolution that the spin behavior of particles takes place the way it does. A particle with upward spin, along the z axis, when passed through a Stern-Gerlach magnet, is deflected upward (ending in the state $|U\rangle$) while a particle with downward spin is deflected downwards (ending in the state $|D\rangle$); but a particle with upward spin along the x axis, as amply discussed in chapter 3 (see Figures 3.7 and 3.8), is simply the linear combination of the two preceding states, and it evolves into a state that does not correspond to any precise localization of the particle, but is really the linear superposition $(1\sqrt{2})[|U\rangle + |D\rangle]$.

Future Predictions

At this stage, at last, the theory's probabilistic character emerges. Recalling the quantization of physical quantities, the most accurate knowledge possible about the system at time t, that is to say, of $|\Psi,t\rangle$, only permits us to know the *probability* of obtaining any possible results from a measurement of any observable quantity. Only in some specific cases, that is, for certain "observables," can these results be foreseen with certainty—that is, can have the probability of 1 (100% certainty) of being verified. When it comes to this important aspect of the theory—that is, its precise rules for prediction—I shall not go into any technical detail, at least for the moment, but shall limit myself to a few general remarks.[6] Once we have the state |Ψ, t> at time t and any observable W, it is possible to give an unambiguous expression (with the appropriate numerical coefficients[7]) of the state, as a superposition of states in which the observable in question assumes a precise value. Once this combination has been explicitly written, the squares of the numerical coefficients yield the probabilities of the outcomes.

In order to clarify this point, it should suffice to look at the examples we have already discussed, especially the analysis developed in the first part of section 3.6 concerning the nonepistemic nature of quantum mechanical probability. In that discussion I stressed the following point: when we know that the polarization state of a photon at time t is in the state indicated as |P> (in formula [3.5]), then the equation

$$|P\rangle = \cos\Theta \, |Q\rangle + \sin\Theta \, |R\rangle \tag{4.3}$$

represents the expression of state |P> as a superposition of states polarized along the orthogonal axes Q and R, with $\cos\Theta$ and $\sin\Theta$ as numerical coefficients. Since the state |Q> corresponds to the fact that a photon in this state would certainly pass the test of polarization in the indicated direction, and the state |R> refers to the fact that it would surely fail to pass the test, it follows that if we know that |Ψ, t> coincides with |P>, then we know the probability that it will pass ($\cos^2\Theta$) or that it will fail ($\sin^2\Theta$) a test of polarization in an arbitrary direction Q.

I would not like to go into any more detail at this time. To sum up what I have been saying: the quantum evolution is perfectly deterministic at the level of the state vector, which is to say, at the level of the mathematical entities used by the theory to describe physical situations. The knowledge of this state furnishes precise information about all the (nonepistemic) probabilities of obtaining any measurement result (permitted of course by the rules of quantization) for any physically observable quantity. Thus the stochastic nature of the theory does not stem from its own laws of evolution but from the relations between the mathematical entities used to describe the various physical situations and the results of measurement processes on the system.

4.8. THE TRUE MEANING OF THE THEORY'S PREDICTIONS

One point in particular needs emphasis now, and although it is something that may already be obvious to some readers, it is extremely important

SUPERPOSITION THEORY

and should always be kept in mind. In the preceding chapters, as the various experimental situations were being detailed, much attention was given (through showing the "detectors" in the diagrams, and pedantic repetition in the text) to stress that the probabilities of the measurements refer to the results of measurements which are actually performed.[8] Because of this characteristic, quantum mechanics becomes a theoretical model allowing only for probabilistic affirmations about the possible outcomes of measurement, and, moreover, affirmations conditional on the fact that the measurements be in fact effectively carried out. And it is owing to this quality, so typical of the orthodox interpretation, that we can legitimately say—along with Schrödinger, Einstein, and Bell—that by its formal structure the theory only speaks of "what we find if we make a measurement" and not what "exists out there." This point, whose epistemological importance cannot be overstated, will constitute one of the foci of the debate that developed in the 1930s about the interpretation of quantum formalism. The debate is still unsettled and provides the central theme for many chapters to follow.

4.9. SOME OTHER ASPECTS OF THE FORMALISM: THE DELAYED CHOICE EXPERIMENTS

In the foregoing discussion there has been more than one anticipation of a kind of Mach-Zehnder experiment that led Wheeler to consider a somewhat embarrassing phenomenon. Starting with an interferometer, as in Figure 4.8, Wheeler considered a case where the two horizontal legs of the journey were made long enough that, in the time it took for a photon to pass through them, it would be possible to add or remove the second half-silvered mirror (H-S2 in the figure). Of course, Wheeler allowed that this insertion or removal could take place through the free choice of the experimenter (showing us, by the way, that he does believe in freedom of the will). The argument runs as follows: a photon is sent into the apparatus in such a way that it hits the first half-silvered mirror (H-S1) at a precisely defined instant. After the photon has passed this mirror, the observer decides by his own choice, whether to leave the mirror H-S2 in place or to remove it. We know, from the discussion of section 4.4, that this action would have a peculiar effect: when the mirror is in place, the photon must inevitably end at detector v, with no chance for it to arrive at detector w (see Figure 4.8). But if the experimenter chooses to remove the mirror, the photon will

CHAPTER FOUR

have a 50% probability of activating either detector v or detector w, indifferently.

Wheeler's puzzle appears when we consider the case of a photon arriving when the half-silvered mirror is in position. The reason why the photon in these conditions is absolutely unable to meet detector w derives from the fact that before we get the information, the photon has passed along both pathways. For it is precisely the destructive interference between the two beams that brings nothing along direction w, and thereby leads to the impossibility of any activation of detector w. On the other hand, in the case of an individual photon that activates detector w while the half-silvered mirror is *not* in place, this photon must follow pathway H-S1 → S1 (since if it followed the other way, it would activate v and not w). But since the decision of the observer is made the instant when the photon has already passed through the first half-silvered mirror, the photon has already had to decide to follow one path only or the two paths together. The experimenter's decision has had a retroactive effect on the behavior of the photon!

Wheeler compared this strange behavior to a "great smoky dragon" (see Figure 4.13, showing the image drawn by Gilbert Field at Wheeler's instructions). The dragon has a recognizably shaped head and tail (corresponding to the fact that at the entrance and exit of the apparatus, the photon has very precise properties: to be propagated along the incident

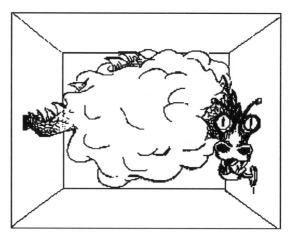

FIGURE 4.13. An eloquent picture of Wheeler's smoky dragon, which symbolizes the irreducible ambiguity, in certain cases, of the physical situation of microscopic systems.

106

ray and then to activate one of the detectors), while the dragon's body, in the place where the two possible pathways are realized, is pictured as a cloud, with no recognizable contours. This image represents the fundamental ambiguity (when seen with "classical" eyes) of microscopic processes, up until the moment when, as Wheeler says, they have been "amplified to the macroscopic level by an act of observation" (i.e., registration by one of the detectors).

To render this example yet more astounding, Wheeler (a noted expert in relativity theory) wanted to translate the process just now described to the cosmic scale by using the gravitational effect of matter upon a ray of light. We imagine the following (see Figure 4.14): a quasar (abbreviation for a "quasistellar object") located billions of light years from earth emits photons in all directions. In reality, if we consider the fact that the processes of emission are individual, this assertion means that the electri-

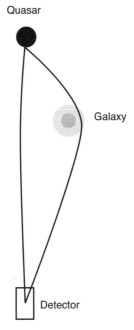

FIGURE 4.14. An analogy of the delayed choice experiment discussed above, transposed to the cosmological scale, by using the gravitational lens effect of a galaxy upon a photon. In the figure we have also drawn as lightly curved the direct quasar-earth path, to point out that it too feels the gravitational effect (although to a much smaller degree).

cal field (the equivalent of the wave function) that guides the photon is extended over an extremely vast region. We consider a case where a galaxy is situated between the quasar and the earth, but not directly on the line that joins them. Suppose we concentrate our attention on the two possible pathways to be taken by the light on its journey from the quasar to our telescope. The first path is directly quasar-to-earth, the other aims toward the zone beyond the galaxy. General relativity states that the gravitational attraction of a significant mass can also affect photons, causing them to deviate from a rectilinear path of propagation.[9] Now it can happen that the path of the light rays going past the galaxy can reach the telescope after being deviated by the gravitational lens. We then suppose that the difference of phase amounts to one-half a wavelength, making the whole situation analogous to a giant Mach-Zehnder apparatus. Now—so Wheeler argues—the moment I decide to insert—or not insert—a half-silvered mirror at the point of intersection of the two paths, I *now* force the photon, which has been traveling for billions of years from the quasar, to decide if it is going to follow both pathways or only one!

It is important to realize that the process Wheeler assumed in constructing his surprising argument is not mere science fiction. It is a reality of nature. Two quasistellar objects classified by astronomers as 0957 + 561 A and B, which appeared to form an angle with each other (the telescope had to be moved from observing one to the other), have been recognized as being one and the same object, but producing two images. The first image would correspond to the direct propagation of light from the quasar to earth, the second to a virtual image brought about by a gravitational lens. Figure 4.15 represents (to the left) an image of the quasar as it appears without any manipulation, while to the right is seen the same image with a digital subtraction of the star's profile. What remains is compatible with the presence of a galaxy producing the effect of a gravitational lens. Independent evidence was afterward found for the existence of a galaxy in this location, so that today there is a remarkable consensus among scientists that the two images on the left refer to a single stellar entity. Thus the hypothetical experiment of Wheeler is quite possible in principle.

Of course, the argument cannot be pressed too literally. The theory tells us, without ambiguity, that before a measurement, the wave function is different from zero on both pathways. Furthermore, the theory does not allow for "retrodictions" without inconsistencies.[10] The purpose of Wheeler's analysis, as he explained, was to show how, according to quan-

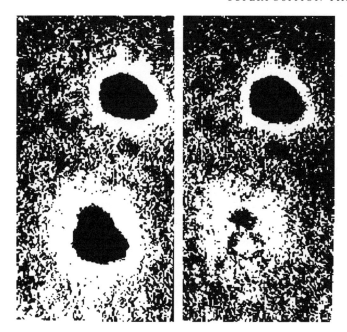

FIGURE 4.15. To the left is shown the image of what appear to be a couple of quasistellar objects (quasars). To the right, the contribution of the lower spot has been removed from the left-side picture, assuming that it was a virtual image — by the gravitational lens effect of a galaxy — of the same object that appears in the upper part. The resulting image is compatible with the presence of a galaxy that would really be responsible for the double image, as explained in the text.

tum thinking, "no elementary phenomenon is a phenomenon until it is a registered (observed) phenomenon." We will later have to turn our attention to the details of this matter, when we discuss the epistemological implications of the theory and the debate that has developed about its proper interpretation.

We have now reached the end of the chapter, and taken important steps toward the comprehension of the new theory. It is quite likely that some readers who were not yet familiar with quantum theory may find themselves a little disoriented. This is completely natural, and they are in good company. All the great physicists have shared the reader's amazement.

Niels Bohr once said, "If you do not get *schwindlig* [dizzy] sometimes when you think about these things then you have not really understood it." And when the theory was still being elaborated, Einstein wrote that

CHAPTER FOUR

"the more success the quantum theory has, the sillier it looks." But even more significant, perhaps, is a statement of the Nobel prize winner Richard Feynman. What Feynman said is all the more remarkable when we consider that he said it many years after the scientific community as a whole had accepted the theory, and the more detailed formulations had greatly extended the range of the theory's validity by bringing amazing new discoveries:

> There was a time when the newspapers said that only twelve men understood the theory of relativity. I do not believe there ever was such a time. There might have been a time when only one man did, because he was the only guy who caught on, before he wrote his paper. But after people read the paper a lot of people understood the theory of relativity in some way or other, certainly more than twelve. On the other hand I think I can safely say that nobody understands quantum mechanics.

CHAPTER FIVE

Visualization and Scientific Progress

> If I had to choose between your wave mechanics and the matrix mechanics, I would give preference to the former, owing to its greater visualizability [Anschaulichkeit].
> —*Hendrik Antoon Lorentz*

IN THIS CHAPTER we will take time out from our step-by-step analysis of the origin and development of quantum mechanics to consider one of the typical problems that arise in connection with the evolution of scientific theories: the question of how to visualize what we understand. How important is this for the development of science, especially when major conceptual changes are involved? The imaginative capacity of individual scientists can be crucial, whether we consider the natural processes under investigation or the hypotheses proposed to solve quandaries. The history of quantum theory provides ample matter for discussion on this topic.

The analysis we are attempting to develop naturally belongs to the field of cognitive science, because it studies the modes of "scientific discovery" with reference not merely to the objective data of the particular science in question, but also to certain other elements that contribute to understanding the "mysteries" of the cognitive process itself. The interest of such study has been expressed very effectively by H. A. Simon: "The cognitive sciences show how it is possible to construct a normative theory, or logic, if you will, of the processes that lead to scientific discovery. They show how phenomena that are normally given the curious labels 'intuitive' or 'creative' can be treated operationally and formally."

In our case I would like to illustrate the influence of two radically different viewpoints on the development of the new quantum theory. The two views reflect two distinct attitudes toward "doing science," and we will see how important it was for some of the founders of the theory to be able to fall back on intuitions and frame their thinking in approved conceptual schemes. Others, however, were stirred by the disconcerting

CHAPTER FIVE

new experimental evidence to venture into the unknown and be guided exclusively by formal developments, without any worries about how well the theory could be intuited. We shall see how personal attitudes and psychological motivations—going well beyond the goal of pure and simple agreement with experimental facts—can play a special role in directing the path of scientific progress.

This theme has been discussed in recent literature with reference to other scientific constructs, and especially by the brilliant physicist and historian of science, Arthur Miller. The detailed framework and penetrating analysis Miller has provided in several works inspires my own, necessarily briefer, exposition.

5.1. NIELS BOHR

We can begin with reviewing the personal evolution in thinking experienced by Niels Bohr, one of the theory's founders. At the beginning of the century, when the scientific community found itself face to face with a real crisis over the impossibility of fitting certain fundamental physical processes (including the behavior of atoms) into the scheme of classical physics, it was still believed possible to find a solution in extending the intuitions of classical theory to microscopic systems, and especially by still retaining the classical position toward phenomena as taking place in space and time and governed by rigid laws of causality. This emerged clearly from Bohr's fundamental work published in 1913, where he presented his atomic model. He insisted on showing how this model permitted a visualization of the atom as a miniature Copernican system and thereby in harmony to some extent with classical conceptions. This position was strengthened during the years preceding 1923. Even though quite conscious of conceptual divergences from the classical framework, and the problems such divergences posed, Bohr presented his vision of the world in a series of lectures at Göttingen, Cambridge, and Copenhagen, using large red and black diagrams showing the electron orbits of atoms of various elements. These pictures fascinated the audience, especially the one showing radium, with each of its eighty-eight meticulously charted orbits (Figure 5.1). Unfortunately the originals do not survive.

Bohr's insistence on presenting a graphic model of a reality that was already known to be extremely problematic can be understood as an extreme attempt to reduce the elusive and embarrassing aspects of microscopic systems into a conceptual framework solidly anchored to a classi-

FIGURE 5.1. Pictorial representation of the 88 electron orbits of an atom of radium, according to Bohr's model.

cal imagination. Bohr was well aware that this process would require future adjustments of no small extent. In fact, problems were not slow in emerging. For, whereas the imagery of classical mechanics permitted the visualization of electrons on stationary orbits like a Copernican system, the transitions of electrons from one orbit to another (associated with the emission and absorption of light) did not permit of any visualization. How was it possible to leap from one state to another without passing through the intermediate states declared impossible by the theory? Bohr himself wavered on this problem of the interaction of atoms and radiation, unable to find a logical pathway between two options: on the one hand, a simple understanding of the atoms as having Keplerian motions, and on the other, the adoption of a position analogous to Planck's, whereby the material constituents of the structures responsible for the emission and absorption of electromagnetic radiation were likened to tiny oscillators.

A clear representation of the dilemma he faced emerges from a comment made by Max Born, one of the researchers who personally experienced the discomforts and hopes of that period: In 1923 Born wrote: "A remarkable and alluring result of Bohr's atomic theory is the demonstration that the atom is a miniature planetary system . . . the thought that the laws of the macrocosmos in the small reflect the terrestrial world obviously exercises a great magic on mankind's mind; indeed its form is rooted in the superstition (which is as old as the history of thought) that the the destiny of man can be read from the stars. The astrological mysticism has disappeared from science, but what remains is the endeavor toward the knowledge of the unity of the laws of the world." But he felt compelled to add: "Alas . . . the possibility of considering the atom as a planetary system has its limits."

Bohr made a further attempt to save at least some element of classical thinking. This was his principle of correspondence, which required that

CHAPTER FIVE

at limits set by large quantum numbers (that is, the cases where the quantum energy levels of the atoms differ so little from one another that they simulate a continuum), quantum predictions can be reduced to classical. This does in fact occur, but it does not remove the underlying difficulties. As Bohr's student and collaborator Arnold Sommerfeld wrote, "It is indeed astonishing how much of the classical wave theory still remains even in spectroscopic processes of a decidedly quantum character," but he added the illuminating remark that the principle of correspondence represented only a temporary solution and that "modern physics is thus for the present confronted with irreconcilable contradictions and must frankly confess its *non liquet* [Latin for 'it is not clear']." The crucial problem, clearly, is still the conflict between the classical description that sees every physical process as developing in a space-time continuum, and the essentially discontinuous nature of quantum processes. Bohr insisted on his attempt to overcome these difficulties through recourse to Planck's schematism, as I mentioned before. This involved explaining the interactions between radiation and matter by visualizing the material particles as miniature oscillators. But this ended in a blind alley: to take account of all the spectral lines, one had to have as many oscillators as there were possible transitions to or from a given quantum state. The visualization of microscopic processes became a lost cause. How can it be possible to conceive an electron as an object with an orbit and at the same time as a collection of oscillators? In Arthur Miller's words, "the honeymoon of the Bohr theory was over."

5.2. Heisenberg: The Conscious Refusal to Visualize

Immediately after Sommerfeld's confession, Pauli wrote to him in approval of his lucid analysis: "Your frank *non liquet* is a thousand times preferable to me than the well-constructed artificial apparent solution of the problem by Bohr, Kramers, and Slater." A little later he would point out the path he thought had to be followed: "I believe that the energy and momentum values of the stationary states are somewhat more real than the 'orbits.'" In this way, he resumed a theme he had already touched on before, when he had harshly criticized the picturesque images of atoms that appeared in a 1923 book: "Even though the demand of these children for *Anschaulichkeit* [German for "visualizability"] is in part legitimate and healthy, still this demand should never count in physics as an argument for retaining systems of concepts. Once the systems of concepts are settled, then will *Anschaulichkeit* be regained."

At this point Heisenberg seized the occasion to turn the problem upside down. The difficulties Bohr was having with multiple-electron systems, the discovery of electron spin (a concept with no classical analogue), the ideas of de Broglie on the wavelike nature of particles, and the observations of Einstein on the indistinguishability of quantum particles (we shall discuss this important point in a later chapter)—all this convinced him of the need for abandoning the classical schemata in their entirety. Instead of staying chained to contradictory mechanical images and to concepts that were insusceptible of being directly experienced (such as electron orbits), the task should be to concentrate attention on observable quantities such as spectral lines, and such characteristics as polarization and intensity.

At the same time, Heisenberg was developing his own formal apparatus, matrix mechanics, which deals more directly with these aspects of atomic processes. Encouraged by his success, he expressed ever more precise reservations about Bohr's model: "It is my genuine conviction that an interpretation of the Rydberg formula [which expresses the line structure of atomic spectra] in the case of circular and elliptical orbits of classical geometry has not the least physical meaning." He thereby reached a precise position that required the renunciation of any visualization in the traditional sense, that is, as based on perception and inspired by familiar theoretical patterns, and favored instead a new visualization founded solely on the formal and mathematical schema at the base of the new theory. This was a deliberate renunciation of visualization, a renunciation that appeared to be an advantage insofar as it brought about a higher level of abstraction and made it possible to conceive new and previously hidden aspects of reality.

5.3. Schrödinger: An Attempt at Visualization in Terms of the Wave Paradigm

As mentioned in chapter 1, although Schrödinger was aware of Heisenberg's theory, the theory had not been formulated and interpreted precisely enough to be considered a complete theoretical model. Consequently, Schrödinger was inspired in his own research by the ideas of de Broglie and Einstein on the wavelike nature of all physical processes. Adopting this perspective would permit a visualization that allied the new description with concepts and physical situations that found a natural place within the recognized classical conceptual schemata. From his first fundamental works of 1926 he would take a clear position on this point and on matrix mechanics: "My theory was inspired by L. de Broglie and by the short but incomplete remarks by A. Einstein. No genetic relation whatever with Heisenberg is known to me. I knew of this theory, but felt discouraged, not to say repelled, by the methods of transcendental algebra, which appeared very difficult to me, and by the lack of *Anschaulichkeit*."

What kind of image *was* acceptable to Schrödinger? He did not like Bohr's orbits, he considered energy levels and transitions to be abstract concepts, and he felt a lively desire to substitute the fundamental discontinuity of matrix mechanics with a scheme that treated material systems as if they were waves, and thus was occupied with continuous processes in time and space, describable by means of a very familiar mathematical apparatus, namely, wave equations.

Einstein immediately expressed his favor for Schrödinger's position[1]: "I am convinced that you have made a decisive advance with your formulation of the quantum condition, just as I am convinced that the Heisenberg-Bohr method is misleading." And of course Heisenberg gave voice to a contrary position: "The more I reflect on the physical portion of Schrödinger's theory, the more disgusting I find it. . . . What Schrödinger writes on the visualizability of his theory . . . I consider trash."

Clearly, not even wave mechanics would have an easy time of it. Schrödinger's original hope of interpreting wave matter as real waves describing the density of charge or matter in space, was quickly shown to be an unfounded hope. Heisenberg himself proved that even a wave function that was well localized in space (and thereby accounting for the fact that an electron appears characterized by a practically pointlike mass and charge) could not be localized for more than an extremely brief moment of time.[2] The image Schrödinger proposes is untenable: the electron that we all know as pointlike (or better, that has been revealed to us as pointlike in all the experiments designed to identify it) is quickly transformed into a kind of evanescent cloud with a charge and mass distributed over macroscopic regions. This criticism is quite pertinent. In fact, to avoid experimental falsification and logical inconsistency, the theory of Schrödinger requires one to replace the interpretation which makes reference to the density of matter or mass with the interpretation of Born, where the wave function squared represents the *probability density* of position.

5.4. CONCLUSIONS

The purpose of this chapter has been to alert the reader to the fact that different scientists can have very different attitudes toward what they believe to be the most adequate method of dealing with a disconcerting scientific problem, and how some are more inclined to put a priority on visualizing the physical processes they study, while others find it preferable, and more stimulating, to trust in mathematical formalism.

The story reveals that in the specific case in question, neither of the two approaches ended up being more effective than the other in the development of the new theory—in fact, both contributed a great deal. This should not surprise us. The microscopic world that was unveiling itself to the eyes of researchers was so strangely new and revolutionary, that the attempts to extrapolate classical conceptions (a procedure that has had some success) at last proved a miserable failure. On the other hand, this very novelty demanded formal approaches so unfamiliar as to be difficult of comprehension. But the interaction between these various modes of confronting the problem (a process, as we have seen, that encountered harsh setbacks) has brought about a very significant result: the birth of a new theory with incomparable predictive capacity, that has opened the door to a whole new way of understanding reality.

I would like to conclude the chapter (which, as I mentioned before, is largely based on Miller's work), with an interesting observation that Miller himself proposed to his readers. Quantum theory has undergone noteworthy developments in later periods as well. The need to make it compatible with the requirements of relativity theory has given rise to new difficulties, but this has also led in turn to significant scientific discoveries (it should be enough to recall that it was just this labor that brought Dirac to predict the existence of antimatter), and ultimately to the formulation of modern quantum field theories. These theories are not soluble in exact terms and thus require the elaboration of suitable methods of approximation to be handled. The most characteristic of these approximations in use is the so-called perturbation theory.

The concept is fairly simple: we try to derive the solution of equations through successive approximations. Every step in this process of approximation produces an ever greater proliferation of terms. Feynman proposes an effective method of representing in graphic terms the various and complex terms of a perturbative series. In origin these diagrams are mere symbols that allow for useful manipulations. But it is interesting

CHAPTER FIVE

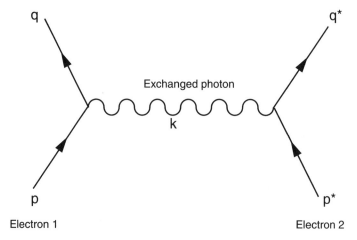

FIGURE 5.2. Feynman's diagram that corresponds to the term that approximately describes the process of diffusion taking place when two electrons collide.

(and a fitting conclusion to this chapter) to observe how such diagrams very naturally become the basis for a new visualization of physical processes.

I will limit myself to a few examples. One term of the series, accounting in a certain approximation for the collision between two electrons, and for their reciprocal repulsion, is pictured in Figure 5.2. In the diagram, according to Feynman's general prescriptions, the continuous lines (when thought of as moving in the direction of the arrows) represent states of electrons that are propagated in the directions indicated, while the wavy line in between represents a photon. The diagram, which was actually introduced as a mere symbol of a complex mathematical expression, can be read in a directly visualizable way: the electron to the left, with momentum p, emits a photon with momentum k, and thus rebounds by changing its direction to q. The second electron, with momentum p*, absorbs the photon and rebounds along the new direction q*. The global effect is the scattering of one electron by another.

Another diagram of interest is shown in Figure 5.3. When we keep in mind that according to Feynman's rules, a straight line when moving in a direction opposite to the arrow, represents an antielectron, the two figures (a) and (b) of Figure 5.3 represent in visual terms, respectively, (a) the process of annihilation of an electron and an antielectron along with

VISUALIZATION AND SCIENTIFIC PROGRESS

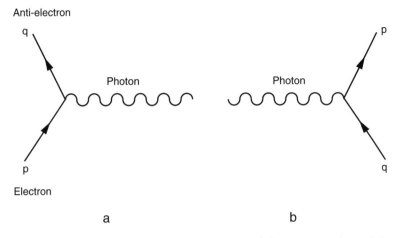

FIGURE 5.3. A representation à la Feynman of the process of annihilation of an electron-antielectron pair, and of the creation of another such pair brought about by a photon. Of course, such processes take place only under very special conditions that need not be discussed in detail in this context.

the production of a photon (light) and (b) the creation, by a photon, of an electron-antielectron pair.

To conclude: in the modern versions of quantum theory—carried out at levels more abstract than the theoretical models of Bohr—it has become common practice, and a rather natural, obvious way of grasping the theory, to visualize complicated mathematical expressions with pictures and diagrams, to which an immediate, intuitive physical sense can be attributed. One could say, in a way, that there has been a late recovery of visualizability, but still as part of the new conceptual perspective required by the most recent versions of quantum formalism.

119

CHAPTER SIX

The Interpretation of the Theory

> Einstein's verdict on quantum mechanics came as a hard blow to me; he rejected it not for any definite reason, but rather referring to an "inner voice.". . . It [the rejection] was based on a basic difference of philosophical attitude, which separated Einstein from the younger generation to which I felt I belonged, although I was only a few years younger than Einstein.
> —*Max Born*

IN THE FOREGOING CHAPTERS we have followed the first phase of the new theory's elaboration. We can now go on to study the second phase, beginning with a glimpse of the prospect shared by the scientific community in the years immediately following 1927. The precise mathematical formulation of the theory was now completed (in 1926 Schrödinger had demonstrated the equivalence of matrix mechanics and wave mechanics), and the peculiarities of microsystem behavior were considered achieved scientific data. Consensus was particularly to be found on a few essential points:

- The dual wave and particle nature of all physical systems
- The discontinuity of natural processes

In addition, with the sole exception of Einstein,[1] the scientific community was also in agreement on the following:

- The fundamentally random nature of physical processes
- The uncertainty principle, which is to say, the impossibility of determining at one and the same time the position and velocity of a particle

An obvious corollary of these two facts was the following:

- The need to grant scientific dignity to a theory that did not in principle permit certain predictions (from the classical perspective, an undeniable demand), and from which, at most, only probabilities of possible outcomes of physical processes were forthcoming

It was only natural that this situation opened a very lively debate about the true meaning of the formalism, driving the scientific community to speculate about the objectives of scientific research. Serious doubts were soon raised about the very object of science, and about the objective reality of natural phenomena. Many scientists lost all hope of constructing a coherent vision of what was really happening at the atomic and subatomic scale. The debate took on precise philosophical connotations, and as we shall see, many physicists, very conscious of the great successes the new theory had to its credit, and influenced by positivist and instrumentalist philosophical currents (see below, section 6.2, for a summary), went so far as not only to give up on the construction of a coherent framework of physical processes, but even to oppose any attempts to do so. The need to construct such a framework began to be looked on as a kind of metaphysical pretense, a superfluity foreign to the spirit of genuine scientific research. Consequently, even to take such a project seriously was considered if not blasphemous, at least "unprofessional." Eventually, the peculiar role the very process of observation played within the formalism became an object of debate in itself, leading some scientists to deny the very possibility of attributing any properties, or even objective reality, to microscopic constituents independently of the processes of observation to which they were being subjected.

Various scientists assumed various positions about these fundamental problems. We will try to comment on them in the following sections, while pointing out as well the subtle differences of viewpoint even within the group (Bohr, Heisenberg, Born, Pauli) who elaborated the so-called orthodox or Copenhagen interpretation, the position on quantum formalism that eventually won the field. We will also discuss the opinions of some firm opponents of this position, Einstein above all.

Before we study the sides taken in this titanic struggle, it will be convenient to set forth in a preliminary and schematic fashion certain basic considerations, which at an intuitive level, and without any special critical explication, have undoubtedly helped orient the scientific community toward adopting an unfamiliar epistemological viewpoint. After these comments, we will be in a position to catalogue and discuss some of the philosophical perspectives of major interest, now so essential for understanding the meaning, motivation, and cultural-historical roots of the second fundamental phase of the theory's evolution.

CHAPTER SIX

6.1. Some Basic Arguments for a Radical Change of Perspective

In this section I will set forth a few basic points that, while not in historical order or in the same language as the arguments of the great protagonists, should nevertheless help to clarify how the various positions emerged and were defended. For even though those positions were not absolutely novel, they were certainly unusual for the scientific community at the time—a community that was very thoroughly grounded in the classical conception of natural processes.

Classical physics treats the objects of our daily experience (leaving aside, at present, the description of thermodynamic processes in terms of their microscopic constituents), and is strictly deterministic, asserting that if the state of a system is perfectly known (for example, in the case of a mechanical system, if one knows all the positions and velocities of all the particles that come into play), then all the quantifiable values of all physical magnitudes can be known with precision: the energy, angular momentum, exact location of the center of gravity, and so forth. These quantities can be measured with instruments and appropriate methods designed to verify that these magnitudes did in fact have all the particular values that had been predicted.

Of course, everyone was perfectly aware that this conception represented a great simplification of the actual state of affairs, and that it was based on some (legitimate) approximations. In particular, the simplification presupposed that, when treating anything short of the entire universe, any physical system could be considered in isolation from its environment. In the second place, it assumed that the procedure of verification satisfied one of the two following conditions:

1. Either the disturbance induced into the measured system can be made as small as possible, making it legitimate to assert that, both before and after being measured, the system still has the property in question—a certain energy value, for example.
2. Or (if we can't avoid procedures that appreciably change the system), it should be possible, at least in principle, by using the universal laws that govern all physical processes (including the system-appparatus interaction), to assess the effect of the measurement, in order to make the necessary corrections. These in turn allow us to infer the actual quantities both before and after the act of measurement.

To provide a concrete illustration of the problem, we can consider how the temperature of a gas can be measured by means of a thermometer. If the gas is not at the same temperature as the thermometer, the process of measurement (that is to say, bringing the thermometer into contact with the gas) activates a process that changes the temperature of both systems, tending to equalize them. Then, at the end of the process, the temperature indicated by the thermometer will be different from what the gas actually had to start with. And so, with reference to the cases described in (1) and (2) above, we can think either that it is possible to construct a thermometer with so small a thermal capacity that it does not exchange heat with the gas to any appreciable extent, or that the (by no means negligible) thermal capacity of a thermometer is exactly known, so that it would be possible, using the equations that govern the tendency to thermal equilibrium, to deduce precisely what the actual temperature of the gas was before it was measured (obviously, the later temperature would likewise be known, and equal to what the thermometer indicates).

It is important to stress how crucial it was for the advance of science, that seventeenth-century scientists had the intuition that under the right conditions it was legitimate to use approximations, and to study a given physical process by way of a very idealized version of the process. This was rightly emphasized by Bell (1987b): "At the beginning, the natural philosophers sought to understand the world around them. In their struggle to do this, they hit upon the great idea of imagining artificially simplified situations, within which the number of factors involved could be reduced to a minimum. *Divide et impera* (Divide and Conquer!) was their motto. This was the birth of experimental science."

Now let us try to look through the eyes of the typical scientist of the revolutionary period we are concerned with. On the one hand, Bohr's and Heisenberg's analysis compelled recognition that the disturbance induced into a microscopic system by the process of measurement could be drastically reduced but never completely canceled (we should recall how every attempt to determine position inevitably changes the velocity of a particle, in a random manner that cannot be precisely determined). On the other hand, the *finite* character of the quantum of action made it impossible to diminish the exchange of energy involved in a process continuously and to the vanishing point. In sum, the scientist could not evade the fact that he was dealing with microscopic systems that were delicate in the extreme.

For example, consider the very simple measurement-of-position experiment illustrated in Figure 6.1, different from the one we studied in chap-

CHAPTER SIX

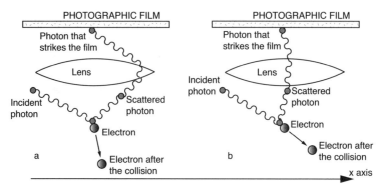

FIGURE 6.1. Measurement of the position of an electron along the x axis, using a lens to observe it. The electron becomes illuminated, scattering photons that leave an impression on the film. At the point of impact on the film, the position of the electron (along the x axis) can be inferred the more certainly, the shorter the wavelength of the light. The figure shows that how if even a single photon is enough to make an impression on the film (since it can reach its final position by following any one of the possible paths through the lens, it is impossible to know how much velocity in direction x it has imparted to the electron at impact. If the photon was scattered at a large angle (case a, to the left), the electron bounces off in a direction almost perpendicular to the x axis, and thus does not acquire velocity in that direction (we suppose that the electron is at rest before it is observed). But if the impact was of the kind illustrated in case b (to the right) the electron recoils in the direction of the x axis and has some nonzero component of velocity in that direction. So, whereas before the measurement it was known that the the electron was practically at rest, but we did not know its position, after the measurement we know its position with a remarkable accuracy but in the meantime we have lost the knowledge of its velocity since we cannot determine the actual path taken by the photon.

ter 4. It is the microscopic observation of a particle (an electron, say) being illuminated by radiation. The whole process of measurement (the illumination of the particle, the diffusion of light quanta, the formation of an image on the photographic plate or on the retina of the human observer) requires that the particle be subjected to a bombardment of photons that must increase in energy for any increase of accuracy desired in the measurement. In fact, in order to increase the accuracy, one has to make the wavelength smaller, i.e., to increase the frequency (or, equivalently, the energy) of the incident quanta of light. For an electron, even a single impact with one of the incident photons is equivalent to a rather strong collision. Of course, this process of measurement is highly invasive! Furthermore, as the caption under the figure explains, the genuinely random nature of the process does not allow us, as it would in the classical situation, to use the theory to provide an exact assessment of the disturbance, and then infer from the result what the property of the particle actually was before the act of measurement.

In this situation it is only natural to ask a question like the following: "If the process of measurement inevitably (and in a way we cannot know) changes the property of the system we are studying, does this not make for a serious crisis about the very possibility of a coherent spatiotemporal description of what happens at the atomic level?" The process becomes so complicated—both the system measured and the instrument of measuring are so inextricably connected—that it is not in the least clear how much we can legitimately infer about the characteristics objectively possessed by the electron (by "objective" is meant "independently of our own action of identifying it"). Pushing this line of thought to its logical conclusion, if we can even say that it is completely impossible to measure with any desired precision both the velocity and the position of a microscopic particle, what sense does it make to say that the particle even possesses these two properties?

As we see, even a rather superficial analysis using the new conceptual framework raises fundamental questions about the knowability of the real, about the role of the measuring process, and about the observer. The present chapter will be devoted to these questions. To understand their importance, and the meaning of the answers various scientists have given, we must illustrate some of the philosophical positions that have been taken up with regard to the problems at hand.

6.2. A Brief Philosophical Interlude: Idealism, Realism, Materialism, and Positivism

The title of this section makes reference to ideas and positions concerning the important philosophical and epistemological problems that have been pursued in a wider context than the strictly scientific one, and much earlier than the historical period we are studying. Above all, I want to emphasize that the relevant changes of perspective required by the new theory have made these issues unavoidable for anyone at all interested in the conceptual implications of science. In this connection I would like to cite a statement that expresses very effectively this strong interaction between philosophy and science—an interaction that does not arise from purely cultural or historical considerations but is imposed, as it were, by the very nature of things: "In my opinion, the twentieth century," (wrote Abner Shimony, a well-known American physicist and epistemologist, in 1989) "is one of the golden ages of metaphysics, probably surpassed by the fourth century B.C. in conceptual innovations, but probably surpassing all previous ages in the control and precision of the best metaphysical thinking." Among the reasons for this situation he assigns a principal role to the fact that "parts of the natural sciences have been deepened to the point where evidence can be brought to bear in a controlled way upon problems traditionally classified as metaphysical." He goes so far as to sketch out what we could, with suitable caution, call "experimental metaphysics."

Such considerations encourage us to study how certain fundamental philosophical themes have become associated with problems raised by the study of atomic systems. Here we will review some specific positions concerning the real and its knowability.

Idealism

According to the American philosopher Anthony Flew, this position is "a name given to a group of philosophical theories, that have in common the view that what would normally be called the 'external world' is somehow created by the mind." Of course, some versions of idealism are more radical than others: for some, representations of the mind have a higher status of being than objects; for others, objects themselves are created by the mind; for still others, nothing exists at all apart from mind. A truly consistent idealist would have only a limited interest in science, since science is primarily concerned with understanding the external world—that is, something whose existence is seriously questioned or even denied outright by philosophical idealism.

Realism

This position involves an affirmation that the objects of the external world have a real existence completely independent of being perceived by anyone. There are also various kinds of realism, the simplest one being known as "naive" realism, which assumes that a direct knowledge of objects and their properties is acquired by the process of perception. This position encounters some difficulties, of course, when problems of error and/or sense illusion are raised. Many realists find it appropriate to acknowledge that the mind plays a mediating or interpretative role in sense perception. But even if this process in a certain sense falsifies the knowledge of the object, the assumption that the object is distinct and independent from perception remains a cardinal point of the realist position. However, to accept a realist position implies no specific position about the problem of mind and conscious perception, especially concerning its degree of dependence on material things, which materialism assumes to be complete. What is essential is the acceptance, contrary to any version of idealism, that something certainly exists "out there" (that is, objects) other than the mind.

Materialism

The kernel of the materialist position consists in the claim that all existence is subordinate to, or can be reduced to, material objects and their interrelations. Matter is the only reality. Evan Squires has rightly emphasized how an articulation of the materialist position that would take into account the evolution of science, and thus be better suited to the present situation, would say that all the laws and fundamental principles of nature coincide with the laws and principles of physics,[2] and are exhausted in them. In particular, materialism tends to negate any specific nature to mental processes: whatever they may be, they are not primary. The materialist position is normally accompanied by a negation of spiritual and religious values, and by an atheistic standpoint.

Positivism and Logical Positivism

As already stated, this position has certainly influenced, and been consciously drawn upon, by some of the protagonists in the debate over quantum theory. In philosophy, positivism is a movement that considers scientific method as the only source of knowledge. The position is noted for its openly declared hostility toward religion and traditional philosophy, especially metaphysics. Its roots go back to empiricism, but the term "positivism" appears for the first time in philosophy at the beginning of the nineteenth century, in the works of Auguste Comte. This thinker maintained the absolute priority of observation and direct verifiability in determining the truth of any statement: all hypotheses and/or metaphys-

CHAPTER SIX

ical or subjective arguments not reducible to direct observation were meaningless.

Aside from the influence it had on Bohr and his colleagues, positivism will be of interest to us especially for its close connections with the general problem of the meaning of scientific research. One of its motivations was the goal of making scientific method more rigorous. The position of positivism not only puts emphasis on the need for theory to take account of experimental data, but also obliged scientists to avoid the risk of assertions, conjectures, hypotheses about entities, events, or properties of things that could not be directly verified. This kept them from wandering into fields now judged less serious such as metaphysics or religion.

It was not by chance that one of the leaders of this movement was Ernst Mach, professor of physics at Prague and Vienna, whose famous book on mechanics won the unqualified praise of Einstein. Mach joined the cause of positivist philosophy but concentrated his attention on physics, a science he felt was the ideal model for all other knowledge. Following positivist logic, Mach denied any value, and in fact any meaning at all, to any assertion that was not empirically verifiable. The target of one of his attacks was the idea of absolute space and absolute time — ideas which could not be subjected to experimental verification. In this respect his thinking exercised an influence on the young Einstein's development of the theory of relativity. Ironically, the very same rigorously scientific approach that motivated the positivists actually played an antiscientific role, and led Mach to join the opponents of Boltzmann in denying any kind of reality to atoms and molecules. Indeed, the kinetic theory of gases was motivated more by a desire to find a unified description of natural processes than by any need to refine the (already excellent) agreement between thermodynamics and experience.

For our purposes it is very important to recall that positivism initiated a school of thought called logical positivism (or logical empiricism) that originated in Vienna around 1920 in the work of a group of philosophers, scientists and mathematicians known as the "Vienna Circle." The leader was Moritz Schlick, professor of philosophy at the University of Vienna, and it counted among its members such men as Rudolf Carnap and Kurt Gödel, the greatest mathematician of the past century. They were inspired not only by Mach but also by Bertrand Russell and Ludwig Wittgenstein. The elaboration of Einstein's theory of relativity had a great influence on this group, by their own admission. The Vienna Circle bolstered the positivist assumption that only scientific knowing represented true knowledge, by stipulating that all arguments be made with an absolutely rigorous logic. This should not seem something that goes without saying, since this stipulation was not broadly intended, but made special reference to important developments at the time in the elaboration of systems of formal logic. Consequently, the essential qualities of knowledge became verifiability and logical deducibility from verifiable premises, that is, tautologies. Metaphysical assertions do not have such characteristics and are therefore senseless. The same holds for ethical, aesthetic, and theological assertions, which are irrelevant from the cognitive point of view, but can possess an emotional meaning. From this standpoint, logical positivism appeared as an attempt to redefine philosophy and its goals, limiting its object to the analysis and definition of scientific language. The philosopher Alfred Jules Ayer maintained in his book *Logical Positivism* that "the originality of the logical positivists lay in their making the impossibility of metaphysics depend not on the nature of what could be known but upon the nature of what could be said." The only task left to philosophy was the analysis of the affirmations that philosophy makes in the light of the aforesaid principles. Such affirmations were linguistic, not factual, and thus philosophy becomes a part of logic. As Carnap put it, "Philosophy is the logic of the sciences."

The ideas of the Vienna Circle dominated the philosophy of science in the crucial years that interest us. This can be a very useful fact to know. For example, to the mind of a positivist, the realist position becomes meaningless by definition. Since there is no way to verify if anything exists "out there" apart from the experience we have of it, even to assert the existence of a reality apart from the perceiving subject (and even posing the very problem) can be neither true nor false—it is simply meaningless. Similarly, logical positivism also maintains that there is no sense in making distinctions or choosing between two rival theories with the same predictive content. The sole motive for privileging one theory over another would be criteria of simplicity.

The review we have now made of the basic positions can help us understand how various philosophers, scientists, and epistemologists of the time took sides in the debate over the interpretation of quantum theory. We need especially to be aware of the two philosophically antithetical positions, realism and positivism, that played so important a role in the formulation of the theory. Both conceptual views required verifiability and logical consistency in physical theories; both accepted as a possible leading criterion the discovery of some kind of simplicity or elegance. But when it came to the ultimate goals of scientific research, they were divided by a great chasm.

Positivism came in an almost natural way to assume instrumentalist positions about scientific knowledge that in various ways approached positions of philosophical idealism.[3] In fact instrumentalism assumes as a basic principle that scientific theories should not be preoccupied with "what exists out there," but should simply be the instruments of experimental prediction. In particular, from this perspective, there is no sense in posing the question whether scientific theories are true or false. Their meaning derives completely and exhaustively from their practical utility. The realist, on the contrary, believes in the existence of an external reality and the primary goal of his research is to understand it. This conviction was expressed very vividly by Einstein when he said, "I want to know God's thoughts . . . the rest are details."

6.3. The Positions of the Protagonists

It should be a fairly simple task now to review the positions of some of the great scientific thinkers on the interpretation of the elusive reality of microsystems, and to find traces of a more or less conscious philosophical formation, and an adherence to one or the other position just described.

Bohr

Bohr's position was remarkably articulate but not always clear. For our purposes it will be enough to identify some of the fundamental characteristics of his thought. Above all, Bohr gave an absolute prominence to language: "Our task is to communicate experience and ideas to others. We must strive continually to extend the scope of our description, but in such a way that our messages do not thereby lose their objective or unambiguous character." This position of his youth became strong and articulate when he entered the arena of science as a combatant. He reflected on his scientific practice and observed that in fact all researchers, no matter from what field, used macroscopic instruments of measurement to acquire data which would then have to be communicated in the familiar language of classical physics. This language thus became a kind of logical prerequisite, even for a theory that treated nonclassical phenomena and processes. Microcosmic events had to be amplified in order to be observed and described, but the amplification involved macroscopic instruments and a description that inevitably fell back on language that had become scientific patrimony from the times of Galileo and Newton—

namely, the language of classical physics. In this emphasis on the absolutely prominent role of language can be seen clearly a trace of the logical positivists' insistence on linguistic and logical analysis.[4] "There is no quantum world. There is only an abstract physical description. It is wrong to think that the task of physics is to find out how nature is. Physics concerns what we can say about nature."

Similarly, clear elements of a positivist position can be found in Bohr's cherished notion of complementarity. The fact that different procedures bring different aspects of reality into the light—aspects that, as Bell noted so acutely, are not so much complementary as completely contradictory to one another—did not disturb Bohr in the least. What reasons led him to justify this peculiar position? It is impossible to show contradictory aspects in the same experiment. Since different and incompatible experiments cannot be brought into confrontation with one another, the question whether their results are consistent with each other has no meaning, and cannot be posed. It is not legitimate to use a description outside the strictly defined range of its applicability.

In the same way, for Bohr the interpretative problems of quantum mechanics have no meaning insofar as every question about its interpretation is posed in the language of classical physics but refers to a field outside the applicability of classical physics. It was no coincidence that the only assertions Bohr considered legitimate were those referring to situations where the principle of correspondence was valid, that is, in a context that largely reenters the domain of classical physics.

From its very birth, then, the orthodox interpretation of quantum formalism found sustenance in logical positivism, and in turn provided it with scientific support.

Heisenberg

We already observed that there were important differences between this man's position and Bohr's. It was not by chance that Heisenberg was very critical of Bohr's attempts to save the visualizability of atomic processes, as discussed at length in the preceding chapter. In fact, the need to distinguish between the observed object and the process of observation appears ever clearer in Heisenberg's thought, together with the conviction that only the process of observation has an effective reality, represented by its results. The atomic world begins to disappear, like Alice's Cheshire cat, or like the body of Wheeler's dragon: "In the experiments about atomic events we have to do with things and facts, with phenomena that are just as real as any phenomena in daily life. But the atoms or the elementary particles are not as real: they form a world of potentialities or possibilities rather than one of things or facts." And he wrote further: "In the past we attributed to electrons 'the same sort of reality' as the objects of our daily world," but "in the course of time this representation has proved to be false." For the "electron and the atom possess not any degree of direct physical reality as the objects of daily experience.... "what the words 'wave' or 'particle' mean, we know not any more."

Once again a positivist position emerges, that largely recalls what Mach had said about the microscopic constituents of a gas: every assertion that makes reference to something beyond our specific and direct experience is devoid of meaning.

Born

For this thinker, I will only recall that he was the one who made it possible to overcome the dilemma Schrödinger found himself in with regard to the meaning of the wave function. Born proposed the probabilistic interpretation. Hence the square of the wave function $|\Psi(r)|^2$ provides the probability density for the position of the particle. But, it may be asked, the probability of *what*? The answer Born gave to this question (and, it should be admitted, the only answer that does not lead to direct logical contradictions with the formalism) is the answer that would become definitively enshrined in the orthodox interpretation: it is not the probability that the particle *is* at point r, but the probability that it *will be found at point r, if a measurement is taken for its position*.

Jordan

Jordan was Born's student and collaborator. He so thoroughly adopted the implications of Heisenberg's position and the probabilistic interpretation, that he maintained with firm conviction that the act of measurement not only disturbs what is measured, it even *produces* it. For example, in a process of measurement of position by a microscope, "the electron is forced to a decision. We compel it to *assume a definite position*; previously it was, in general, neither here nor there; it had not yet made its decision for a definite position.... [I]f by another experiment the *velocity* of the electron is being measured, this means: the electron is compelled to decide itself for some exactly defined value of the velocity. ... [W]e ourselves produce the results of the measurement."

Pauli

Some of Pauli's views were mentioned in the previous chapter. Here I shall only cite one famous statement of his, made to rebuff Einstein's repeated attacks on the Copenhagen orthodoxy. Once again, his definition of what statements can legitimately be made is clearly reminiscent of neopositivism: "As O. Stern said recently, one should no more rack one's brain about the problem of whether something one cannot know anything about exists all the same, than about the ancient question of how many angels are able to sit on the point of a needle. But it seems to me that Einstein's questions are ultimately always of this kind."

In what follows, we shall see how this statement must be considered incorrect, since it has recently become possible — by a kind of experimental metaphysics as suggested by Abner Shimony — to give a clear answer (alas, an answer that Einstein would not have liked) in a laboratory to some of those questions of Einstein that Pauli (and probably all the scientists who unconditionally adopted the new paradigm) would have considered metaphysical, and thus illegitimate. At the moment, however, these particular developments are not of leading interest to us: the foregoing remarks are intended only to show how the positivist climate of that period contributed in no small way to preparing the ground for the interpretation of the theory that would eventually become dominant.

A few observations need to be made, now that we have concluded our brief review of the thought of the great exponents of the "orthodox" interpretation of quantum mechanics. Above all, it must be admitted that it is perhaps to some extent misleading to use a single label — "orthodoxy" — to indicate a whole series of positions that have important shades of difference. The many quotations of Bohr so often cited by authors are not always clear or even mutually consistent. Bohr's positions differ appreciably from those of the other members of the group. For example, I do not believe that Bohr would have subscribed to Jordan's assertion that the observer actually creates the measured properties themselves. The most practical test for the complexity of positions held by these scientists (even if to a certain degree they were in harmony with one another) is to ask any given physicist or epistemologist today what he or she means by "the Copenhagen interpretation of quantum mechanics." It will be difficult to get the same answer twice. But apart from these important differences there are certain points that can be said to be shared by all those who elaborated the interpretation. Above all, there is the view that the primary objective of science is to foresee the outcomes of our measurements: the goal of scientific research is not to understand what exists but to understand what we get when we perform measurements. Consequently, the interpretation lends an absolutely primary role to the observer. It has been said many times that quantum mechanics has in a certain sense undone the Copernican revolution by placing the human being at the center of the universe. In sum, it is extremely important to emphasize the conceptual structure of the orthodox interpretation from an epistemological perspective: the theory only makes probabilistic assertions about the results of measurements on physical systems, but such assertions are conditional on the effective carrying out of the measurements. We can already anticipate some of the problems to be encountered in connection with the desire to adopt the formalism for the description of all physical processes, especially the evolution of the entire universe. Problems of this sort are still open for discussion and have become the subject of a lively debate. Many chapters of the present book are dedicated to this.

At this point we can briefly study the positions taken by the opponents of this interpretation of quantum formalism.

De Broglie

De Broglie was always profoundly dissatisfied with the dominant interpretation of the theory he had done so much to create. In particular, he had the conviction that it was wrong to abandon the classical idea of the trajectory of material particles. The quantum wave plays an important role, to be sure, in determining these trajectories, but the reality of the trajectories is not to be doubted. Later we shall see the importance of this position for the development of the so-called hidden variable theories, a line of thinking that has initiated controversy and new theoretical researches, and even a series of experiments that have brought a great advance in our understanding of microscopic phenomena. Certain aspects have come to light that were entirely unsuspected in the period we are studying.

Schrödinger

As I have already emphasized several times, Schrödinger is one of the founding fathers of the new theory, and his thinking has many points in common with de Broglie and Einstein. He was primarily inspired by a firmly rooted belief in the need for saving spatiotemporal continuity in the description of physical reality. In one of a whole series of fundamental works written in 1926 dealing with the problem of the wave-particle duality, he stated as follows:

> No special meaning is to be attached to the electron path itself . . . and still less to the position of an electron in its path. . . . The wave not only fills the whole path domain all at once, but also extends far beyond in all directions. . . . This contradiction is so strongly felt that it has even been doubted whether what goes on in an atom can be described within the scheme of space and time. From a philosophical standpoint, I should consider a conclusive decision in this sense as equivalent to a complete surrender. For we cannot really avoid our thinking in terms of space and time, and what we cannot comprehend within it, we cannot comprehend at all.

In October of the same year he was invited by Bohr to Copenhagen, to join Heisenberg and himself in discussing the problems of the new theory, but he did not change his opinion. On this occasion, speaking in reference to his opponents' position about the discontinuous measurement process (the so-called reduction of the wave packet we shall discuss in what follows) he went so far as to say: "If we have to go on with these damned quantum jumps, then I'm sorry that I ever got involved." Later we shall see how this instinctive refusal of the discontinuity implied by the formalism, of the impossibility of fitting the phenomena into a coherent spatiotemporal framework, led him to be one of the few who could appreciate the true and profound import of the fundamental work done in 1935 by Einstein, Podolsky, and Rosen (to be discussed at length in chapter 8). But before concluding this brief discussion of Schrödinger's critical position, it is worthwhile to emphasize that, unlike Einstein (with whom he shared many convictions), Schrödinger never considered quantum indeterminism as a possible source of problems. In fact, from 1922 on, in agreement with his teacher Franz Exner, he was ready to accept it as a fact that the fundamental laws of nature might have a statistical character: "Once we have discarded our rooted predilection for absolute Causality, we shall succeed in overcoming the difficulties."

Einstein

As I have already emphasized more than once, Einstein's position underwent important changes as the years went by, and detailed study is needed to understand the implications of these changes. All too often, both technical and popular works have betrayed the thought of this great genius in his struggle to understand quantum mechanics, by referring almost exclusively to his supposedly firm and unwavering opposition to indeterminism. The endless citation of his famous statement, "God does not play dice," has certainly not contributed much to a widespread knowledge of his actual opinions. For he also once said, "I have spent at least one hundred times more time in thinking about quantum problems than about general relativity." Such persistent and dedicated effort would not have led merely to the naive prejudices that are usually put forward as his primary motives for opposing "orthodoxy."

To be sure, Einstein, at least in the first phase of the theory's elaboration, mounted a detailed attack against it, and showed his intolerance of the idea that natural processes could be fundamentally random. Born's comment, quoted at the head of this chapter, was made in reference to the more extensive version of Einstein's statement: "Quantum mechanics is certainly imposing, but an inner voice tells me that it is not yet the real thing. The theory says a lot, but does not really bring us any closer to the secret of the 'old one.' I, at any rate, am convinced that *He* is not playing at dice." To take this as the synthesis of all his thinking does not do justice to the depth and richness of Einstein's position: it amounts to an accusation of dogmatism entirely foreign to his style of thinking. To prove this, it should be enough to report some of his statements which show that he did contemplate seriously the possibility of a stochastic nature of natural processes. As early as 1920 he wrote, "That business about causality causes me a lot of trouble, too. Can the quantum absorption and emission of light ever be understood in the sense of the complete causality requirement, or would a statistical residue remain? I must admit that there I lack the courage of a conviction. However I would be very unhappy to renounce complete causality." This was far from any kind of a priori dogmatism. It became a real misrepresentation of his true position, espoused at first by some of the scientists whom he had disagreed with from the beginning, but later passed along and exaggerated by careless and superficial popularizers. On this point it is interesting to recall that Pauli, a fierce critic of Einstein's views, accused Born of erecting a mythical Einstein, a "straw man" for his own attacks. As Pauli wrote, "In particular Einstein does not consider the concept of 'determinism' to be as fundamental as it is frequently held to be (as he told me emphatically many times) . . . he *disputes* that he uses as a criterion for the admissibility of a theory the question, 'Is it rigorously deterministic?'"

For Einstein, as for Schrödinger, the necessity of saving spatiotemporal continuity played a fundamental role and its absence was a factor in his dissatisfaction with indeterminism. In a letter to Born he wrote,

> I cannot make a case for my attitude in physics which you would consider at all reasonable. I admit, of course, that there is a considerable amount of validity in the statistical approach which you were the first to recognize clearly as necessary given the framework of the existing formalism. I cannot seriously believe in it because the theory cannot be reconciled with the idea that physics should represent a reality in time and space, free from spooky actions at a distance.

However, as much as I want to correct a one-sided view of Einstein's thought, I do not want to fall into the opposite error. In order to counteract that impression, and at the same time underline the profound interests and honest struggle of this great mind, I shall refer once again to his letter to Born: "We have become Antipodean [i.e., are at opposite poles] in our scientific expectations. You believe in the God who plays dice, and I in complete law and order in a world which objectively exists, and which I, in a wildly speculative way, am trying to capture. I firmly *believe*, but I hope that someone will discover a more realistic way, or rather a more tangible basis than it has been my lot to find."

These words seem to me a kind of summing up of Einstein's long philosophical journey. He passed from a position of denying indeterminism—that is, the "God who plays dice"—in the first debate with Bohr (as we shall see in the next chapter), to a firm profession of faith in the objective existence of the world. This will be the leitmotif of Einstein's argument about the incompleteness of the theory, which will be taken up in chapter 8.

But before we begin our study of the debate, and its interesting ramifications, we shall have to make more explicit the "orthodox" position on the effect of performing measurements on a physical system. According to this view, the measuring process is what makes the actual properties emerge from the limbo of formal potentialities. The effect in question—Schrödinger's "damned quantum jump"—taking place in connection with the act of measurement, is known in technical parlance as the "reduction of the wave packet." The following section is devoted to its analysis.

6.4. The Process of Measurement and the Reduction of the Wave Packet

When we explained (in section 4.7) how quantum formalism describes the preparation and evolution of a physical system, and what kinds of

CHAPTER SIX

predictions it allows us to make, we put off a detailed discussion of the effect of the measurement process on the state vector of the system. We must now attend to this most delicate issue, in order to ease the reader's comprehension of later chapters. The analysis will progress in stages, by looking first at the simplest cases, where the measured quantities are quantized, and the possible outcomes are few. Typical examples of this would be measuring the polarization of single photons or the spin components of electrons along specified directions, where the possible outcomes are "the photon passes (or does not pass) the test," or "the electron has upwards (or downward) spin." From these cases we move to more complicated ones, such as the measurement of physical quantities like position, that are not quantized. In these cases, the possible outcomes cover a continuous range of values.

It will also be a good idea to define more precisely what it means to decompose any given state into the sum of other states corresponding to definite values for certain quantities. This plays a crucial role in connecting the theoretical description of measurement with the experimental procedure that actually brings the measurement about.

One point of particular importance gives us a precise indication of the effect of measurement, and how it is expressed in formal language. The idea is rather simple. The formalism should incorporate the legitimate (and in fact necessary) request that we are going to explicate and that finds a perfect confirmation in the experiments. If a measurement is made of a certain observable W, one of the values permitted by the theory is obtained. Typically, if the observable is quantized (as are the energies of atomic systems) the result of the measurement, say ω_k, will be one among a certain collection of possible values $(\omega_1, \omega_2, \omega_3, \ldots)$. Now, if we repeat the same measurement of W immediately afterward, it would be only natural to assume that the same measurement would have to provide, again with certainty (by which I mean, "with probability equal to one") the result obtained beforehand, that is, ω_k.

To clarify the argument, we can begin by considering the experiment of measuring the polarization of a photon along a certain direction Q, assuming that the incident photon (that is, before being measured) is characterized by a state of plane polarization in plane P, forming an angle θ with plane Q (the reader should consult Figure 2.14). We already know that two mutually exclusive outcomes are possible: either the photon passes the test, or it is absorbed by the filter. The occurrence of one or the other result is genuinely random. It is characterized by the probabilities defined by the decomposition of the state of plane polarization P into the

INTERPRETATION OF THE THEORY

states (as standardized in Section 2.4) of polarization along Q and along Q's perpendicular R (as in Formula 3.5, which we repeat here for convenience):

$$|P\rangle = \cos\theta \,|Q\rangle + \sin\theta \,|R\rangle. \qquad (6.1)$$

As for the probabilities of these outcomes, we already know that this expression implies (according to the prescriptions of the theory) that such probabilities are equal to $\cos^2\theta$ and $\sin^2\theta$, respectively. Let us now identify the observable W: it is what corresponds to the fact that the photon passes or does not pass the test of polarization. To simplify our treatment of the problem, let us also suppose that the amplification process of the microevent takes place via an instrument equipped with a meter indicating the value +1 on a scale, when the photon does pass the test, and –1 when it does not (see Figure 6.2). We can therefore say that the observable W can assume only one of the two values: +1 or –1. In what follows we shall indicate with symbols $\mathcal{P}(\Omega = +1|P)$ and $\mathcal{P}(\Omega = -1|P)$ the probabilities of the two outcomes, when immediately before the measurement, the state of the system is |P>.

Suppose now that we successfully carry out the measurement in question, and the photon passes the test. Our device will register that W = +1. If we repeat the same act of measurement on the system immediately afterward, as explained above, we must require that we will obtain the result +1 with certainty. We now ask ourselves: what does this requirement imply about the state of the photon right after the first test was passed? Since we know that any state of plane polarization (we can call it |T>) can be expressed as a combination of the states |Q> and |R> with coefficients a and b:

$$|T\rangle = a|Q\rangle + b|R\rangle \text{ with } |a|^2 + |b|^2 = 1, \qquad (6.2)$$

and since the numbers $|a|^2$ and $|b|^2$ represent, respectively, the probabilities of the outcome W = +1 and W = –1 in a measurement of polarization along direction Q, after the measurement we should have

$$|a|^2 = 1, \ |b|^2 = 0, \qquad (6.3)$$

because only in this case will the expected result, that is, W = +1, occur with certainty. It follows from this that, after the first measurement, in the case when the photon passes the test, its state would be |Q>.[5] The effect of the measurement can therefore be described in the present case by saying that the system (which, before being measured, was in the state |P> = $\cos\theta$ |Q> + $\sin\theta$ |R>), is transformed instantaneously (and thus

a). Before the measurement

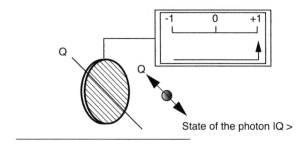

b). After the measurement: the case when the photon passes the test

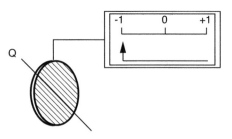

c). After the measurement: the case when the photon does not pass the test

FIGURE 6.2. The measurement of the observable "polarization along plane Q" of a photon already characterized by a polarization plane along P. The crucial element of the apparatus is a polarized filter but the fact that this filter absorbs or does not absorb the photon is amplified by an apparatus that uses a macroscopic gauge to indicate the outcome of the measurement.

INTERPRETATION OF THE THEORY

discontinuously—performing one of Schrödinger's "damned quantum jumps") into the state |Q>. Formally, we would write

$$|P\rangle = \cos\theta |Q\rangle + \sin\theta |R\rangle \xrightarrow[\Omega = +1]{\text{measurement}} |Q\rangle. \quad (6.4)$$

In other words, the measurement abruptly changes the state,[6] by eliminating the component $\sin\theta |R\rangle$ that refers to the other possible outcome (W = −1) and by restandardizing the state vector (that is, by substituting $\cos\theta$ by 1).

It is worthwhile repeating that the two outcomes are mutually exclusive and that they exhaust all the possibilities. In fact, as discussed in chapter three, the photon either passes the test or does not pass, with no other alternatives. In the present case, for the specific tool of measurement we are using (namely, a polarized filter), if the outcome had been negative (W = −1), the photon would have been absorbed by the filter and disappeared. However, this is not inevitable: it is possible to construct so-called ideal measuring instruments, which would not destroy the system but simply keep to the business of registering the outcome of the measurement. Under such conditions, instead of limiting ourselves to the process we have summarized (in Section 6.4, for example), we have to take into account the possibility that after the measurement (and with the appropriate probabilities) one of the following two changes of the state vector occurs:

$$|P\rangle \cos\theta |Q\rangle + \sin\theta |R\rangle \xrightarrow[\Omega = +1]{\text{measurement}} |Q\rangle, \quad (6.5)$$

$$|P\rangle \cos\theta |Q\rangle + \sin\theta |R\rangle \xrightarrow[\Omega = -1]{\text{measurement}} |R\rangle. \quad (6.6)$$

The sudden change of the system's state as an effect of the process of measurement, a random change that depends on the outcome of the measurement itself, is technically referred to as the "reduction" or "collapse" of the wave packet. After being measured, the state ought to coincide with a state that guarantees that the system will give the same result when subjected again to the same process of measurement.

. . . . .

The above discussion permits us to focus on an extremely important matter: the problem of how to express the state before the measurement in terms of the specific states corresponding to the certain occurrence of the alternative outcomes (|Q> to W = +1 and |R> to W = −1). In fact, we should recall that the relation be-

141

CHAPTER SIX

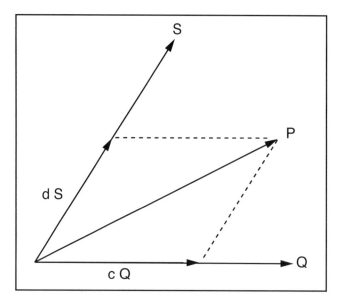

FIGURE 6.3. The expression of a vector P of a given length (let us say, equal to 1 in some appropriate unit) as the sum of two vectors cQ and dS. These are obtained by reducing—by the factors c and d—two unit-length vectors lying in the same plane as P and forming any nonzero angle to it.

tween the three states |P>, |Q>, and |R> was brought about, as shown in chapter 3, through reference to the decomposition of the electric field vector (described in chapter 2) along two perpendicular directions. However, such a vector can always be expressed (according to the parallelogram rule) as the sum of two vectors even not perpendicular, provided they are not aligned (Figure 6.3).

This suggests that it might be possible (and in fact it is possible) to express the state |P> as a combination of the states of polarization planes |Q> and |S> along two nonperpendicular directions Q and S:

$$|P> = c|Q> + d|S>. \qquad (6.7)$$

Nevertheless, this decomposition is not meaningful for describing the actual situation of a measuring process intended to determine whether or not a photon passes a test of polarization in direction Q. In fact, since the state |S> corresponds to a polarization in plane S that is not at right angles to Q, a photon that is found in this state can very well pass another test for polarization along Q. In other words, the two states |Q> and |S> do not each correspond to one of the two states that give a certain, precise, and distinct outcome in a measurement of polarization along Q. And here we can see the whole interpretative structure collapsing:

INTERPRETATION OF THE THEORY

supposing the first measurement gives the outcome S, the second can give a different one, namely, Q. In short, if we want to avoid logical and interpretative inconsistencies, we must assume something that is automatically guaranteed by the way the formalism describes the physical observables: given an observable W, and the set of possible outcomes of a measurement, these outcomes must be exhaustive and mutually exclusive. The change of state that results as a consequence of the measurement of a specific observable must be such as to assign, in an immediate repetition of the same measurement, a probability of 1 to the outcome just attained and a probability of 0 to any other potential outcome.

With these preliminaries we can now proceed to discuss the phenomenon of the collapse of the wave packet in the case of a typical nonquantized observable, i.e., one that can assume any value. For convenience we can refer to the position observable along a certain axis.

So, then, we consider a particle, an electron for example, characterized (as in Section 3.7) by a state that corresponds to a wave function $\Psi(x)$ as represented in Figure 6.4. We know that $|\Psi(x)|^2$ represents the probability density (according to Born) of finding a particle at point x when a measurement is taken of its position. However, because this variable position is not quantized and thus can assume any one of the values for which $|\Psi(x)|^2$ is different from zero, we immediately realize that there is no meaning in asking if the particle can be found at a definite point.

The reasonable question that can be asked here concerns the probability of finding the particle in an interval Δ, which can be designated as small as we want, as long as it is not reduced to one point. If we need to visualize the process of measurement, we can compare it to placing a screen with a slit of width Δ in the path of the light waves (see Figure 6.5), and induce a process of amplification that, as before, brings about different macroscopic situations for the dial of some

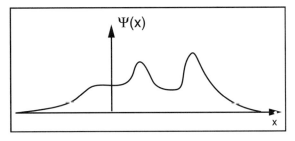

FIGURE 6.4. The dependence of the electron's wave function on the position variable x, before the process of measurement.

143

CHAPTER SIX

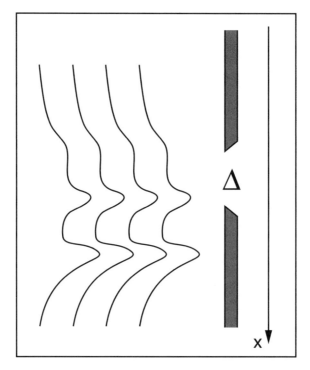

FIGURE 6.5. The wave function of Figure 6.4, propagated virtually unchanged (the figure shows a very short interval of time) meets a screen. The process in the figure is that of a measurement made to ascertain whether the electron is found inside or outside the interval Δ.

apparatus. The process of measurement is thus adapted to respond to the question: will the particle be found inside the interval Δ or outside it? For a correct analysis of the problem we would argue as follows:

1. First, we recall that the probability of finding a particle in any given interval is obtained by the area subtended by the function $|\Psi(x)|^2$ in the interval under consideration (referring also to Figure 3.11). From this it follows in particular that, since the particle is certainly to be found in some place when it is measured, the area subtended by the function on the entire x axis ought to be equal to 1, that is, if we consider the interval to be $(-\infty, +\infty)$, then the particle will certainly be found inside it. This characteristic of the wave function—that is, the fact that the square of the modulus subtends an area equal to 1—is essential for the probabilistic interpretation. It represents the property of a standardized state, and is technically referred to as *normal-*

INTERPRETATION OF THE THEORY

ization. From now on I will use this more correct expression to denote this characteristic of the states.

2. For the special problem at issue, once the interval Δ has been fixed, we write the wave function Ψ(x) as the sum of the two functions $\Psi_\Delta(x)$ and $\Psi_{\text{not in }\Delta}(x)$, defined as follows (see Figure 6.6):

$\Psi_\Delta(x) = \Psi(x)$ if x is in Δ, $\Psi_\Delta(x) = 0$ if x does not belong to Δ,

$\Psi_{\text{not in }\Delta}(x) = 0$ if x is in Δ, $\Psi_{\text{not in }\Delta}(x) = \Psi(x)$ if x does not belong to Δ.

Two observations need to be made. First, the functions being considered $\Psi_\Delta(x)$ and $\Psi_{\text{not in }\Delta}(x)$ are identically equal to zero outside the interval and inside the interval, respectively. Secondly, they are not normalized because it is the *sum* of the areas subtended by their squares that is equal to 1. But now (see Figure 6.7) we can define two new functions $\hat{\Psi}_\Delta(x) = N\Psi_\Delta(x)$ and $\hat{\Psi}_{\text{not in }\Delta}(x) = M\hat{\Psi}_{\text{not in }\Delta}(x)$ proportional to the previous ones but which are normalized, that is, the area subtended by the square of each function is equal to 1.

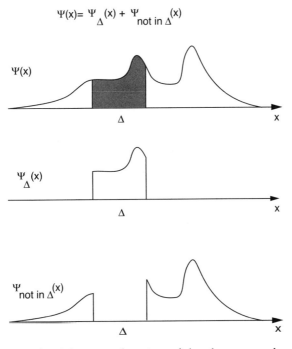

FIGURE 6.6. Graph of the wave function of the electron as the sum of the two functions appropriate for understanding a process of measurement undertaken to determine if the particle is in the interval Δ or outside it.

CHAPTER SIX

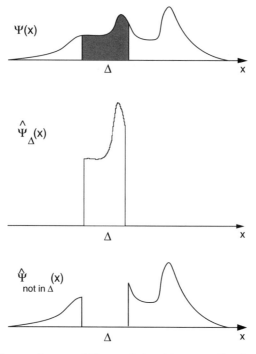

FIGURE 6.7. The analogue of Figure 6.6, with normalized versions of the functions of Figure 6.7.

3. We can then write the following identity:

$$\Psi(x) = \frac{1}{N}[\hat{\Psi}_\Delta(x)] + \frac{1}{M}[\hat{\Psi}_{\text{not in }\Delta}(x)], \qquad (6.8)$$

which represents the expression of the function in terms of the two functions $\hat{\Psi}_\Delta(x)$ and $\hat{\Psi}_{\text{not in }\Delta}(x)$ associated with certain outcomes for the process of measurement. In fact, the function $\hat{\Psi}_\Delta(x)$ is different from zero only in that interval, and its square subtends an area equal to 1, in virtue of which we can guarantee that if the particle is in this state, it will certainly be found within the interval when a measurement is taken. Alternatively, the function $\hat{\Psi}_{\text{not in }\Delta}(x)$ is different from zero only outside the same interval and it subtends an area equal to 1, in virtue of which it guarantees that if the particle is in this state, it will certainly be found outside interval Δ when measured. Especially to be noted is how the two certain outcomes associated with these two states are once again mutually exclusive and how one of the two will certainly occur in the process of measurement.

4. It follows from the formula above (6.8) that $1/N^2$ and $1/M^2$ (really, noth-

INTERPRETATION OF THE THEORY

ing other than the area subtended by the square $|\Psi(x)|^2$ of the original function within the interval Δ, and the area outside it) represent the probabilities of the two outcomes: the particle will be found inside the interval, or outside it, in an act of measurement.

5. The effect of the measurement for the case in question should be obvious. Since we require that by repeating the measurement the same result will occur, we have

$$\Psi(x) = \left[\frac{1}{N}\hat{\Psi}_\Delta(x) + \frac{1}{M}\hat{\Psi}_{\text{not in }\Delta}(x)\right] \xrightarrow[\text{the particle is in }\Delta]{\text{measurement}} \hat{\Psi}_\Delta(x),$$

$$\Psi(x) = \left[\frac{1}{N}\hat{\Psi}_\Delta(x) + \frac{1}{M}\hat{\Psi}_{\text{not in }\Delta}(x)\right] \xrightarrow[\text{the particle is not in }\Delta]{\text{measurement}} \hat{\Psi}_{\text{not in }\Delta}(x). \tag{6.9}$$

The above two formulas express the modalities of the process of the reduction of the wave packet for a system of the type we have been considering.

All the examples considered hitherto have to do with measurements that only give two mutually exclusive outcomes. To avoid any misunderstanding, it would be well to point out that usually there are more outcomes. To mention just one example, we can consider a case of measurement of position analogous to the one just discussed, but where the particle has a choice of being found in interval Δ_1, Δ_2, or outside both intervals, so that there would be three possible outcomes. The argument would follow the same steps as just now shown, and the analogous graph is shown in Figure 6.8.

Finally, we can also go back to the discussion of quantized angular momentum (as in Section 3.4) to understand how the measurement of the component along the z axis can yield one of the possible values.

I think the discussion has gone far enough at this point to allow the reader willing to put forth the needed effort to understand fully the logic that underlies both the predictions of the theory and the effects of measurement, two aspects that I would like once more to emphasize, knowing full well I risk the charge of being too repetitive:

When we are interested in some observable, we must consider the possible and mutually exclusive outcomes that the theory says can occur in a measurement process of the observable in question. Then we can identify appropriate states in such a way that each one of these assigns a probability of 1 to only one of the permitted outcomes, and a probability of zero to all the others. Next, we must express the state of the system before the measurement as a linear superposition (with appropriate coefficients) of the states in question. The formal structure of the theory guarantees that all these requirements are satisfied. Now, the theory asserts that when a measurement is carried out on the observable under inspection, one of the

CHAPTER SIX

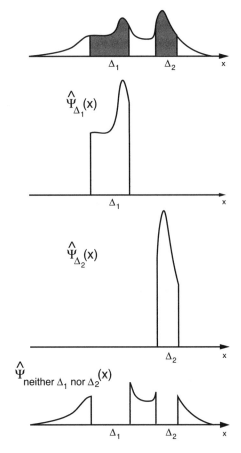

FIGURE 6.8. The analogue of Figure 6.7, but showing a measurement of position with three possible outcomes.

possible and mutually exclusive outcomes will be obtained, with a probability equal to the square of the coefficient that multiplies (in the superposition) the normalized state for which that outcome is certain. Finally, the measurement induces the phenomenon of the collapse of the wave packet: the state *after* the measurement is no longer what it was *before* the measurement. It has been instantaneously transformed into the normalized state corresponding to the outcome actually obtained.

· · · · ·

With these specifications, we have the formal preparation necessary to follow our next theme: the exciting developments that came after the process of the theory's formulation and interpretation.

CHAPTER SEVEN

The Bohr-Einstein Dialogue

> No more profound intellectual debate
> has ever been conducted—and, since they
> were both men of the loftiest spirit, it
> was conducted with the noble feeling on
> both sides. If two men are going to
> disagree, on the subject of the most
> ultimate concern to them both, then that
> is the way to do it.
> —*Charles Percy Snow*

MORE THAN ONCE I have emphasized how Einstein's position on quantum mechanics, and especially on the right way to interpret it, underwent important changes over the years. His increasingly subtle attacks forced the defenders of the orthodox position to reply to his criticisms point by point. The exchange brought a deeper understanding of formalism in many crucial points.

Bohr was always the natural opponent of Einstein (see Figure 7.1). More than any other member of the School of Copenhagen, Bohr was animated by a particular personal interest in the philosophical and epistemological aspects of the theory, finding in surprising aspects of the microscopic world the stimulus to advance daring hypotheses about reality and its knowability, such as his theory of complementarity. But the two giants of scientific thought had a profound respect for one another, and each paid careful attention to the other's penetrating insights. The study of this debate goes beyond historical interest: as we shall see, Einstein's attacks often provoked reactions on Bohr's part that required him to revisit and reelaborate crucial elements of the formalism and its interpretation. This very articulate process in which other famous scientists also participated, from Heisenberg to Born, and from Schrödinger to John von Neumann, led to a progressively more exacting focus on certain especially problematic points of the theory. Even though it was not always clear where the debate was headed, in the long run the way was paved for important developments. The process had several stages, which we shall review in detail in this and the following chapters.

CHAPTER SEVEN

FIGURE 7.1. Bohr and Einstein on a walk through Brussels during one of the Solvay Conferences; the photograph is by Paul Ehrenfest.

First Phase

In this phase, Einstein tried to disprove the uncertainty principle by suggesting Gedankenexperimente (thought experiments) that would allow for the simultaneous determination, to any desired accuracy, of incompatible variables such as velocity and position. According to all the official records, Bohr came out of this debate as the winner. But there is more to it: in my view, the records do not sufficiently reveal the price Bohr had to pay for the victory. For without admitting it in any explicit way, he was forced to modify certain assertions he had formerly made concerning the status of macrosystems and the role they played in measuring microscopic variables. In particular, as we shall see, he had to acknowledge that such systems could not be assumed always to be governed by the laws of classical physics. This fact, even though it did not become immediately clear, contributed in an important way to the dawning realization (a few years later) by the more attentive members of the scientific community that one of the more serious problems the new theory would have to face was its inability to demarcate with mathematical precision just where microscopic processes leave off and where macroscopic processes begin. This ambiguity has played an important role in the debate since 1930. Many of the "solutions" proposed to rescue the formalism (from Bohr to von Neumann, and Wigner to more recent attempts) are based on the possibility of setting this line of demarcation (to a large extent) at any point one pleases.

Second Phase

This phase is intertwined in terms of historical period and in terms of logic with the first and third phases. However, for clarity of exposition, and in order to bring out the influence it had on certain important developments, it makes sense to describe this phase separately, and keep it distinct from the others. For here is where Born's probabilistic interpretation comes especially to the fore. While Einstein recognized that it was practically impossible to determine simultaneously the values of incompatible magnitudes, he was not willing to grant, with the Copenhagen orthodoxy, that we could not even *think* there were precise values really there to be known. In other words, as we have mentioned before, Einstein was very reluctant to accept the nonepistemic nature of quantum probabilities, and insisted that the theory (which he saw as otherwise very valuable[1]) did not tell the whole story. While it furnished an appropriate description at a certain level, it said nothing about what lay underneath: "I have the highest respect for the achievements of the physicists of the last generation, the achievement that goes by the name *quantum mechanics*. I think that such a theory represents a profound level of the truth, but I also think that its need for laws of a statistical nature will render it obsolescent." And further: "There is no doubt that quantum mechanics has seized hold of a beautiful element of truth and that it will be a touchstone for a future theoretical basis in that it must be deducible as a limiting case from that basis, just as electrostatics is deducible from the Maxwell equations of the electromagnetic field or as thermodynamics is deducible from statistical mechanics." Nevertheless, he was firm in his conviction that "assuming the success of efforts to accomplish a complete physical description, the statistical quantum theory would, within the framework of future physics, take an approximately analogous position to statistical mechanics within the framework of classical mechanics."

I would like to emphasize that this was the first time that the possibility of a completion of the theory was contemplated, though not explicitly stated as such. We shall see how this line of thinking would bring David Bohm to formulate the first explicit example of a theory of hidden variables. This was in large measure due to the influence of his frequent discussions with Einstein at Princeton.

Once again, the position taken by Einstein provoked a reaction that led to later developments. The reaction became concretely realized in von Neumann's theorem that it was impossible to complete the theory in a deterministic way. This was a peculiar outcome to which we shall return to discuss in detail, since it brings certain irrational processes of scientific progress into the open, and shows how a community that has already adopted (in Thomas Kuhn's words) a new scientific paradigm also accepts uncritically all the arguments that appear to reinforce it. Strangely enough, this "mistaken" theorem played a decisive role in signaling the definitive triumph of the Copenhagen interpretation, and more than twenty years had to pass before it was understood to be inconclusive.

Third Phase

In collaboration with Boris Podolsky and Nathan Rosen, Einstein now developed a thought experiment that would become famous as "the EPR paradox." On the basis of natural hypotheses of realism and using the relativistic principles universally recognized as valid, this argument implies that quantum mechanics is fundamentally incomplete as a theory. As we shall see, Bohr's reaction was not entirely convincing, but once again the scientific community, impressed by the many successes of the new formalism, did not show the respect that should have been forthcoming for so acute a criticism, put forward by someone of such high caliber as Einstein. On this occasion, very few of the illustrious protagonists in the debate managed even to understand the true meaning of Einstein's criticism. Pauli disposed of it in a few words, while Born completely misunderstood it. Once again the orthodox interpretation appears to have won the latest battle, but in truth, Einstein's defeat (and defeat it surely was) represents one of the highest moments in our century for scientific research, since it brought attention to an element of utmost importance for the understanding of reality: quantum nonlocality. Many years would pass before it was recognized that Einstein was indeed defeated not so much by the superficial arguments of his opponents, as by nature herself, who now was revealing aspects no one ever suspected or imagined. And this proved to be one of Einstein's great achievements: to have forced everyone to confront those unsuspected aspects of reality that are at the center of scientific debate even today.

Fourth Phase

In his last works, Einstein articulated his position one last time, and drove to the heart of the epistemological problems of the formalism: no longer did the uncertainty principle torment him, nor even the incomprehensible character of the real at the microscopic level; rather, it was the fact that the theory, if assumed to be complete, required the denial of even minimal requirements of realism at the macroscopic level. This was a price he, like so many others, thought too high to pay.

In this sense it can be said that the ultimate victory Einstein was expecting was in a certain sense realized: "I hope that someone will discover a more realistic way, or rather a more tangible basis than it has been my lot to find." Today, there is a vast consensus among the experts in the field about the fact that the Copenhagen orthodoxy (including the postulate of the collapse of the wave packet, which must inevitably be preserved to keep the whole structure intact) fails to yield a sensible and acceptable picture of physical reality. Hence new interpretations have emerged in recent years, and even proposals to modify the equations at the base of the theory. The contemporary debate is focused on these attempts.

CHAPTER SEVEN

The fundamental role of Einstein's repeated, ever renewed critiques should emerge clearly in the pages to follow. In this chapter we shall limit ourselves to an analysis of the first phase of the debate, most of which took place before 1930, during the Solvay Conferences.[2]

7.1. Is It Really Impossible to Reveal the Wave and Particle Aspects of a Physical Process in One and the Same Experiment?

Einstein's first serious attack on the orthodox conception that was becoming more and more accepted took place at the Fifth Conference on Physics of the Solvay Institute in 1927. His presentation stirred up a series of lively debates among the participants. Einstein asked if it was possible to use the (established) laws of the conservation of momentum (that is, the product of mass times velocity) and energy to obtain some additional knowledge about the state of a particle that is involved in a process of interference. To understand his argument, as well as Bohr's reply to it, we can refer to Figure 7.2, which recalls the experiment discussed in chapter 3. A precisely directed beam of light (i.e., such that its particles can be assumed to have a velocity perpendicular to the x axis) with a sufficiently large extension, meets screen S1 with a very small slit in direction x. After passing through this aperture, the beam diffracts and falls on a second screen S2 that has two slits. The subsequent propagation of the light causes the formation of an interference pattern on the final screen F. As before, we consider a situation where only one photon at a time passes through the apparatus.

Recalling Section 3.8, and Figure 3.14, the reader should remember how, at the passage of the two slits of S2, the wave aspects of light play an absolutely essential role. We also pointed out how any attempt to reveal the particle aspects (which in the present instance would require finding out through which of the two slits each particular photon passes) inevitably destroys the wave aspects: the interference pattern disappears, and is replaced by a diffraction line which simply confirms our knowledge of which slit the light passed through.

At this point, Einstein directs our attention to the first screen and argues as follows: since the incident particles have a velocity practically perpendicular to the x axis, since it is only their interaction with the screen that can give them a vertical component of velocity, and since the

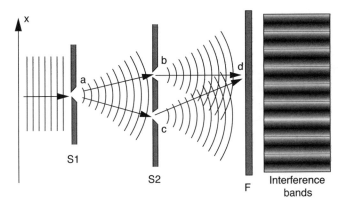

FIGURE 7.2. A monochromatic beam shines on the first screen (S1) and diffracts; the associated wave hits the second screen (S2) with two slits, causing an interference figure to form on the film (F). As always, it is assumed that one particle at a time passes through the apparatus, so that the figure of interference is formed over time through the accumulation of successive elementary processes.

laws of the conservation of momentum require that the sum of the momenta of two interacting bodies remains constant, then, if the particle is deflected upward, the screen ought to be pushed backward, and vice versa. In real conditions the mass of the screen is so great that it would remain almost perfectly stationary, but in principle it should be possible to measure even a very tiny amount of movement. If we imagine carrying out this measurement after the passage of every particle, we would be able to know, for those particles whose deviations bring them to enter one or the other of the slits in S2, whether they went through the upper slit or the lower slit. But since measuring the direction of movement of screen S1 after the particle has moved on to the second screen can have no influence on the successive development of the process, the interference pattern will still be there on screen F. Now, as we have seen repeatedly, the interference pattern requires that the state of the beam at the passage of the two slits is the *superposition* of the states that correspond to a passage through one or the other slit, and this means that wave aspects are governing the process at that moment. Nevertheless, for each particle we will be able to know (by the result of measuring the slight disturbance of the first screen) whether it has passed by the upper or lower slit of the second screen. This ideal experiment allows us (in principle, at least) to observe the particle aspects (i.e., knowing which of the two slits has been passed through) without destroying the wave aspects (the superposition) at the moment when the light goes through the second screen. And thus the same experiment allows us simultaneously to determine the position and velocity of a particle with greater precision than the uncertainty principle allows us to do.

And what was Bohr's response? Using two illustrations incorporating Einstein's points (Figures 7.3 and 7.4), Bohr replies as follows. The second figure shows how the first screen should be modified (ideally) so as to be able to detect an upward or downward movement. This is why Einstein's idea does not hold: an extremely precise knowledge of the vertical movement of the screen is presupposed by the argument. In fact, if the velocity of the screen along the x direction before the passage of the particle were not known with greater precision than the change caused by the particle (that is, if the screen is already moving upward or downward with an unknown velocity, and greater than what it has as a result of being hit), the determination of its movement after its passage would not give us the desired knowledge. However, Bohr argued, a precise knowledge of the transverse movement of the screen, when the uncertainty principle is applied, brings an inevitable imprecision about its position in direction x. Before the process, the position of the screen would be indeterminate to at least a certain amount (perfectly defined by the formal-

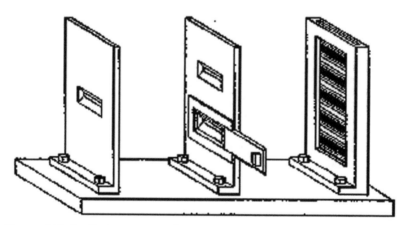

FIGURE 7.3. Preliminary setup for Einstein's proposal of showing simultaneously wave and particle aspects at the passage of the second screen. The moving shutter is designed to emphasize, as in chapter 3, that the attempt to know which slit the photon passes through destroys the figure of interference (compare Figure 3.14).

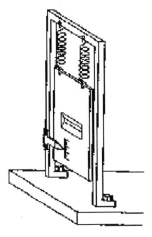

FIGURE 7.4. In order to carry out Einstein's experiment, the first screen must be replaced with a screen that can move vertically, and whose "bounce" can be measured.

ism). Let us consider, for example, the point of Figure 7.2, where destructive interference appears. Clearly, a displacement of the first screen would change the lengths of the two pathways a-b-d and a-c-d, making them different from what is actually shown. If the difference between the two paths changed by one-half a wavelength, there would be constructive interference at d rather than destructive. The ideal experiment requires, then, a kind of averaging over all the possible positions of screen S1, and, for every position of S1, a different type of interference would appear, somewhere between fully destructive and fully constructive. The effect of this averaging process would be a uniformly gray appearance on screen F: once again, our attempt to reveal *particle* aspects at S2 has destroyed the possibility of an interference pattern on F, which essentially depends on having *wave* aspects at S2.

The argument is correct and convincing. Nevertheless, as Bohr himself acknowledged, for understanding the phenomenon

> it is only decisive that, in contrast to the proper measuring instruments, these bodies together with the particles would in such a case constitute the system to which the quantum-mechanical formalism has to be applied. As regards the specification of the conditions for any well-defined application of the formalism, it is moreover essential that the *whole experimental arrangement* be taken into account. In fact, the introduction of any further piece of apparatus, like a

CHAPTER SEVEN

mirror, in the way of a particle might imply new interference effects essentially influencing the predictions as regards the results to be eventually recorded.

And later, when seeking a way out of the ambiguity about which macroscopic parts of the system ought to be treated in quantum terms and which not, he said,

> In particular, it must be realized that—besides in the account of the placing and timing of the instruments forming the experimental arrangement—all unambiguous use of space-time concepts in the description of atomic phenomena is confined to the recording of observations which refer to marks on a photographic plate or to similar practically irreversible amplification effects like the building of a water drop around an ion in a cloud chamber.

A few observations are called for. Above all, we should note that Bohr's argument—about the impossibility of using the experimental setup proposed by Einstein as a disproof of indeterminism—itself made use of the fact that a macroscopic system (screen S1) would have to obey quantum laws. On the other hand, Bohr always maintained that, if we wanted to reveal microscopic aspects of reality, we have to initiate a process of amplification involving macroscopic instruments, whose fundamental characteristic is to obey classical laws of physics, and to be described in classical terms. How can we concede that at least one part of the macrocosm, with all its definiteness, must, so to speak, be reabsorbed into the cloudy world of quanta? And more than anything else, what criteria are needed (since simply being macroscopic in itself is not enough) to identify which parts of a system obey quantum laws and which obey the classical?

In 1927 the powerful force of this question was still not fully felt. However, as we shall see, a few years would be enough for Schrödinger to make it clearly explicit. If Einstein had taken full advantage of the fact that his objection forced Bohr to engulf a portion of the macroscopic world in the world of quanta, he could have articulated his own argument further, and compelled his opponent to admit that ever more significant parts of the real must be recognized as being governed by quantum laws. I personally am of the opinion that perhaps subconsciously Bohr suspected that he was running the risk of being caught in a blind alley, since he could not propose any general principle for locating the exact boundary between classical and quantum. So he tried to avoid this insoluble problem by shifting the emphasis away from the classical/quantum separation, which up until that time had been identified with the microscopic/macroscopic separation, and placing emphasis instead on reversible versus irreversible processes. The outcome of a measurement, the formation of a band of light on a photographic plate, or of a droplet in a cloud chamber, were examples of processes of amplification which Bohr then saw as essential for obtaining information about microsystems, but which now acquired the new, crucial characteristic of "irreversibility."

Once again, this point marked a subtle adjustment which would have the effect of changing the terms of the debate, and in the years that followed, many defenders of the orthodox position were now supplied with a new, ambiguous escape route from the conceptual difficulties of formalism.

7.2. The Time/Energy Uncertainty Relation

Chapter 3 discussed the uncertainty principle for the two variables of position and velocity of a particle. The uncertainty relation can be expressed more suitably in terms of the two variables of position and momentum. Momentum is designated as p = mv, and at the level of analysis we are currently interested in, mass represents an unchanging aspect of any body; therefore a possible uncertainty of momentum can only be derived from an uncertainty of velocity (that is, $\Delta p = m\Delta v$), so that the uncertainty relation of Section 3.7, can be written more simply as follows:

$$\Delta x \Delta p \geq h. \tag{7.1}$$

It is a good time to recall that the wave nature of physical processes implies the existence of a further uncertainty relation that connects time with energy.[3] In order to understand this relationship we can look at a simple experiment. We prepare a wave with a limited spatial extension. As shown in Figure 7.5, we suppose that a beam of light with a good longitudinal extension meets a screen with a slit, and the slit has a gate that is only open for a brief interval of time, Δt. On the other side of the opening we will have a wave propagating to the right but with a limited spatial extension.

We must now keep in mind an important fact, known very well to all students of electronics, who can observe it clearly whenever they analyze an electric signal on an oscilloscope by looking at the signal itself and its Fourier transform. This process shows what frequencies enter into a signal, and is exactly analogous to a test of a musical instrument's timbre, which identifies the various harmonics that compose a single note played on the instrument.

CHAPTER SEVEN

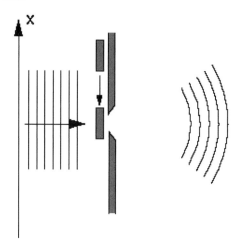

FIGURE 7.5. A light beam with longitudinal extension passes through a hole that is open only briefly, for interval Δt. Therefore, on the other side of the opening, there is a wave with only a limited spatial extension in the direction of propagation.

For those who are unfamiliar with electronics we can also argue as follows: a perfectly monochromatic wave, such as represented in Figures 1.3 and 1.4 in the first chapter of this book, must be extended infinitely. If we want a wave with a limited spatial extension (technically known as a wave packet), various waves of differing frequencies have to be superposed, distributed continuously in a certain interval of frequencies surrounding an average value, which we can call v_0. Now, it happens that at a certain instant of time there is a certain region of space (displaced with the passage of time) in which the various contributions of the various fields of the superposition are constructively summed together. However, according to a precise mathematical theorem, when we consider a region at some distance from this region, the phases of the various fields at any predetermined point become randomly distributed, causing destructive interference. The region where the wave is nonzero is thus spatially limited. It is easy to show that if the wave has a spatial extension Δx, which in our example would mean that the gate is open for time $\Delta t = \Delta x/v$, v being the velocity of the wave, the wave would "contain" (or "be a superposition of") various monochromatic waves whose frequency would cover an interval Δv, satisfying the following relationship:

$$\Delta v \geq 1/\Delta t. \qquad (7.2)$$

Keeping in mind that according to Planck's universal relationship the frequency and the energy are in the proportion

$$E = h\nu, \qquad (7.3)$$

it follows immediately (from the inequality [7.2] above) that the particle associated with the wave will have to possess an imperfectly defined energy (since various frequencies enter into the superposition) characterized by an uncertainty in energy of

$$\Delta E = h\,\Delta\nu \geq h/\Delta t. \qquad (7.4)$$

From this it immediately follows that

$$\Delta E \Delta t \geq h, \qquad (7.5)$$

which is just the uncertainty relation between time and energy. As with position and momentum, this is an inevitable consequence of quantum formalism.

7.3. 1930: Einstein's New Attack

At the Sixth Solvay Conference in 1930, the uncertainty relation just analyzed was the object of a serious new attack by Einstein. His idea was simple enough: he imagined an experimental setup that Bohr would later sketch in minute detail (see Figure 7.6) in order to bring out the essential elements of the argument and emphasize the points of his reply.

Einstein imagined a box containing electromagnetic radiation, and a clock that controlled an opening in one of the sides of the box with a gate. The gate is opened for a certain interval Δt, which can be arbitrarily chosen. While the gate is open, we imagine that one of the photons from inside the box escapes it. In this way, a wave is produced, having a spatially limited extension, as described in the previous section. Naturally, if we want to have recourse to the time/energy uncertainty relation, we will have to find a way to determine with an appropriate precision the energy carried by the photon. At this point Einstein resorted to the celebrated relation between mass and energy of special relativity: $E = mc^2$. From this it follows that the knowledge of the mass of an object furnishes us with precise knowledge about its energy. The idea is simplicity itself: if the box is weighed before and after the opening of the gate, and if the box emits a certain amount of energy, it will be lighter afterward. The change in weight multiplied by c^2 would provide us with a precise knowledge of

CHAPTER SEVEN

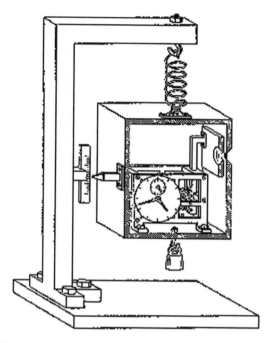

FIGURE 7.6. Einstein's 1930 "thought experiment" intended to show a violation of the time-energy uncertainty relation.

the emitted energy. Obviously, since in principle the weight of the box could be determined to any arbitrary degree of exactness, the emitted energy could be determined as accurately as desired. Thus the product $\Delta E \Delta t$ can be made smaller than what is implied by the uncertainty principle.

Like all the ideas Einstein advanced, this one too was particularly acute and seemingly unassailable. We can relive the episode in the vivid description made of it a few years later by Leon Rosenfeld, one of the Congress participants:

> It was quite a shock for Bohr . . . he did not see the solution at once. During the whole evening he was extremely unhappy, going from one to the other, trying to persuade them that it couldn't be true, that it would be the end of physics if Einstein were right; but he couldn't produce any refutation. I shall never forget the vision of the two antagonists leaving the club [of the Fondation Universitaire]: Einstein a tall majestic figure, walking quietly, with a somewhat ironical smile, and Bohr trotting near him, very excited. . . . The next morning came Bohr's triumph.

The expression "triumph" to describe this event is fitting for two reasons. First, because once again Bohr succeeded in showing that Einstein's subtle argument was not conclusive, but even more because, in order to succeed he had recourse to one of Einstein's own great ideas, namely, the principle of equivalence between gravitational mass and inertial mass. The argument runs as follows: in analyzing the process, it would be necessary to take account of the inevitable uncertainty of the needle's position on the gauge (otherwise, thanks to the uncertainty principle, the gauge could have a transverse velocity that would nullify the whole experiment). This uncertainty would clearly be translated into an uncertainty in determining the weight, and thereby the emitted energy. On the other hand, since the whole system is embedded within a gravitational field that varies with position q, by the principle of equivalence the uncertainty in the position of the clock implies an uncertainty about its temporal accuracy[4] and to the value of the interval Δt. A precise determination of this effect leads to the conclusion that in fact, the relationship $\Delta E \Delta t \geq h$ cannot be violated.

I would like to conclude this chapter with a few reflections. First, once again, in order to refute Einstein's objections, Bohr was constrained to treat macroscopic parts of the apparatus, such as the scale and the clock, in quantum terms, and as such subject to restrictions following from the uncertainty principle. Second, I want to emphasize that the particular relevance the scientific community saw in this event was that Bohr triumphed by using one of Einstein's most profound discoveries. But this is based on a misunderstanding. Since Einstein's argument did not make any use of the principle of equivalence between gravitational mass and inertial mass, its refutation should not have involved the use of that principle. Some years ago, discussing this point with one of my students, we were together able to show that Einstein's argument could be refuted anyway, even without having recourse to the principle of equivalence. Somewhat later at a conference, while discussing the point with some colleagues, Professor Asher Peres called my attention to the fact that W. G. Unruh and G. I. Opat had reached the same conclusion in a work published in 1979 in the *American Journal of Physics*. The article, which is elegantly written and quite interesting from a historical perspective, sheds new light on this crucial moment in the debate.

Finally, before continuing our story, I would like to add that this episode and the year it occurred (1930) marked a decisive change in Einstein's attitude toward the theory. He was now convinced that there were limitations in principle to the possibility of determining with any desired precision both the velocity and the position of a particle—that is to say, from now on he accepted the uncertainty principle. But he retained his firm conviction that these properties could be considered as belonging to the systems in question and not as created by the act of measurement. From this time forward his attacks will no longer be against what he saw as the theory's inconsistency, but against its incompleteness. It could be said that his critique no longer revolved around the idea of "playing dice with the world," which the theory seemed to imply with reference to the outcomes of measurement, but rather around the fact that the theory required us to accept that in reality there was *nothing other than* this "playing at dice," that we are not allowed to think of properties objectively possessed by physical systems independently of our procedures of measuring them. This will be the theme of our next chapter.

CHAPTER EIGHT

A Bolt from the Blue: The Einstein-Podolski-Rosen Argument

> Paradox indeed! But for the others, not for EPR. EPR did not use the word "paradox." They were with the man on the street in this business. For them, these correlations simply showed that the quantum theorists had been hasty in dismissing the reality of the microscopic world. In particular Jordan had been wrong in supposing that nothing was real or fixed in that world before observation.
> —*John Stewart Bell*

1935 WAS AN EXTRAORDINARY YEAR. In that year, two fundamental works were published that reignited the debate about the interpretation of the formalism, forcing the scientific community to concentrate its attention on a few facts not sufficiently pondered in the discussions of previous years. Formally speaking, both works were based on the same specific characteristic of the theory: the so-called quantum entanglement between the constituents of a composite system. The original German expression used by Schrödinger, Verschrankung, has become known in the scientific literature as "entanglement," and we will conform to this usage. There is no better way to begin than with the words of Erwin Schrödinger: "I consider [entanglement] not as one, but as *the* characteristic trait of quantum mechanics, the one that enforces its entire departure from classical lines of thought."

It is the absolutely peculiar correlations that in general accompany entanglement that led Einstein, Podolski, and Rosen (in the first of the two papers mentioned) to present their subtle argument that quantum mechanics is fundamentally *incomplete* as a theory. They argued with reference to a system of two elementary particles in an entangled state. In an analogous way, that very same year, Schrödinger (in the second paper mentioned, featuring his famous cat) used entanglement between a microsystem and a

CHAPTER EIGHT

macrosystem to show that the theory enters into an incurable conflict with certain unavoidable requirements concerning macrosystems.

In this chapter, we inspect the EPR argument in detail, the debate it started, and the misunderstandings to which it gave rise. Some of its developments will then be treated in the following chapters. Schrödinger's work, on the other hand, will have to wait until chapter fifteen, when the thorny problem of quantum measurement will be taken up.

Before beginning our analysis of EPR's subtle critique, certain premises need to be established about the quantum description of composite systems. Among other things, this will enable the reader to appreciate fully the meaning of entanglement. In view of the great importance of the argument, we proceed step by step, at the risk of seeming overly pedantic.

8.1. THE PROBLEM OF THE PROPERTIES OF A QUANTUM SYSTEM

A quantum system is described at every instant by a state vector. According to the theory, this represents the maximum of information that can be had about the system. To illustrate, we can consider the polarization states of a photon. For example, we take a photon associated with a state vector indicated by |45>, as discussed in previous chapters. What exactly does this information tell us about the properties of the photon? In fact, as we have often emphasized, knowing the state vector gives us information solely about the outcomes of possible measurements carried out on the system. In the present instance, we know that if we test the photon for vertical polarization, it will have a 1/2 probability of passing the test and a 1/2 probability of failing it. Although the theory usually gives us only probabilistic information about the outcomes of hypothetical measurements, for some tests it can assign the value of 1 (one) or 0 (zero) to the probability of obtaining some particular outcome. And so, in our case, the theory tells us that our photon has the probability of 1 of passing a test for 45° polarization, and the probability of 0 of passing a test for 135° polarization. In this instance, then, with precise and exclusive reference to the observables (polarization at 45° or 135°) for which we know the outcome of the measurement a priori, we can assert that the photon really possesses the property in question, that it is polarized at 45° or rather, that it possesses the property that guarantees that it will pass such a test with certainty.

When we consider a quantum system in its entirety, there always exists at least one observable for which it is possible to predict the outcome of a measurement with certainty. In the case of polarization states of a single photon, this means that, whatever its state vector may be, there exists an appropriate polarization test (elliptical in the most general case) that the photon will pass with certainty, if the test is actually carried out. If the perspective just described is assumed—that we attribute an objective property to a system in cases where we can predict the outcome of a measurement without in any way disturbing that system—we can sum up the lesson of quantum mechanics (without forgetting, of course, the uncertainty principle) by saying that the study of microsystems reveals something remarkably different from the classical situation. In classical terms, any system whatsoever always has precise values for all conceivable observables (in other words, the knowledge of the state—the maximum of information about a system on the classical level—allows us to predict the outcome of any test on it with certainty). In quantum terms, by contrast, a single system indeed possesses some properties, but such an assertion could not be made with regard to other properties, apart from probabilistic predictions about the results of possible measurements, provided they are actually carried out. In a way, the theory has taught us to think that a system cannot be thought of as having "too many" properties—in particular, that some properties will be incompatible with others, so that if one of them is possessed, then the other would certainly not be possessed. Thus a photon polarized at 45° in fact does not possess any property with reference to vertical or horizontal polarization.

I wanted to repeat some points, even if it seems tedious, because the reader will need to be familiar with them in order to understand composite systems. Furthermore, the analysis just presented is essential for understanding one of the ingredients of the EPR argument. For these authors make the following natural assumption (R) about realism that is essential to their argument:

> ***Assumption R.*** If, without in any way disturbing a system, we can predict with certainty (i.e., with a probability equal to 1) the value of a physical quantity, then there exists an objective element of physical reality corresponding to this physical quantity.[1]

8.2. COMPOSITE SYSTEMS: THE CASE OF FACTORIZED STATES

Consider the following situation: Two photons are emitted from a light source S, and are propagated in two opposing directions. At a certain in-

CHAPTER EIGHT

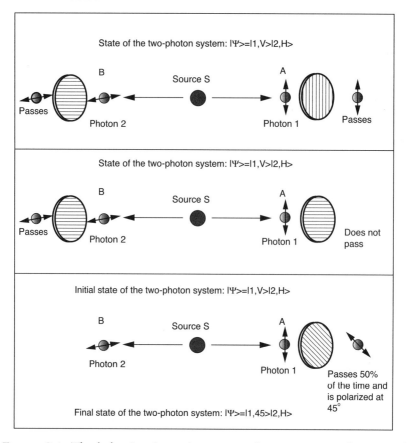

FIGURE 8.1. The behavior, in various measuring processes, of a pair of separated photons in the case of the factorized state |1,V> |2, H>.

stant, one will be in region A to the right of the source (see Figure 8.1), and the other will be in region B, symmetrically located on the other side of S. Omitting a description now of the spatial characteristics of the process (this means we will not explicitly consider the wave packets that describe the propagation of the two photons) we concentrate our attention only on their polarization characteristics. We indicate by 1 the photon to the right,[2] and we suppose that it has vertical polarization. As before, we indicate its state by the vector |1,V>. In an analogous way, we suppose that the photon to the left, which we designate 2, has horizontal polarization: its state will be indicated as |2, H>.

The theory tells us that the system of the two photons is described by the state |1,V> |2, H>. Technically, this state is known as the direct product of the two states in question, but such formal aspects should not interest us now. On the other hand, we are interested—along with Roger Penrose—in the fact that in the case in question the product can be thought of as the quantum equivalent of the conjunction "and" between its two constituents, in the sense that the two systems 1 and 2 are simultaneously present, and that the state

$$|\Psi\rangle = |1,V\rangle |2, H\rangle \qquad (8.1)$$

is a single quantum state corresponding to the fact that "(a photon is in A with vertical polarization) *and* (a photon is in B with horizontal polarization)." In the preceding sentence, everything between the quotation marks—the conjunction, in other words, of the two propositions in parentheses—represents a true assertion for the composite system, provided that the principle we adopted earlier is in force, namely, that we identify the statement "has vertical polarization" with the assertion that it "will certainly pass the test for vertical polarization," and analogously for the other photon as well.[3]

The state under consideration can be called factorized, because it is the product of the state of one photon and the state of another photon. Its properties are clear: each of the two photons behaves completely independently of the other when tested for polarization or when sent through a birefringent crystal. For example, if, given the state $|\Psi\rangle = |1,V\rangle |2, H\rangle$, a test for vertical polarization is carried out on the photon to the right and one for horizontal polarization on the photon to the left, we know that both photons will pass their tests with certainty. Similarly, if a test for horizontal polarization is made on the photon to the right and the same test on the photon to the left, the first will certainly fail and the second will certainly pass. Finally, we consider the more general case where the photon on the right is subjected to a test of 45° polarization. What will happen? The answer is not surprising: the photon to the right will pass the test in one-half of the cases, and will become polarized at 45°, and in the other half of the cases will fail the test and be absorbed. The photon to the left, meanwhile, which is not now being subjected to any test, will remain horizontally polarized. The three possibilities are pictured in Figure 8.1.

In order to clarify the third possibility, it should help to review the formulas of Section 2.4. We must especially keep in mind that the state |1,V> ought to be written in the form that is now familiar to us, since we are interested in a measurement of 45° polarization:

CHAPTER EIGHT

$$|1, V\rangle = \frac{1}{\sqrt{2}}(|1, 45\rangle) + \frac{1}{\sqrt{2}}(|1, 135\rangle), \quad (8.2)$$

which, substituted into the expression of the state $|\Psi\rangle$, gives us

$$|\Psi\rangle = |1, V\rangle |2, H\rangle = \frac{1}{\sqrt{2}}(|1, 45\rangle |2, H\rangle) + \frac{1}{\sqrt{2}}(|1, 135\rangle |2, H\rangle). \quad (8.3)$$

According to this formula, a measure of 45° polarization in A can lead to photon 1 passing the test with equal probability (since $1/\sqrt{2}^2 = 1/2$), in which case the system will be represented by the (normalized) state of two superposed states corresponding to the outcome of the polarization test (according to the rule of the reduction of the wave packet discussed in Section 6.4):

$$|\Psi\rangle = |1, V\rangle |2, H\rangle = \frac{1}{\sqrt{2}}(|1, 45\rangle |2, H\rangle) + \frac{1}{\sqrt{2}}(|1, 135\rangle |2, H\rangle)$$

$$\underset{\substack{\text{45° measurement} \\ \text{passes test}}}{\Rightarrow} |1, 45\rangle |2, H\rangle. \quad (8.4)$$

Analogously, if the measurement had not been destructive (for example, if it had been carried out with the help of a birefringent crystal) and the photon was discovered to have had a polarization of 135°, the final state would be the one that corresponds to the second term of the state $|\Psi\rangle$ (except for the factor $1/\sqrt{2}$).

The argument can now be generalized, assuming any arbitrary state of the two photons. If the state is factorized, each photon will have the properties that belong to it in virtue of its state, and the probabilities that it has of passing or failing any tests for that state.

A case that will be particularly important in what follows is when two photons have the same polarization, that is to say, are both associated with the state

$$|\Phi\rangle = |1, V\rangle |2, V\rangle, \quad (8.5)$$

the properties of which should be obvious to the reader and are illustrated in Figure 8.2. Analogously, another state of interest to us is where the two photons again have equal polarization, but this time they have horizontal instead of vertical polarization:

$$|\Lambda\rangle = |1, H\rangle |2, H\rangle. \quad (8.6)$$

The reader can be spared another discussion of the outcomes of polarization measurements.

EINSTEIN-PODOLSKI-ROSEN ARGUMENT

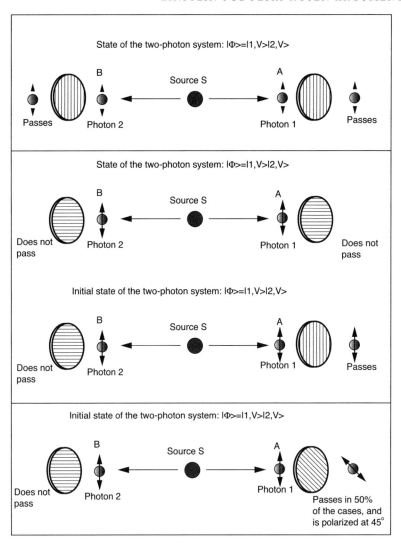

FIGURE 8.2. The analogue of Figure 8.1, in the case where both photons begin with vertical polarization. This case is of interest for understanding the entangled state discussed in the next section.

171

CHAPTER EIGHT

Before concluding this section, one last point needs to be made. In the case of factorized states of a composite system, each of the two constituents still has at least one objective property (which belongs to it in virtue of the state associated with it), and the correlations between measurements made on the two systems are those that characterize independent events; that is to say, the joint probabilities are simply the product of the individual probabilities. If, for example, I am interested in the joint probability that photon 1 passes a test of polarization (say, Q) and photon 2 passes a test of polarization (say, R) , and I know that the relative probabilities are P(1,Q) and P(2,R), respectively, then the probability that both will pass the test is simply P(1,Q) × P(2,R). The situation from this perspective is perfectly analogous to that of classical probabilities for independent events. If I toss a die and flip a coin and ask myself what the joint probability is for getting a three on the die and heads for the coin, we all know the result would be 1 in 12: (1/6) × (1/2) = 1/12.

8.3. Composite Systems, Entangled States

We must now face the most characteristic trait of quantum formalism: entanglement. Like almost all the other peculiar aspects of the theory, this follows from its linear character—that is, from the principle of superposition.

Once again we contemplate a system of two photons as in the preceding section. We observe that the state |Φ> of the equation (8.5) is clearly a possible state for the system. The same can be said for the state |Λ> of the equation (8.6). However, if this is the case, the theory assures me that even the superposition of the two states (suitably normalized), i.e., the state

$$|\Psi> = \frac{1}{\sqrt{2}}[|\Phi> + |\Lambda>] = \frac{1}{\sqrt{2}}(|1,\ V>|2,\ V>) + \frac{1}{\sqrt{2}}(|1,\ H>|2,\ H>),$$

(8.7)

is a possible state for a system of two photons. In fact, this state is not only possible but is also relatively easy to prepare. What are its properties?

Right away we see that neither of the two photons has the property of vertical or horizontal polarization, since the probability of photon 1 passing (for example) a test of vertical polarization is characterized by the coefficient (1/√2) of the state in which it has just such a polarization, and the square of this coefficient is equal to 1/2. Thus if we test photon 1, half of the time it will pass the test, half the time it will fail, in a random and unpredictable way. The same holds for a horizontal test and for the other photon.

But something else is also at work. Suppose we are interested in measures of polarization at 45° and 135°. The attentive reader already knows what we need to do: we will have to express the states of vertical and horizontal polarization for the photons under consideration as superpositions of polarization states, at 45° and at 135°. We recall the formulas (they clearly will be the same for both photons):

$$|V\rangle = \frac{1}{\sqrt{2}}[|45\rangle + |135\rangle],$$
$$|H\rangle = \frac{1}{\sqrt{2}}[|45\rangle - |135\rangle]. \tag{8.8}$$

Now, let us substitute these expressions (for photon 1 and for photon 2) into the last term of the preceding formula. Following out the calculations explicitly, we come up with a product as follows:

$$|\Psi\rangle = \frac{1}{\sqrt{2}}\left\{\frac{1}{\sqrt{2}}[|1,45\rangle + |1,135\rangle]\frac{1}{\sqrt{2}}[|2,45\rangle + |2,135\rangle]\right.$$
$$\left. + \frac{1}{\sqrt{2}}[|1,45\rangle - |1,135\rangle]\frac{1}{\sqrt{2}}[|2,45\rangle - |2,135\rangle]\right\}$$
$$= \frac{1}{2\sqrt{2}}\{|1,45\rangle|2,45\rangle + |1,45\rangle|2,135\rangle + |1,135\rangle|2,45\rangle$$
$$+ |1,135\rangle|2,135\rangle + |1,45\rangle|2,45\rangle - |1,45\rangle|2,135\rangle$$
$$- |1,135\rangle|2,45\rangle + |1,135\rangle|2,135\rangle\}$$
$$= \frac{1}{\sqrt{2}}\{|1,45\rangle|2,45\rangle + |1,135\rangle|2,135\rangle\}. \tag{8.9}$$

The result is surprising: the state turns out to be the superposition of the state of two photons polarized at 45° and of two polarized at 135°! The two new perpendicular directions have taken the place of the V and H directions in the earlier formula. This implies that every photon has a 1/2 probability of passing a test of this kind: exactly the same probability it has of passing a vertical or horizontal one.

. . . . . .

CHAPTER EIGHT

We should not be surprised that the same state can be expressed in different ways. This holds true in general and has nothing to do with the fact that the state is entangled. It is a consequence of the linear character of the theory, or, as it is technically expressed, a consequence of the fact that the states constitute a linear vector space. The possibility of giving different expressions of the same state is simply the analogue of the possibility, already discussed several times, of decomposing a vector into the sum of any two vectors, so long as they are not collinear (compare Figures 1.6 and 2.9). On the other hand, the peculiar fact that the state (8.7) assumes exactly the same form when it is expressed in terms of the states |V> and |H> or the states |45> and |135> and in general for any pair of polarization states along any two perpendicular directions (see equation [8.10] below) depends on its peculiar form. If in equation (8.7) the coefficients of the two terms were not equal, then in equations (8.9) and (8.10) four terms would correspond to the fact that each of the photons can be found in one of the two states of plane polarization. The peculiarity of the state in question derives from the characteristic of the state |Ψ> to be symmetrical under rotations, so that no one direction is privileged. The vertical and horizontal orientations are perfectly equivalent to any two perpendicular directions.

. . . . .

We can now continue this line of questioning and look into other possible measurements of plane polarization along any arbitrary direction. The patient and diligent reader can make the calculations using the formulas of Section 6.4, which express a state of assigned plane polarization in terms of the states corresponding to two polarizations along two perpendicular planes of any arbitrary orientation. If the reader does not have time or inclination to carry out this very helpful exercise, he or she can simply take my word for the results: the state |Ψ> always has the same form whatever direction is chosen. This means that it is the superposition of two states, in the first of which both photons are polarized in the chosen direction, which we can call n, in the second of which they are both polarized in the perpendicular direction $n_\perp$:

$$|\Psi> = \frac{1}{\sqrt{2}}(|1, n>) + \frac{1}{\sqrt{2}}(|1, n_\perp>|2, n_\perp>). \qquad (8.10)$$

Taking a moment to reflect on this result, we see that, according to the rules now becoming familiar to the reader, each of the two photons has a 1/2 probability of passing a test for polarization along any arbitrary direction.[4] And so here is the first extraordinary implication of entanglement: the theory has taught us that we cannot attribute too many properties to a physical system, but it has also taught us that some property is always possessed. Now, the constituents that enter into our state no longer have any individual properties in so far as there is no polarization test pertaining to any one of them the result of which we could predict with certainty. Of course, the system as a whole still has some properties (which we are not discussing now) but despite their distance from one another and their lack of interaction, the constituents represent an inseparable unity: they really are entangled with each other.

8.4. The Effect That Measurement of a Constituent Has on the State as a Whole

Suppose an observer decides to carry out a measurement of polarization on photon 1, along direction n. What happens? If the photon passes the test, according to the principle of the reduction of the wave packet as applied to equation (8.10), we have

$$|\Psi\rangle = \frac{1}{\sqrt{2}}(|1, n\rangle |2, n\rangle) + \frac{1}{\sqrt{2}}(|1, n_\perp\rangle |2, n_\perp\rangle) \xrightarrow[\text{passes the test}]{\text{measurement of } n} |1, n\rangle |2n\rangle$$

(8.11)

and the final state ends up being factorized! Unexpectedly, photon 2, which before the measurement had no property of polarization, has acquired a precise property as a consequence of the measurement of photon 1 (no matter how distant): it is polarized along the direction of polarization of the first photon, if the first one has passed the test (needless to say, if the first photon did not pass the test, there would be reduction on the second term of the superposition and thus the second photon would wind up as polarized in a direction perpendicular to n). I trust that every reader is beginning to sense that the analyses are leading us to something unexpected and surprising. We are now ready to confront the central point of the chapter.

8.5. The EPR Argument

We now have at our disposal all the elements we need to follow the argument that led Einstein, Podolski, and Rosen to affirm the incompleteness of quantum mechanics. They developed their analysis with reference to an entangled state of two distantly separate particles for which a situation holds that is conceptually very similar to what we analyzed for the state |Ψ> in the two preceding sections. But it must be clarified that the observables they were interested in (i.e., what they were measuring at one end of the apparatus in order to infer something about properties to appear at the other end of it) were not the polarizations of a pair of photons but the observables of position and momentum of a pair of elementary particles. By the appropriate choice of initial state (which once again can easily be prepared in a laboratory) the argument of the original work is logically the same as what we will present for the two entangled photons.

In fact, the original argument of EPR was reformulated by David Bohm in his book *Quantum Theory* (1951) with reference to the measurements of spin components of a system composed of two particles of spin 1/2, in an entangled state strictly analogous to what we considered for the photons. This new presentation was advantageous because it remarkably simplified the formal character of the argument: he worked with variables that can assume discrete values only, and in a finite number (recalling that the sole possible outcomes concerning spin components, for a particle with spin 1/2 are, in units of h/2π, equal to +1 and −1). Bohm's reformulation played an important role as well. It was in reference to this that Bell was able to derive the famous "inequality" that bears his name.

Our choice of a pair of photons was dictated by two reasons: first, our example is formally identical with the example Bohm used, provided one takes into account the analogies we discussed (in chapter three) between polarization states of photons and the spin states of particles with 1/2 spin. Second, the revolutionary experiments undertaken by Alain Aspect to show quantum nonlocality in laboratory conditions involve entangled states of two photons and measuring procedures very similar to our own. In fact, in recent years the developments in quantum optics, lasers, light guides, and other technology have brought photon systems into use in many experiments on the fundamental principles of quantum mechanics. In practice, of course, there is a difference between the experiments we are using and the one discussed by EPR, but now that the reader has been made aware of this we can return to a discussion of the logic of the argument.

To take the next step we need to establish another link. This consists in adopting with EPR a very natural hypothesis, directly inspired by the theory of relativity. It incorporates Einstein's idea of the locality of physical processes, and can be designated as *E.L.*

EINSTEIN-PODOLSKI-ROSEN ARGUMENT

Assumption E.L. The elements of physical reality objectively possessed by a system cannot be influenced instantaneously at a distance.

With these premises the argument can be articulated as follows:

1. At time t two photons of a composite system are in regions A and B, spatially removed from one another and in an entangled state of polarization $|\Psi\rangle$ as seen above:

$$|\Psi, t\rangle = \frac{1}{\sqrt{2}} |1, V\rangle |2, V\rangle + \frac{1}{\sqrt{2}} |1, H\rangle |2, H\rangle. \qquad (8.12)$$

2. At time t, the photon in region A is subjected to a measurement of plane polarization along the vertical direction. We suppose the outcome of the measurement is that the photon passes the test. As discussed in the previous section, according to the postulates of quantum mechanics, the effect of the measurement is to reduce the wave packet, producing a *damned quantum jump* into the state that corresponds to the outcome obtained. In other words, immediately after the measurement, say at time $t + dt$, the state of the system becomes

$$|\Psi, t + dt\rangle = |1, V\rangle |2, V\rangle. \qquad (8.13)$$

3. At this point, the observer in A who carried out the measurement is able to predict without performing any further action and without disturbing photon 2 in any way, that photon 2 will certainly pass a test for vertical polarization. Referring to assumption R, the observer can assert, immediately after the measurement at A on photon 1, that photon 2 possesses an element of physical reality, namely, it has vertical polarization.

4. According to the assumption *E.L.* about locality, we must say that it could not have been the measurement of photon 1 in region A that created this element of physical reality for photon 2. Consequently, we must conclude that photon 2 possessed this property (of being able to pass the test for vertical polarization with certainty) even before, and independently of, the measurement of photon 1.

We must now take note that two different strategies are available for concluding the argument. The first one (we can call it α) represents the argument of EPR, and contemplates the possibility that at time t, a hypothetical polarization measurement is made, now along a different direction, and incompatible with the first one. For example, photon 1 could have been put to a test of 45°. The second strategy (we can call it β) jumps directly to the conclusion about the incompleteness of the theory without

177

CHAPTER EIGHT

citing any incompatible measurements. We follow both lines because Bohr made a direct reference to the first argument, whereas the second is better suited to the spirit of the debate on formalism.

5α. At time t the observer in A might have decided to carry out the polarization test at 45° by obtaining a certain result: say, for instance, the photon does pass the test. Now (recalling the last expression of [8.9]), the same observer would have concluded that photon 2 is polarized at 45°. Alternatively, if the photon had not passed the test, the conclusion would have been that photon 2 is polarized at 135°. Joining any one of these two alternatives with the conclusion reached in point 4, it must be asserted that photon 2, before its measurement in A, had its own property of "passing with certainty a test for vertical polarization" and of "passing with certainty a test of 45° polarization." This means it simultaneously has properties that are claimed incompatible by the formalism.

6α. At this point the EPR argument does not, in fact, say that it is possible to measure simultaneously incompatible observables, i.e., that the uncertainty principle is violated, nor does it say that the theory is logically defective. The conclusion is something else: since natural and obvious requirements have led us to the conclusion that photon 2 has simultaneously incompatible properties, this means that, in fact, even if it is not possible actually to determine these properties with the desired precision at one given moment, they nevertheless must be considered as objectively possessed by the system. But quantum mechanics denies this possibility and furthermore the state $|\Psi, t\rangle$ (of formula [8.12]) contains no formal element that refers to or in any way specifies these properties. The conclusion? Quantum mechanics is fundamentally incomplete as a theory, since it cannot describe or in any way account for elements of physical reality which must be recognized as belonging to a physical system.

We now consider the other line of thought, β:

5β. At point 4, we had been brought to the conclusion that the single photon under examination (photon 2) had the property of vertical polarization, before the measurement. This privileged one spatial direction (vertical). On the other hand, the theory asserts that before the measurement the state of the system was the state $|\Psi, t\rangle$, which, as already observed, is symmetrical under rotation, and thus no spatial direction is privileged. Now the completeness of the theory is a key point of the orthodox interpretation: the knowledge of the state vector represents the maximum information that one can have about a system. In the present case, that means that the theory can-

EINSTEIN-PODOLSKI-ROSEN ARGUMENT

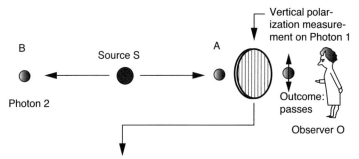

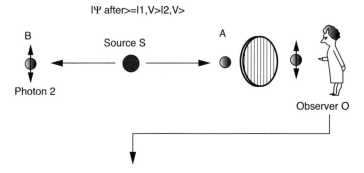

FIGURE 8.3. Schematic representation of the EPR incompleteness argument.

not in any way describe the privileged direction that characterized photon 2 before the measurement. But then not all the elements of reality recognized as present in a system can find expression in the formal apparatus of the theory, and this amounts precisely to asserting that the theory is incomplete.

To conclude this exposition of the EPR argument (summed up schematically in Figure 8.3), I would only like to add that nothing in the pre-

CHAPTER EIGHT

ceding arguments depends on the actual result obtained in the experiment on the first photon. If the outcome had been the opposite, the argument would have taken us to the same conclusion.

8.6. Bohr's Reaction

The EPR argument was shattering. It was simple and based on very reasonable assumptions. Not surprisingly, it represented another formidable challenge for Bohr, whose reply was published five months later in the same journal (The *Physical Review*), and had the same title as the EPR article: "Can the Quantum-Mechanical Description of Physical Reality be considered Complete?" Once again it will be worthwhile to relive the moment through the words of the eyewitness Rosenfeld:

> This onslaught came down upon us like a bolt from the blue. Its effect on Bohr was remarkable. . . . [A]s soon as Bohr had heard my report of Einstein's argument, everything else was abandoned: we have to clear up such a misunderstanding at once. We should reply by taking up the same example and showing the right way to speak about it. In great excitement, Bohr immediately started dictating to me the outline of such a reply. Very soon, however, he became hesitant. "No, this won't do, we must try all over again . . . we must make it quite clear. . . ." So it went on for a while, with growing wonder at the unexpected subtlety of the argument. . . . The next morning he at once took up the dictation again, and I was struck by a change in the tone of the sentences: there was no trace in them of the previous day's sharp expressions of dissent. As I pointed out to him he seemed to take up a milder view of the case, he smiled. "That's a sign," he said, "that we are beginning to understand the problem."

Despite this optimistic declaration and despite the fact that the greater part of the scientific community at the time and theoretical physicists until recent times took it for granted that Bohr was victorious once again in his battle with Einstein, it has to be openly admitted that Bohr's reply was remarkably obscure and could not be considered as a definitive refutation of EPR. Great scientists, including Bell and Einstein himself, stated explicitly that they did not understand Bohr's position on the question. Instead of simply reporting the opinions of these great masters we can see even more directly how inconclusive Bohr's reply was, by quoting a passage from his article that he himself considered conclusive, since he would repeat it verbatim years later (in 1949), when asked to contribute to the volume *Albert Einstein: Philosopher-Scientist*, edited by Paul Arther Schilpp on the occasion of Einstein's 70th birthday.

Bohr reacted to assumption R (see above) as follows:

> From our point of view we now see that the wording of the above-mentioned criterion of physical reality proposed by Einstein, Podolsky, and Rosen contains an ambiguity as regards the meaning of the expression "without in any way disturbing the system." Of course, there is in a case like that just considered no question of a mechanical disturbance of the system under investigation during the last critical stage of the measuring procedure. But even at this stage there is essentially the question of *an influence on the very conditions which define the possible types of predictions regarding the future behavior of the system*. Since these conditions constitute an inherent element of the description of any phenomenon to which the term "physical reality" can be properly attached, we see that the argumentation of the mentioned authors does not justify their conclusion that quantum-mechanical description is essentially incomplete. On the contrary, this description, as appears from the preceding discussion, may be characterized as a rational utilization of all possibilities of unambiguous interpretation of measurements, compatible with the finite and uncontrollable interaction between the objects and the measuring instruments in the field of quantum theory.

The obscurity of this passage speaks for itself. It contains a series of points that in their sum are incomprehensible, as Bell lucidly explained. What precisely is meant by the adjective "mechanical" used to describe those "disturbances" Bohr did not think should be taken into account? In the passage that Bohr himself wanted to stress, how are we supposed to understand the expression "an influence on the very conditions . . . ," if not in the sense that diverse measurements in A furnish diverse kinds of information on the system at B? But this fact is not only openly admitted, it provides one of EPR's most forceful arguments. Finally, as Bell observed, what can an expression like "unknowable interaction between the object and the measuring apparatus" mean, once we keep in mind that the central point of the EPR argument lies in the hypothesis that if locality is accepted, only the system in A can be disturbed by the measuring process, and nevertheless, this process furnishes precise information about the system in B? Was Bohr contemplating the possibility[5] of instantaneous action at a distance? It should be obvious that if we abandon assumption E.L. (see above), the entire argument of EPR will fall, but can we understand this to be Bohr's real meaning, since he didn't say so explicitly?[6]

Personally I think that on the one hand, Bohr did not fully understand the subtle implications of Einstein's argument[7] that called upon unsuspected aspects of reality such as nonlocality, and on the other hand, he did not succeed in identifying (and in fact could not have) any conclusive arguments against Einstein. He almost seems to have allowed himself to be guided only by his principle of complementarity (we see a trace of this in his vague phrasing). This could be put to immediate use to show how even Einstein's indirect and ingenious method (i.e., the line of thought α) for obtaining information about the system in B requires having recourse to procedures in A that cannot coexist. This is because information is needed about incompatibles (such as position versus momentum in EPR's original argument, or vertical versus 45° polarization as in our example). In particular, the specific experimental setup in A that furnishes information about the position of the constituent in B is incompatible with what would allow any information to be gathered about its momentum in B. This observation is certainly very important and perfectly correct. In my view, it may even have inspired Bohr's obscure expression, "an influence on the very conditions which define the possible types of predictions regarding the future behavior of the system." And yet, as correct as it was, the observation was quite inappropriate for refuting the EPR argument. Those men knew perfectly well that the choice of measurements of position or momentum were mutually exclusive, and nothing in their argument required that they be carried out at the same time on the same system.

8.7. Misunderstandings of the EPR Argument

We said that the scientific community was immediately convinced (rather uncritically) that Bohr had once again defeated Einstein. What is more, most of the scientists of that time and in the following years proclaimed Bohr the victor without fully understanding the meaning either of Einstein's criticism or of Bohr's response. The situation is actually extremely complex and requires further comment. For not only was the EPR analysis judged incorrectly, it was often misunderstood and taken to nonsensical conclusions. Now, the purpose of my book is not only to tell the story of quantum theory, but especially to enable the reader to understand its subtle and very profound conceptual implications. The analysis of these so-called misunderstandings and recognition of their weak points is needed not simply for reasons of completeness; it will also help us get a better understanding of the crucial points of the theory.

We begin with one of the great figures of those years, Max Born. This profound thinker encountered difficulties especially when it came to understanding the true significance of the EPR argument. A few years later, when his correspondence with Einstein was published, he expressed his point of view in the following terms: "The root of the difference between Einstein and me was the axiom that events which happen in different places A and B are independent of one another, in the sense that an observation on the state of affairs at B cannot tell us anything about the state of affairs at A." It would be difficult to invent a more extreme misunderstanding. Einstein had no difficulty admitting that distantly separated events present strict correlations and thus that information obtained in one area can in turn furnish a more precise knowledge of the state of affairs in another place; what he denied was that an action carried out in one region could immediately influence the physical situation in another region.

This point deserves a special comment. Various writers (I will not list them here) have claimed that Einstein's argument does not cause any problems since it is known, even at the classical level, that knowledge acquired about one part of a system can very well bring an increase of information about the whole system and thus about some other of its parts, no matter how far away. To illustrate this point, reference was frequently made to the following example (Figure 8.4). There are two boxes. We know that one has a white ball and the other has a black ball, but we do not know which box has which color ball. We take one of the boxes and remove it as far away from the other as we please, let us say, to the other end of our galaxy. Now an observer, before observing the color of the ball in the box near him, can say only that it has a 1/2 probability of being white. At this moment he opens the box and sees that, indeed, it is white. Because of the correlation (the fact that the two balls have opposite colors) he can immediately infer with certainty that the ball in the box on the other side of the galaxy is black. According to some writers, the situation EPR considered is exactly the same—involving an increase of local information in order to have an increase of information about far distant parts of the same system.

I think all my attentive readers will already have understood how far off the mark it is to reduce the EPR argument to something of this kind. The crucial point of their argument has nothing whatever to do with the fact that acquiring information locally tells us something about distant situations. That was not what disturbed them, as Born mistakenly thought. The conceptually crucial point of the EPR analysis lay in the fact

183

CHAPTER EIGHT

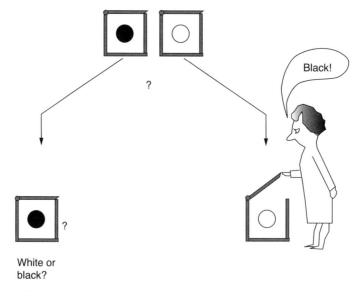

FIGURE 8.4. A typical misunderstanding of the EPR argument, that underestimates its value. According to those who hold this point of view, the situation is perfectly analogous to a classical one where an observer knows that if two boxes contain balls of different colors, he will know the color of the ball distantly removed, by discovering the color of the ball near him.

that, whereas in the classical case just discussed, there is no contradiction, and it is correct and appropriate to assert that even if unknown to us, *the distant ball was black before, and independently of, the observation*, but in quantum mechanics, and with reference to entangled states, such an assertion would be absolutely illegitimate. Further: the assertion that even before the measurement, the pair of photons would be both vertically polarized or both horizontally polarized can be easily falsified in the laboratory.[8]

This naive reading of the EPR argument was criticized with subtle humor by Bell, who presented the famous example of "Dr. Bertlmann's socks" (Figure 8.5).

As Bell said,

> The philosopher in the street, who has not suffered a course in quantum mechanics, is quite unimpressed by Einstein-Podolsky-Rosen correlations. He can point to many examples of similar correlations in everyday life. The case of Dr. Bertlmann's socks is often cited. Dr. Bertlmann[9] likes to wear two socks of dif-

FIGURE 8.5. The amusing example Bell used to illustrate a situation like that of Figure 8.4, to satirize those who had a naive understanding of the EPR argument.

ferent colours. Which colour he will have on a given foot on a given day is quite unpredictable. But when you see [Figure 8.5] that the first sock is pink, you can be already sure that the second sock will not be pink. Observation of the first, and experience of Bertlmann, gives immediate information about the second ... and is not the EPR business just the same?

After this preface, Bell explains the situation of the EPR thought experiment in detail, and concludes:

It is in the context of ideas like these that one must envisage the discussion of the Einstein-Podolsky-Rosen correlations. Then it is a little less unintelligible that the EPR paper caused such a fuss, and that the dust has not settled even now. It is as if we had come to deny the reality of Bertlmann's socks, or at least of their colours, when not looked at. And as if a child has asked, How come they always choose different colours when they *are* looked at? How does the second sock know what the first has done? Paradox indeed!

And he continues with the statement I put at the head of this chapter, in which he pointed out that EPR did not intend to reveal a paradox, but to draw extreme conclusions from the conceptual structure of the theory, and show its incompleteness.

A similar misunderstanding was shown by the great philosopher of science, Sir Karl Popper. In a book of some of his collected writings, *Quantum Theory and the Schism in Physics,* Popper presents his own criticisms of the orthodox interpretation, attacking in particular the tra-

CHAPTER EIGHT

ditional position about the reduction of the wave packet, saying, "No doubt, the reduction of the wave-packet can happen very suddenly; even with super-luminal velocity, as I explained in section 75 of *The Logic of Scientific Discovery*; for it simply is not a physical event—it is the result of the free choice of new initial conditions." It should be noted that this wording, with its explicit reference to superluminal events, suggests a position like that of the ball in the box or the shoes on Dr. Bertlmann. Just how these new initial conditions, defined as an action that takes place in A, can, to use the language of Popper, render immediately actual some potentialities and not others present in B at the moment of measurement, does not seem to interest him. This work dates from the 1950s.

A few years after writing the preface of that book, Popper fell into an opposite, and equally serious error about an "EPR situation." On this occasion, contrary to the preceding one, there is an over- rather than an underevaluation of the EPR analysis. On p. 27 of the same book, Popper proposes an experiment that constitutes a variant of the EPR argument, asserting that if the Copenhagen interpretation is correct, the experiment just analyzed would allow for sending signals faster than the speed of light. This work is one of a lengthy series we will discuss later, in which it is maintained that quantum formalism would permit us to use the reduction of the wave packet to violate one of the postulates at the basis of relativity (i.e., that the speed of light cannot be exceeded). Now, despite the peculiarity of the situation addressed by EPR, this conclusion is fundamentally erroneous and arises from an incorrect use of quantum formalism.

I recall a spirited discussion I once had with Popper at the International Center for Theoretical Physics at Miramare in 1983. Professor Abdus Salam informed me that on the occasion of Popper's visit (for delivering a lecture on the foundations of quantum mechanics), he would be very pleased if the Center would have on hand some competent person in the field, and asked me to take part in the discussion. I knew Popper's work well and told Professor Salam that my intervention could be critical. Salam's reply was simple: "I have full confidence in you, and if you think you are right, you should explain your position without any fear." Popper presented his thought experiment (a variant of the EPR argument), which, according to him, left us with only two alternatives: either the orthodox interpretation was correct, and it would then be possible to send signals faster than the speed of light, or there would not be any action at a distance and the experiment would constitute a falsification of quantum theory. At the end of the conference I explained to him in simple, but mathematically precise terms, the reasons why his point of departure was erroneous: he had not correctly applied the rules of the theory and in fact, the *impossibility* of sending superluminal signals would confirm the theory rather than falsify it—the exact opposite of what he maintained. At the end of my intervention he only said that he could not answer my objection since he did not have a mastery of the mathematics of the formalism, but was still convinced that the theory implied the possibility of superluminal signals. This strange, and, as we shall see, fundamentally erroneous idea has been supported by various researchers in various scientific works, and published in prestigious reviews.

We now come to the last example of a serious misunderstanding of the meaning and importance of the EPR objection. In 1982 a book was published that is one of the most beautiful, complete, and serious biographies to date: *Subtle Is the Lord,* by the great physicist Abraham Pais. The author studies Einstein's life as a whole, providing detailed and documented biographical information, along with a clear and exhaustive account of Einstein's enormous scientific output. The book is extremely valuable, and an inexhaustible source of precise information. The one aspect of the book that seems lacking to me is where Pais accounts for Einstein's response to quantum mechanics. There are numerous imprecisions, revealing a prejudice that arises from the author's unconditional adherence to the Copenhagen orthodoxy. This does not allow him to see the subtle shadings and meanings in many of Einstein's observations. I cannot omit to mention one of these, which has reference to the theme of the present chapter. According to Pais, the interest of the work of EPR consists completely and exclusively in the fact that it contains two statements that reveal Einstein's true position—or at least what Pais thinks was Einstein's true position. To quote Pais directly: "The only part of this article that will ultimately survive, I believe, is this last phrase, which so poignantly summarizes Einstein's views on quantum mechanics in his later years. The content of this paper has been referred to on occasion as the Einstein-Podolsky-Rosen paradox. It should be stressed that this paper contains neither a paradox nor any flaw of logic. It simply concludes that objective reality is incompatible with the assumption that quantum mechanics is complete." Up until this point, the statements of Pais are correct and to the point. But the conclusion, expressed in what follows, contains a very weighty judgment, which, considering the year it was written, is surprising for any researcher with even a minimum of competence in the field: "This conclusion has not affected subsequent developments in physics and it is doubtful that it ever will."

The study of what happened in the 1950s (with the work of Bohm) and in the 1960s (with the work of Bell) should make it clear to the reader that Pais failed to understand the richness, profundity, and fertility of the acute analysis by EPR. It was a seed that would grow into a qualitative increase in our comprehension of the implications of formalism and of physical reality.

Perhaps the best way to conclude this chapter in the light of the assertions just made is to review the direct implications on the running debate of the EPR analysis, and to point out the unsuspected characteristics of the microcosm that that analysis brought to light.

8.8. A Preliminary Evaluation of the Work of EPR

At this point in our presentation, that is, before we undertake to explain the developments triggered by the EPR paper in the years that followed, I would like to focus the attention of the reader on certain important concepts, and how revolutionary they were at the moment of their publication. I will list them and comment briefly on each one.

The Uncertainty Principle

By 1935 almost the entire scientific community had adopted the orthodox point of view on the interpretation of quantum formalism. In particular, there was a wide consensus with Heisenberg, who wanted to explain the limitations imposed by his analysis on all possible measuring processes, by insisting that every such process involves interactions that disturb the system under observation in unknown ways. This perspective on uncertainty relations was known as the "disturbance interpretation." In the first stage of the debate, at least, the accent was placed more upon the impossibility of measuring incompatible observables to the desired precision than upon the primary question, whether or not physical systems actually possessed the properties being measured. Later, there would be an evolution leading to the subsequent positions of Heisenberg himself and of Jordan which we have mentioned in the previous chapter, who denied even the conceptual possibility that a system could objectively possess incompatible properties.

The EPR argument had a lot to say on this crucial point: at the least, if one is not going to deny the assumption of locality *E.L.* (as above, section

5 of this chapter), the argument shows above all else that a physical system has to possess incompatible properties objectively, even though we cannot know them. Furthermore, the argument signaled the end of the disturbance interpretation of indeterminacy relations, by showing how precise and instantaneous knowledge could be acquired of a property of one system (in B) by means of a measuring process that exclusively involves a quite distant physical system (in A), and no interaction with the other system. Even the most authoritative protagonists in the debate on formalism (that is, all those who did not agree with Einstein that the theory was incomplete), had no difficulty in conceding with Bohr, that "of course, there is in a case like that just considered no question of a mechanical disturbance of the system under investigation during the last critical stage of the measuring procedure." In conclusion, the work of EPR brought the definitive end to the disturbance interpretation. The indeterminacy relations are henceforth understood to be simply the inevitable consequence of the formal apparatus.

Quantum Nonseparability

Next, by forcing researchers to confront the peculiar properties of systems associated with entangled states, EPR set clearly in the light a certain fact that had not been sufficiently noticed before: that it is one thing for a composite system to be described as "entangled," and quite another for it to be "factorized." In the latter case, the constituents of the composite system (see section 2 above) preserve their individuality in any measurement, even if only to the limited extent that the formalism allows for any system: that is, they possess some properties objectively. On the other hand, if the constituents are entangled (as shown in section 3 above), they lose their individuality in the sense that, in general, no property is there that could be thought of as being objectively possessed by them as individuals. Hence the peculiar phenomenon known as quantum nonseparability: in the case of a composite system, in general, even when the constituents are very distant from one another and do not interact in any way, they cannot be conceived as separate parts of the system to which they belong; only the whole has what may be defined as "an objective existence," i.e., some objective property.

It is important to note that even if at a certain instant two systems are associated with a factorized state, when they interact with each other, they almost inevitably wind up in an entangled state. The simplest and most illuminating example would be the collision of two microsystems.

CHAPTER EIGHT

Let us suppose that an elementary particle which at a certain instant is characterized by a certain velocity v (it is usually better to refer to the momentum, $p = mv$), collides with another particle (the "target") with its own momentum P, something that occurs every day in our laboratories (and also in the real world, when two molecules collide as a result of thermal agitation). At the beginning, that is, when the two particles are distant from each other and not interacting, the state of the composite system, projectile + target, is factorized, that is to say, has the form $|1, p\rangle|2, P\rangle$ and describes two constituents that are being propagated in suitable directions (and their momenta are the properties objectively possessed by them). If their directions of propagation are such as to lead them into proximity with each other at distances where the forces they exercise on each other (for example, nuclear forces in the case of a proton-neutron collision) become appreciable, the effect of the collision (that is, the quantum laws of its evolution) require that the particles will undergo various deflections[10] that are strictly correlated, the various possibilities being characterized by appropriate probabilities. As is well known, this represents another way of saying that the final state is entangled. To simplify our language here, instead of a continuum of possible pairs of final impulses, which occur in practice, we can suppose that only a single series of pairs of correlated values p_i and P_i are possible. The final state will be

$$|\Psi\rangle = \Sigma_i \, c_i |1, p_i\rangle \, |2, P_i\rangle. \tag{8.14}$$

This is precisely the type of entanglement that allows EPR to infer the momentum of (for example) the target particle by a measurement of the momentum of the projectile (and analogously, for the corresponding pair of observable positions,[11] incompatible with the former).

And so, what lesson do we draw from this example? That in general the interaction between two constituents produces an entangled state even if the state was factorized before they interacted with one another. Practically every interaction brings with it a loss of identity of the systems that are interacting. But since in the long run everything in practice interacts with everything, what emerges is a vision of the universe as an "unbroken whole," an undivided unity whose parts no longer have any identity. The theory implies a fundamentally holistic vision of the universe.

Now clearly, this argument is not valid if the theory is taken to be incomplete, which is how Einstein took it. In fact, Einstein's idea can be reduced to the assertion that since one is forced to admit that even before any measurement the constituents have objective properties that the theory is not able to describe, the theory itself must be considered incomplete. It is clear that the completion would involve the introduction of formal elements that describe these properties (even quantum-incompatible ones), and thus the constituents would reacquire an individuality. But Einstein's analysis constrained the supporters of quantum completeness to accept quantum nonseparability. It is very likely that if it hadn't been for Einstein's profound contribution, this important issue would not have been raised until much later.

It should also be mentioned that the orthodox conception, which includes the postulate of the reduction of the wave packet in the formalism, concedes that entangled constituents can reacquire individuality as a result of a measuring process (recall that in section 5 above this very fact was used to conclude that after the measurement on an entangled state, an objective property suddenly emerges, and thus individuality comes back to the constituents). But this is only an evasion—a solution, as we shall see, that poses more problems than it solves. If we suppose that even the process of measurement obeys the theory's evolution equations, then, as emphasized by Schrödinger the very same year (1935) as EPR (a point we shall discuss in detail toward the end of this book), a "sinister" entanglement must occur between the measured system and the measuring apparatus. The analysis brings an extension of the holistic view of reality even at the macroscopic level and in practice for all the physical systems of the universe. It was not by chance that David Bohm and Basil Hiley entitled their recent book *The Undivided Universe*.

The Problem of Completing the Theory

To say that a theory is fundamentally incomplete (in the sense that it lacks any formal counterpart to physical properties which must be considered as objectively present) is equivalent to suggesting that it can be completed. In this sense Einstein can be regarded as the instigator of an important branch of research, known as hidden variable theories. The basic idea here is that the assignment of the state vector does not constitute the most complete specification possible and that the specification can be enriched through additional parameters, the knowledge of which (which might be impossible to attain in practice) would furnish precise information on all the properties of the system, turning quantum probabilities into epistemic probabilities. The idea is to elaborate "a deterministic completion of the theory" such that its probabilistic structure would derive from our ignorance of the knowledge of appropriate parameters, that could furnish certain predictions, if we only knew what those parameters were. The theory would then, at this fundamental level, take on a certain resemblance to statistical mechanics with respect to classical physics. The statistical quantities are the averages of the quantities precisely defined when, for example, the positions and velocities of all the molecules of a gas are known. It is only the ignorance of these variables (which, we should recall, only Maxwell's little demon could know perfectly) that makes the theory of gases fundamentally probabilistic.

The fascinating story of this type of research will be told later. Here I would like to limit myself to considering the role Einstein played in suggesting this line of thought. I sense the need to say something on this point, because in recent years a dispute has arisen between the historian Max Jammer, on the one hand (author of the interesting book *The Philosophy of Quantum Mechanics*) and J. S. Bell on the other. This particular dispute I believe to be unwarranted, and find myself completely on Bell's side.

Jammer accuses Bell of having misled the public by suggesting, in the famous work in which Bell presented his theorem, that Einstein was a proponent of hidden variables. According to Jammer, Einstein's true aspiration was to overcome the quantum description in the spirit of classical mechanics, and especially to attempt the elaboration of a formalism in line with classical field theories, using, in particular, for its fundamental elements the "continuous functions of the four-dimensional continuum." In support of his thesis, Jammer cites Einstein's reservations about the specific model of the hidden variable theory presented by David Bohm in 1952 (to be discussed in the next chapter).

Quite appropriately Bell maintained the following:

First the idea that this line of thinking implied making recourse to classical field theory is not in any way incompatible with the idea that there are hidden variables. At most it shows that Einstein had one kind of variables in mind and not another.

Second, it is true that Einstein was not enthusiastic about Bohm's proposal, and that he was in fact rather critical of it, but that cannot be taken as a conclusive proof that Einstein did not contemplate the possibility of a theory of hidden variables. Bell observes quite rightly what Born said about Einstein's statement to him in a personal letter. When Einstein said, "Have you noticed that Bohm believes (as de Broglie did, by the way, 25 years ago) that he is able to interpret quantum theory in deterministic terms? That way seems too cheap to me," Born commented, "Although this theory was quite in line with his own ideas. . . ."

Finally, we may be able to discern an explicit reference to the need to elaborate a hidden variable theory in the following words of Einstein:

Assuming the success of efforts to accomplish a complete physical description, the statistical quantum theory would, within the framework of future physics, take an approximately analogous position to the statistical mechanics within the framework of classical mechanics. I am rather firmly convinced that the development of theoretical physics will be of this type; but the path will be lengthy and difficult.

I am, in fact, firmly convinced that the essentially statistical character of contemporary quantum theory is solely to be ascribed to the fact that this (theory) operates with an incomplete description of physical systems.

And even the EPR article ends with the following words: "While we have thus shown that the wave function does not provide a complete description of the physical reality, we have left open the question of whether such a description exists. We believe, however, that such a theory is possible."

I would like to add a further citation that I came upon recently. Writing to Tatiana Ehrenfest, the widow of the great physicist Paul Ehrenfest who had done so much to settle the controversy between Bohr and Einstein, Einstein said, "In the future we will see the emergence of a theory that will be free from statistical aspects but will have to introduce a remarkable number of variables."

In conclusion, it is very significant that Bohm's achievement, which we discuss in the following chapter, revealed the great influence upon Bohm of his discussions with Einstein at Princeton. In a few months he moved from unconditional adherence to the Copenhagen orthodoxy to a formulation of the first (and to date, the most interesting) hidden variable theory.

Nonlocality

The work of Einstein, Podolsky, and Rosen was based in an absolutely crucial way on the assumption of locality in Einstein's sense, defined above as $E.L.$[12] Obviously, the authors did not consider leaving aside this requirement, which as we have seen was considered absolutely indispensible by them as well as by their opponents. As I have already suggested, the crucial point to determine the destiny of EPR's work is that nature is so mysterious and unpredictable that we are forced to modify deeply rooted convictions that had seemed self-evident. But there is no doubt that it was really Einstein's work that for the first time put the problem of locality clearly before everyone's eyes, and forced those who held that the theory was complete to face this new conceptual dilemma. Since we will

193

soon be treating these issues at length, I will not delay the discussion here, but only provide a sketch of the main events. As already mentioned, in 1952, Bohm succeeded in presenting a theory of hidden variables equivalent from the point of view of predictability to quantum mechanics, but perfectly deterministic. However, his theory (apart from other peculiarities as we will see) is explicitly nonlocal. Bell was profoundly struck by Bohm's work, which seemed to show explicitly, as he said, how to realize what had seemed impossible before. He then tried to purify it of this defect, by elaborating a model of the same characteristics in all respects but this one. He tried—but without success. At this point he was suddenly struck by a new idea: could it not be, that the predictions of the theory, and not its interpretation, are what enter into an incurable conflict with the locality requirement?[13] And thanks to this happy intuition, Bell succeeded in deriving his celebrated inequality, which shows that nonlocality constitutes an undeniable trait of physical reality, if the correlations between distant systems (as predicted by the theory) are verified. Bell's inequality has been defined by the Nobel Prize recipient Brian Josephson as "the most important breakthrough in recent physics." If he had lived a little longer, Bell would certainly have received the Nobel Prize. I recall Abdus Salam telling me that for several years Bell was being seriously considered for the prize.

I do not believe there is any chance of going wrong in saying that the work of Einstein, Podolski, and Rosen played an essential role in the development of the theory we are soon going to analyze. In the light of these facts, it should be better understood why I consider the statement of Pais I quoted above to have been particularly unfortunate. Unfortunate, yes, but not surprising, since Pais is one of the few physicists who apparently did not appreciate Bell's work. This, and his oversimple observation about EPR, only confirm something else: without wishing to detract in any way from the profundity of his thought and the important contributions he has made to science, it must be said that, through uncritical prejudice, Pais failed to understand the subtleties of the debate about the fundamental concepts of the theory. This situation is indeed ironic when we consider that in his book Pais continually wonders how a genius like Einstein could fail to fully understand quantum mechanics, owing to philosophical prejudice!

CHAPTER NINE

Hidden Variables

> I hope this book will be a fitting testimony to this radical and original thinker [David Bohm] who rejected the view of conventional quantum mechanics not for ideological reasons but because it did not provide a coherent overall view of nature, a feature that David felt an essential ingredient of any physical theory. It was like Escher's "The Waterfall," a fascinating picture in which region by region appeared to be carefully constructed and consistent, but when one stepped back to perceive the whole, a contradiction was there for all to see.
> —*Basil J. Hiley*

IN THIS CHAPTER we will analyze what happened after Einstein, Podolski and Rosen posed their question. For, once their conclusion is accepted—that quantum mechanics is an incomplete theory—we cannot help asking whether or not it can be completed, and how such a completion could be accomplished. Now I should explain at the outset that for the pioneers in this line of research, any completion of the theory would make it rigidly deterministic, so as to satisfy the following requirements: we should be able to determine precisely the value of any observable of a system, through assigning hypothetical parameters that completely specify the state of any physical system whatsoever (even though such parameters may very likely be experimentally inaccessible—hence the name "hidden variables").[1]

Obviously, this cannot be any arbitrary value. It has to coincide with one of the values permitted by quantum formalism for the observable in question. This specification is essential, for it recalls the phenomenon of the quantization of physical magnitudes so typical of the theory: it implies, in fact, that the greater part of the observables can only assume one out of a discrete set of definite values, and not any arbitrary value as occurs in classical physics. If the conditions just now stated were not rigidly respected, the hidden variable theory would already be in direct contradiction to quantum mechanics. In that case, the theory would not represent a completion (i.e., would not be equivalent to it on the level of predictability) but would constitute an alternative theory. Consequently, in the hypothetical theory we are now configuring, the assignment of hidden variables (and thereby the complete characterization of the state) will precisely determine, for example, the value of an atom's energy, within the set of permitted values. Similarly, in the case of an electron and with reference to its angular momentum (see Section 3.4 above), the knowledge of the hidden variables ought to determine which one value it will assume out of the range of values $\sqrt{\ell(\ell+1)}$ $(h/2\pi)$ ($\ell, = 0,1,2,\ldots$), and which value, out of the range of values $m(h/2\pi)$ ($m = -\ell, -\ell, + 1, \ldots, \ell -1, \ell$), will be assumed by its component along any previously fixed direction. But the physical procedure of preparing the state of the system, which in the standard form associates it with a very precise state vector $|\Psi\rangle$, does not allow for complete control over hidden variables. The distribution of these variables will be such that the probability of a certain result for any observable will coincide with the corresponding probability determined in the standard theory by the state vector in question. This is the exact sense in which we must interpret the requirement that the theory is predictively equivalent to quantum mechanics: the physical preparation of the system, and its subsequent evolution, bring about a distribution of hidden variables at every moment, so that the outcomes of any process of measurement must coincide—in terms of possible values and of the associated probabilities—with the values implied by the standard formalism for the same physical process.

Consequently, quantum probabilities will acquire an epistemic character, and the indeterminism of the theory will be entirely attributable to our ignorance of these parameters. In other words, the need to fall back on a probabilistic description is not derived from reasons of principle (i.e., not from the fact that "God plays with dice") but exclusively from our incapacity (or total inability) to get access to the hidden variables.

At this time it would seem a good idea to say that the history of the attempts to complete quantum theory in a deterministic sense is extremely interesting and instructive from both the historical and scientific points of view. We can now take up the exposition of this real-life drama in several acts, starting with the first: von Neumann's theorem. This theorem asserted the impossibility of reaching the desired object through the rigorous demonstration that no theory can possibly exist that is at once perfectly deterministic and also predictively equivalent to quantum mechanics.

9.1. THE PECULIAR STORY OF VON NEUMANN'S THEOREM

John von Neumann was undoubtedly one of the greatest mathematicians of the twentieth century. Unlike almost all other mathematicians then and now, he had a lively interest in physics. In particular, he paid much attention to the theoretical model that had just been born and which had already seen such incredible triumphs, so fascinating both for its fine mathematical structure and for the profound philosophical questions it raised. In this way von Neumann became, so to speak, "the" mathematician of the new scientific community then adopting the new paradigm. In his 1932 book, *Mathematische Grundlagen der Quantenmechanik* (*The Mathematical Foundations of Quantum Mechanics*) von Neumann presented a lucid formulation of the theory that would become a firm point of reference for all who considered logical and mathematical rigor an indispensable quality for any theory aspiring to the dignity of science. He espoused the orthodox interpretation and contributed to its refinement, especially by his thorough treatment of the collapse of the wave packet. He devoted no small part of his work to the problem of measurement. It was von Neumann's peculiar merit (in comparison with his great colleagues from Bohr to Heisenberg and Born) to have become fully aware, in the course of his formal polishing up of the theory, that something is not quite right when one faces the problem of measurement from an authentically quantum mechanical point of view. He is probably the first scientist to have understood clearly that the postulate of the reduction of the wave packet enters into an incurable conflict with another idea: that processes of measurement (and all similar processes in which a microsystem sets off a process of amplification at the macroscopic level) are physical processes like any other, and would thus be governed by Schrödinger's equation with its linear character. To overcome this problem, von Neumann took up a position on the peculiar role of the observer within a theoretical scheme. This position reinforces the orthodox perspective and leads directly to extreme positions such as Wigner's, which we shall discuss in more detail below.

As mentioned before, in the book cited above, von Neumann provided a mathematical demonstration that a deterministic completion of the theory was not possible, that is, that no theory predictively equivalent to quantum mechanics could assign precise values (even if unknown) to all the observables. His theorem is correct, and although it was so refined from a technical point of view that surely only a very tiny minority of physical theorists of the time were in a position to evaluate it critically, it quickly assumed the status of dogma (thanks to the huge prestige of its author) used by the knights of orthodoxy against the "heretics": it was pointless to look for something that von Neumann's *ipse dixit* had decreed to be impossible. Any researchers who were inclined to find new ways to describe natural processes were completely discouraged, and if they were so stubborn as not to abandon their ideas, they were marginalized by "serious" scientific discourse. The project to find hidden variables was abandoned for about twenty years.

What then is the meaning of my assertion in the preceding chapter, that this theorem was "mistaken" and that therefore it appeared to prevent research for so many years in a direction that would certainly have been worthwhile to explore? Surely, John von Neumann could not have committed an error in deducing a conclusion from his premises! And yet, in the 1950s, Bohm would succeed in presenting an explicit example of a theory that realized precisely what von Neumann had shown to be impossible. Was something amiss? If the argument was correct, but the conclusion wrong, there must be some incorrect element higher up, as it were, in the very hypothesis of the theorem. Now, a few years after the appearance of von Neumann's book, Grete Hermann asserted that the author's argument was circular, that is to say, that it presupposed what it was trying to prove. Her criticism, calling for further reflection on the real value of the theorem, was completely ignored at the time. It was finally Bell who would give the needed emphasis to the fact that Neumann's demonstration had only a limited conceptual relevance, since it falls back on an assumption that, while natural and valid in both classical and quantum mechanics, is logically unnecessary. Once the role of this hypothesis was clearly identified, anyone who understood the requirements of a theory of hidden variables would immediately be able to conclude that *this hypothesis and this hypothesis alone* was what was preventing the realization of the hidden variables program, and that it would have to be subjected to a strict critical analysis. The problem can be explained in detail without recourse to technical subtleties, and I believe an explicit illustration will help the reader understand.

9.2. A Detailed Analysis of Von Neumann's Hypothesis

To understand von Neumann's hypothesis, we can consider a hypothetical theory of hidden variables. We use λ to synthetically denote the hidden variables (which can include the state vector, if this is required by the hypothetical theory, provided the variables are not reduced to the state vector alone). The specification of λ completely determines the system's state, and thereby all the values of any observable pertaining to it. So then, looking at a given observable A, the value $A(\lambda)$ that this observable assumes is a determinate value, and coincides with one of the specific values allowed by quantum mechanics for A. We can now precisely formulate von Neumann's hypothesis: if the observable A turns out to be a linear combination of other observables, let's say B and C, with real numbers b and c:

$$A = bB + cC, \qquad (9.1)$$

then the defined values $A(\lambda)$, $B(\lambda)$, $C(\lambda)$ that the observables assume once the λ variables have been specified should be connected by the same relationship that connects the observables themselves, and that means

$$A(\lambda) = bB(\lambda) + cC(\lambda) \qquad (9.2)$$

Given these premises, it is relatively easy to understand that this is equivalent to declaring the whole hidden variables program to be impossible a priori! In fact, by the phenomenon of quantization, A, B, or C can only assume certain determined values, and within quantum formalism such values cannot, in general be connected by the relationship in question (the appendix to this chapter provides a simple demonstration of this, with reference to a component of angular momentum in any given direction). It is well known that such a component can be expressed as a suitable linear combination of the components along three Cartesian axes, but the quantized values that the theory assigns to any component are not connected by the same relationship. The demonstration clarifies how von Neumann's tacit assumption (never to be satisfied in a hidden variables theory) is perfectly natural for the observables of classical mechanics and for the average values of standard quantum theory.

The conclusion we have now reached leads us to reconsider the meaning of von Neumann's assumption. On the one hand, it is clear that, to a certain degree, the charge of circularity is warranted.[2] Since the assumption of the quantization of observables is incompatible with the additivity assumption, the argument bears a remarkable resemblance to asking ourselves a question of the following kind: "Is it possible to construct a theory according to which bodies are not affected by the earth's gravity?"—all the while assuming, as one among our hypotheses, that gravitational attraction is universal! It would be ridiculous to waste any time "proving" that the answer to the question would be "no." But then we should ask ourselves, "Is the hypothesis of universal gravitation well founded?" In our case, this would mean asking if it were logically necessary and physically plausible for the theoretical scheme whose existence we are supposing, to require that the additivity assumption be satisfied at the level of each individual physical system characterized by precise values of the hidden variables. And surprisingly, the answer follows straight from the position of the father of orthodoxy, Bohr himself, who held that it was not legitimate to study a microscopic system while ignoring the measuring apparatus. Actually, what hidden variable theories require is that knowledge of these variables determines the measurement outcomes with certainty. But, as Bohr himself taught us, the most profound lesson to be gained from our researches into microscopic systems consists in the fact that our theory *implies the impossibility of any sharp separation between the behavior of atomic objects and the interaction with the measuring instruments which serve to define the conditions under which the phenomena appear.* And thus we cannot ignore that the observables in play are incompatible and thus that the physical processes necessary to determine the values will be different and mutually exclusive. Consequently, we cannot see why the relative outcomes should be connected by a linear relationship.[3] With reference to the particular problem at hand we would correctly conclude that von Neumann's stipulation is neither logically necessary nor appropriate.

9.3. Enter David Bohm

Bohm was the first to establish a consistent, deterministic model for a theory of hidden variables equivalent to quantum mechanics. We can now follow this story in detail, as Act Two of our drama. Bohm was an American, who began his work in the 1940s at the University of California at Berkeley as the student of Robert Oppenheimer. In 1947 he followed Oppenheimer to Princeton. In these years Bohm was working on his very important book *Quantum Theory*, which unlike almost all other books on the subject discusses the conceptual and interpretative problems of quantum formalism in detail. His presentation of the various points made it clear to everyone that he was a complete adherent of the orthodox interpretation, that he sided with Bohr in the Bohr-Einstein debate. In particular, in his discussion of the EPR paper he reformulated the argument in terms of spin variables rather than the position and impulse variables of the original. Not only was this version simpler, it made the problem more open to experiment.

In 1947 Bohm was at Princeton, where he remained until 1951. Of course, since he was deeply interested in the conceptual and epistemological implications of formalism, he had, as he said, "many stimulating discussions with Einstein." We shall soon see how this involvement would lead to what could be called a conversion from orthodoxy to heresy. But to continue the tale: it was the height of the Cold War, and the beginning of the McCarthy era. In 1949 Bohm was sought out by the Committee on Un-American Activities and was asked to testify in the spring of that year. He appealed to the Fifth Amendment, and was charged with contempt of Congress. Although he was acquitted of that charge, the president of Princeton University refused to renew his contract. Bohm could not get another position in the United States, so, in 1951, with the help of a strong letter of reference from Einstein, he was offered a position at the University of São Paolo in Brazil. He moved there, but after a few weeks, was summoned by the American Consulate in Brazil and given a new passport that said "valid only for reentry to the United States." It was feared that he might transfer to the Soviet Union. Like any scientist in such a position, Bohm was convinced such drastic limitations on his movements would isolate him from the scientific community, and he circumvented the problem his own government had caused by applying for (and obtaining) Brazilian citizenship. In 1955 he was able to spend two years at the Technion Institute in Haifa, Israel, before moving to England, where, after a few years at the University of Bristol, he settled definitively at Birkbeck College in London. In 1991 he was awarded the Elliot Cresson Medal by the Franklin Institute. The U.S. government made public acknowledgment of their errors, and offered to reinstate him as a U.S. citizen, but Bohm turned down the offer.

But let us look a little more closely at the history of Bohm's scientific thought. His text, which showed him to be a convinced adherent of Bohr's position on EPR, appeared in 1951. But already in 1952 he published, while in Brazil, two works that revealed the profound influence that (as Bohm himself admitted) Einstein had exercised on him. In particular he was led to reconsider the conceptual basis of the theory. In these works he offered a certain model we shall soon consider in detail, known as the "pilot wave" model, inspired by ideas suggested years earlier by Louis de Broglie. It is the first example of a hidden variable theory. Interestingly enough, the David Bohm of São Paolo was ready to expound a theory that the David Bohm of Princeton had declared impossible, in the textbook he had just published in which he discussed von Neumann's theorem and considered it significant.

In the former of the two 1952 articles he clarified his intentions as follows:

> Most physicists have felt that objections such as those raised by Einstein are not relevant, first, because the present form of quantum theory with its usual probability interpretation is in excellent agreement with an extremely wide range of experiments . . . and secondly, because no alternative interpretations have as yet been suggested. The purpose of this paper . . . is, however, to suggest just such an alternative interpretation. . . . It permits us to conceive of each individual system as being in a precisely definable state, whose changes with time are determined by definite laws, analogous to (but not identical with) the classical equations of motion. Quantum mechanical probabilities are regarded (like their counterparts in classical statistical mechanics) as only a practical necessity and not as a manifestation of an inherent lack of complete determination in the properties of matter at the quantum level.

So then, what is the theoretical scheme he proposed on this premise?

9.4. The Pilot Wave Theory

The hidden variable theory propounded by Bohm in 1952, known as the "de Broglie–Bohm pilot wave theory," requires a choice of hidden variables that will probably surprise the reader (but as we shall see, is very appropriate): the hidden variables of the model are simply the positions of all the particles that constitute the physical system under examination.

To understand how this particular solution was arrived at, and explicate the model, it will help to keep in mind Schrödinger's description of the states of a physical system in terms of the wave function (which is simply a specific mode of representing the state vector). To keep matters simple, we can limit ourselves for now to the case of a particle in one dimension—that is, a particle moving only along a certain direction. The relative wave function is simply a function $\Psi(x)$ of the position variable x of the particle somewhere along the line. As we know, the square (of the

HIDDEN VARIABLES

modulus) of the wave function $|\Psi(x)|^2$ represents, according to Born's probabilistic interpretation, the *probability density* of finding the particle at position x, once a measurement is carried out. When treating the uncertainty relationship we have laid emphasis on the fact that the theory permits the consideration of wave functions that are as localized as desired, which means that the indetermination of the particle's position can be made less than any arbitrarily chosen Δx value. Normally we can refer to a function like that pictured in Figure 9.1 (which is supposed to satisfy the condition of normalization, which requires that the area subsumed by the function $|\Psi(x)|^2$ is equal to one). This function is different from zero only in the interval $(a - \varepsilon/2, a + \varepsilon/2)$ centered around the point $x = a$. Since the function is identical to zero outside this interval, any measurement of position would certainly find the particle within the interval. We can therefore say that the uncertainty of the position of the particle itself is less than ε. As already emphasized, the theory places no conceptual limit on how small we make ε, but of course we would not be able to make it equal to zero, since the position, by not being quantized, can assume any single value along a continuous interval; thus no procedure could fix its value with an infinite precision.[4]

The pilot wave model is based on the following essential point: the process of physical preparation of a system leads us to a wave function

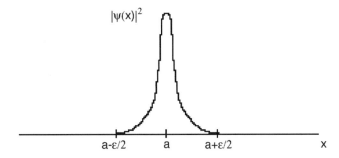

FIGURE 9.1. The square of the modulus of the wave function of a particle represents, as is well known, the probability density of finding the particle on a point of the x axis. From this it follows that the area under the curve in any interval represents the probability of finding the particle somewhere within the interval. In the present case, the particle is securely confined in the interval of amplitude ε shown in the figure, and the normalization condition (which requires that the area under the curve is equal to 1) guarantees that the particle will certainly be found within the interval in question in a position measurement.

203

that, to the extent that its localization permits, is compatible with the fact that, when measured, the particle associated with such a wave function can be found in a measurement of its position in any point of a nonzero interval on the axis. The hypothesis at the base of the theory is that, in fact, the particle in question is really and objectively in a very precise position among the positions compatible with the wave function that describes it (that is, it cannot be in a point x for which $|\Psi(x)|^2 = 0$). But that is not all: if we consider many systems, all described by the same wave function (that is, if the same process of preparation for a system is repeated many times over), then the objective distribution of the positions of the individual particles constituting our ensemble reproduces the distribution of outcomes that would be obtained according to standard quantum mechanics, by subjecting all the particles of the ensemble to a measurement of position. The following points in particular need to be stressed: in the case of a single system it is possible (with the right kind of experimental procedure) to prepare it in such a way that its wave function is perfectly defined and coincides with a function we have arbitrarily chosen, $\Psi(x)$. At the end of the preparatory phase, the system has a precisely defined position which we are not able to control (indeed, the position *is* the hidden variable of the theory). If we then move from an individual particle to an ensemble of particles that have been prepared in identical fashion, the definite positions of the members of the ensemble reproduce the distribution associated with $|\Psi(x)|^2$. In other words, if we consider an interval in which this function assumes an appreciable value (let's suppose the area covered by $|\Psi(x)|^2$ in such an interval turns out to be 70% of the total area), then many members of the ensemble (exactly 70%) will actually have a position that belongs to this interval, whereas in an interval in which $|\Psi(x)|^2$ turns out to be small (or nothing), then only a few (or none) of the particles will be there.

This is the theory's point of departure. Having analyzed it now, perhaps to an excessive degree of pedantry, we can review the conceptual framework as a whole.

First, once the system has been prepared, we say that, at the time conventionally denoted t = 0, the system is associated with a certain specific wave function $\Psi(x,0)$ *which we know perfectly,* and it is furthermore at point x (one among the points compatible with the function in question) *which we do not know at all.*

Second, the wave function $\Psi(x,0)$ and position x together determine (by a simple mathematical prescription we do not now need to explicate) the velocity of the particle at time t = 0. Of course, since we do not know where the particle actually is, this mathematical prescription will precisely assign a velocity to it, as dependent on the specific position of the particle among all the positions allowed.

Third, Schrödinger's equation, as we know, determines the wave function unambiguously at any succeeding instant; we therefore suppose that we know this function at any arbitrary time t > 0, and indicate it as $\Psi(x,t)$.

Fourth, by another precise mathematical procedure, we can define a potential in terms of this function, the so-called quantum potential $V_Q(x, t)$, which describes quantum-mechanical forces. These forces, in conjunction with the classical forces to which the particle happens to be subjected, unambiguously determine the movement for any fixed initial position (and velocity), in exactly the same way as in classical physics the initial position and velocity, plus the potential, unambiguously determine the evolution of some material point. In all of this the following idea is above all to be clearly emphasized: the particle is being subjected to forces that correspond to the physical situation in which it is found (for example, by having charge, the particle is subjected to the force of an electrical field present in the region it happens to be traveling through), and which must be taken into account both in classical physics and in standard quantum physics. In conjunction with these, even further forces are considered that are supposed to have an effect on the particle, being determined at every point and in every instant by the system's wave function. In the model, the crucial role of the wave function (evolving according to Schrödinger's equation) is precisely and exclusively to determine these forces.

Finally, a particle that is initially in a well-defined position will move from there along an unambiguous trajectory (in dependence on the initial velocity and position) determined by Newton's equation for the law of moving bodies in the world of classical physics, even though account is being taken within this equation both of standard forces and of forces resulting from the addition of the quantum potential. Consequently, the particle itself will always possess an absolutely precise position, at every instant following the initial one.

Before we study this model any further, I would like to anticipate the striking quality that characterizes it and makes it predictively equivalent to quantum mechanics: the quantum potential is so ingeniously constructed that, when we consider an ensemble of identically prepared particles (and all thereby associated with the same wave function), but which inevitably have diverse initial positions, distributed according to

205

CHAPTER NINE

the law $|\Psi(x,0)|^2$, the evolution implies that the distribution of their positions at time t exactly reproduces what is given by $|\Psi(x, t)|^2$. Now because, as so many scientists, including Feynman, have observed, all measurements of physical magnitudes can be reduced by definition to measurements of position,[5] it is clear that the property now stated guarantees an agreement between the model and the standard theory, when the same experiment is repeated on a number of systems that have been identically prepared.[6]

Naturally, the proposed model can immediately be generalized to a real three-dimensional situation, and to the description of a system of any number of particles: the relation between the wave function and the quantum potential is formally more complicated but follows the same steps as the case just considered, and all the above-mentioned properties are in general completely valid.

To conclude: in the perspective of the pilot wave theory, "quantum-mechanical force" is the only difference between the quantum and classical worlds. The model is perfectly realist and determinist, and the description that it provides of a physical process is quite similar to the classical one. Nevertheless, it produces outcomes that agree perfectly with all the accurately verified statistics of quantum theory. The nonepistemic quantum probabilities become epistemic, and completely attributable to our inability to know the hidden variable (again, nothing other than the initial position) that characterizes an individual physical system.

9.5. A Few Examples

To understand this model fully, we need to study some individual cases. We begin (in Figure 9.2) with the process of diffraction already illustrated in Figure 3.10. We suppose that the particle is not subjected to any "classical" kind of force (electric, magnetic, or gravitational) while passing through the region between the slitted screen (which "prepares" the wave function at its initial moment) and the detector screen.

What, then, is Bohm's vision of the process? At the beginning, the wave function is shown in the figure right underneath the screen with the slit: it has a constant value in the interval corresponding to the slit's extension, and is zero outside of that interval. However, according to Bohm, the particle (which, according to the theory, would have a velocity perpendicular to the slit) is located in a very precise position (although unknown to us) which we call point a. Since there are no forces that can af-

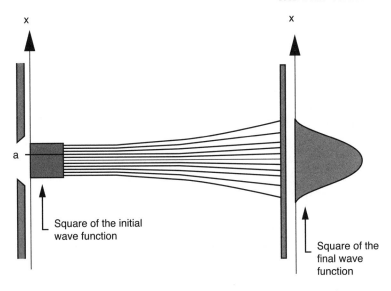

FIGURE 9.2. View of the process of diffraction from one slit, in terms of the pilot wave theory.

fect the particle between the two screens, according to classical physics the particle would be propagated in a straight line and would strike the detector screen at a point directly opposite a. Now, at this stage of the analysis we have omitted the wave function which, evolving according to Schrödinger's equation, changes its form in the intermediate region. This would correspond in general to a quantum force that acts on the particle itself within the region. Explicit calculations show that this force determines a deflection of the particle that depends on its initial position and makes particles starting higher move slightly upward and particles starting lower move slightly downward. These trajectories are illustrated in the figure, starting from various initial positions and forming a figure of diffraction on the detector screen, a geometrical enlargement of the slit on the first screen, with the familiar bell shape.

Something much more interesting and yet conceptually analogous would emerge from the analysis of an interference experiment. In this case, the initial wave function (i.e., immediately after passing through the two slits) is the quantum superposition of two wave functions, each one differing from zero only in correspondence with one of the slits. Once again, the theory asserts that in fact the particle is located at a precise point among those possible for it, some point in either slit A or slit B. But the presence of the wave function of the term corresponding

CHAPTER NINE

to the region of the slit where the particle is not plays a role (alongside the term corresponding to the slit where it is) essential for determining the global wave function in the various points between the initial screen and the detector screen. This wave function produces in its turn a quantum force field that determines the evolution of the particle that was at the initial moment at a certain point of slit A with a certain velocity, a field that guides it in the intermediate region, and brings it at last to one of the precise points compatible with the figure of interference that we know is being formed on the screen when the experiment is repeated numerous times. Of course, the repetition of the experiment with a second particle requires, in general, that from the beginning this second particle has a different position, a different velocity, and thereby a different trajectory, than the first. Now the crucial and very interesting characteristic of the theory is this: if we take all possible initial positions into account (and the correct probability assigned them by the initial wave function), the ensemble of the trajectories will guarantee that the distribution of positions on the final screen will exactly reproduce the quantum mechanical figure of interference. Figure 9.3 illustrates the graph of the trajectories of an electron (every trajectory depending essentially on its initial condition) passing through a double-slit appara-

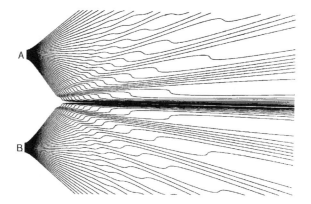

FIGURE 9.3. Trajectories of the particles in a diffraction experiment by two slits. Note now that each particle follows a very precise path completely determined by its initial position. However, by the effect of the quantum mechanical potential, the trajectories tend to be distributed in such a way as to reproduce the interference shape that forms on the screen, when the experiment is repeated numerous times.

tus. The trajectories have been calculated with reference to the quantum mechanical potential.

A few observations are in order. First of all, the trajectories regroup themselves so as to reproduce the alternation between the light and dark bands of the experiment. Secondly, it should be noted how the figure makes clear the role of the quantum mechanical potential, and illustrates how the theory, to the extent that it is "classical," shows some very peculiar aspects, representing, of course, the deterministic counterpart of the quantum world's peculiarities. According to the theory, it is thought that a particle at the initial instant is indeed and objectively present at one point within slit A, and has a certain velocity. Now let us set this situation against another situation, wherein the same particle is at the same point and with the same velocity, but with slit B closed, so that the wave function in that region is zero. The subsequent evolution of the particle is completely different in the two cases. In the second case (see Figure 9.4), the particle would follow a particular trajectory (which it has thanks to its initial position) among the possible trajectories shown in Figure 9.2. In reality, however, it follows a trajectory of the type shown in Figure 9.3. This observation has a special conceptual and pedagogical interest since it allows us to understand, finally, that

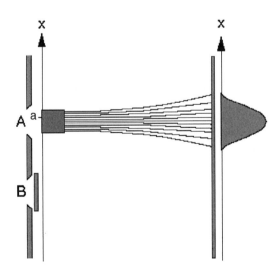

FIGURE 9.4. The trajectories of the particles that pass through slit A are drastically altered by the fact that slit B is closed, as can be seen clearly in a comparison with Figure 9.3.

the theory presents us with some surprising characteristics. The particle in slit A seems able to "know" if slit B is open or closed. Of course, this happens in the standard quantum model as well, which is absolutely equivalent to this one when it comes to measurements of position. But we cannot forget that whereas in quantum formalism the very idea of a trajectory is not even contemplated, here we are moving in the atmosphere of a classical, deterministic framework, which wants to give every particle a precise trajectory. And now, the fact that closing a slit can have a significant effect on a particle that passes through the other leads us to intuit the existence of the nonlocal characteristics of the formalism, raising some rather embarrassing questions for the classical perspective.[7] But this will form the theme of subsequent chapters. Before concluding the present chapter, we need to analyze another peculiarity that is true not only for this theory but for all deterministic hidden variable theories. It forces us to examine carefully how the various hidden variable programs satisfy the fundamental demands that brought them into being in the first place. The factor in question is known as "contextuality." Despite the rather unsettling word, it is easy to understand the meaning of this characteristic, and it may prove interesting to illustrate it with reference to the pilot wave theory.

9.6. The Unavoidable Contextuality of Deterministic Hidden Variable Theories

What, then, are the essential points underlying any attempt to complete quantum mechanics in the spirit of hidden variables theories? The postulates are as follows:

1. That, once the state of a physical system has been fixed, all the observable quantities have precise values
2. That the process of the state's preparation causes a dispersion of the hidden variables in such a way that the statistics for the outcomes of all these observables coincide at every instant with the results that would be obtained by applying quantum rules

As we shall see in the next chapter, postulate 2 leads to an incurable conflict with Bell's locality requirement, an articulation of Einstein's idea that what happens here cannot have any instantaneous influence on what happens there. But even postulate 1 enters into conflict with some hypotheses that seem perfectly obvious, thus raising a few further issues that need to be addressed. We can ask ourselves, "How can postulate 1 be contradictory in itself?" If someone were to inquire how it could be satisfied in practice, we could simply list all the observables and assign to each one of them a precise value within those permitted by quantum mechanical formalism. The problem arises when we try to claim that the theory satisfies some further requirements which were so obvious that nobody even thought they needed mentioning before the work of Bell in 1966 and of Kochen and Specker in 1967.[8] These hypotheses are simply the embodiment of the claim that it is possible to carry out "reliable measurements," that is to say that if the formalism says a certain value is objectively possessed by an observable, then this coincides with the value one gets by carrying out the measurement, and that, if one particular observable turns out to be a function of other observables among those compatible with it, the value of this physical quantity coincides with the function of the values assumed by those other quantities. It seems rather clear to me that this would have to be accepted by anyone wanting to complete the theory in the objectivist spirit that characterizes and motivates the hidden variable program. A little later on we shall evaluate the real impact of the so-called Kochen-Specker paradox on that program. At the moment, however, we should look at a very simple physical system and consider the odd conclusion to which we are led, namely, that every hidden variable theory must inevitably imply some form of contextuality.

The first indispensable step is obviously that of clarifying just what we mean by this expression "contextuality." We should recall that we are interested in hypothetical theories that permit the attribution of values objectively possessed by all the observables of a physical system. Let us pick one, which we can call A, and then designate (as λ) the variables that perfectly determine the state of the system, and with $A(\lambda)$ the value that A assumes in the state under consideration. We say that there is contextuality if the truth or falsity of the assertion "A assumes the value $A(\lambda)$" depends not only on the hidden variables λ, but on the whole physical context.

In Appendix 9B, we give an illustration of this fact with reference to the squares of the spin components of a particle with spin 1. The demonstration is remarkably elegant, and involves the interesting problem of the coloration of a sphere, but I do not think it worthwhile to slow down the argument with it at this point, since the reader will certainly be able to grasp the meaning of contextuality in the next section, where it is illustrated in detail with reference to spin variables in the pilot wave theory.

Only in the precise sense just specified—that is, only when one accepts the inevitably contextual character of physical observables—does it turn out to be possible to make a deterministic completion of quantum mechanics and turn the relative probabilities into epistemic ones. At any rate, it will be appropriate to ask a new question here: "Considering the motive of those who propose hidden variable theories, doesn't the recognition of the 'inevitable contextuality' of at least some observables conflict with the claim of 'objectivity' for a system's properties?" What if the truth value (i.e., the fact that it turns out to be true or false) of the assertion "A assumes the value $A(\lambda)$" depends on whether some other observer decides (through an act of free choice) to measure observable B or observable C (both compatible with A but incompatible with each other)? And if—as we shall see in the next chapter—this measurement (in cases of entanglement) can involve a very distant part of the system, in what sense can the assertion in question be considered to have an objective value?

It is of utmost importance to emphasize that not all observables are contextual for any given hidden variable theory. In fact, it is possible, for every physical system, to construct a hidden variable theory for which a sufficiently rich array of compatible observables (technically speaking, a "complete system") are noncontextual. These observables thus come to acquire a state of particular objectivity.[9] Almost all the other observables are fundamentally contextual and are therefore in a certain sense nonobjective. The choice of which compatible variables will be noncontextual is to a remarkable extent arbitrary. For example, in the Bohm–de Broglie theory, as we have repeatedly pointed out, all the particles of any physical system always have completely and objectively defined positions[10] (the positions in this case coincide with the hidden variables themselves and represent the noncontextual observables of the theory). Bell himself displayed the appropriate attitude to take toward contextual and noncontextual variables with reference to the pilot wave theory: "The moral of this story is that in physics the only observations we must consider are position observations, if only the positions of instrument pointers. It is a great merit of the de Broglie–Bohm picture to force us to consider this fact. If you make axioms, rather than definitions and theorems, about the 'measurement' of anything else, then you commit redundancy and risk inconsistency."

We will return somewhat later to these interesting issues. For now, it may help to reconsider the problem of contextuality with reference to the pilot wave theory, taking into consideration the spin variables of a particle with spin 1/2—in this theory, these are contextual variables.

9.7. CONTEXTUALITY OF SPIN VARIABLES IN THE PILOT WAVE THEORY

Particle position enjoys a very privileged status in the de Broglie–Bohm theory. This is because position represents a property objectively possessed by all particles of the system as a whole, even of the entire universe, if that is our object of interest. Spin, the nonclassical degree of freedom that characterizes many elementary particles, enters into the formalism in a rather peculiar way. To be exact, spin does this only in so far as its presence influences the evolution of the system and thus the wave function, which in turn determines the trajectories of the particles. In terms of the pilot wave theory, spin in a certain sense merely represents the property a particle has of undergoing certain deflections (and thus of following certain trajectories rather than others) in certain experimental situations. Assertions so typical of quantum formalism like "the particle has upward spin along the z axis" are simply used to indicate that the particle will undergo an appropriate deflection in dependence on its position and initial velocity, and this deflection is determined by several things, including spin.

In the de Broglie–Bohm formalism spin is a contextual observable, even when it is only spin 1/2.[11] This trait offers us the opportunity to give the reader a detailed discussion of an experiment in which a spin component is measured. Such an analysis should enable a more direct and intuitive understanding of contextuality, and will have the added benefit of introducing the discussion in the next chapter about nonlocality in hidden variable theories.

We can begin with the standard case already discussed in chapter 3, with special reference to Figures 3.6, 3.7, and 3.8. We described what happens if we introduce a neutral particle with spin 1/2 (in particular, a particle whose spin and magnetic momentum have the same orientation—such as an antineutron) into a region with a nonhomogeneous magnetic field in the direction of the z axis (in such a way that the field increases as z increases). The quantum state in which the particle will be found at its exit from the Stern-Gerlach magnet will depend on the spin state it had before it was measured. In particular, if the spin is upward with respect to the z axis (|z up>), the particle will follow a trajectory (or rather, the relative wave function will be nonzero in only a small, cylindrical space) that will cause it to strike the detecting screen close to point U (up), whereas a particle with downward spin (|z down>) would end up

CHAPTER NINE

near point D (down). It goes without saying that for the two cases under consideration, the pilot wave theory provides predictions that agree perfectly with quantum predictions. The one conceptual difference, of course, derives from the fact that, since the quantum mechanical wave function will be nonzero in a very small (but not infinitely small) radius around U or D, in the standard quantum scheme it is not legitimate to give a fully defined location to a particle before the particle hits the screen. On the contrary, the theory we are now considering maintains that every individual particle follows a perfectly definite trajectory, and that the totality of the possible trajectories in a great many repeated experiments will reproduce exactly the same smudge on the screen as predicted by quantum mechanics.

Let us now move to a more interesting case, where quantum theory predicts the superposition of two different trajectories. As discussed in connection with Figure 3.8, this occurs when the particle is introduced into a region in which there is an inhomogeneous magnetic field with spin "perpendicular" to the vertical, for example, in state |x up>. Now the linearity of the theory implies that the wave function before detection is at every instant the quantum superposition of two wave functions, and that these functions will be appreciably different from zero only inside a very small radius around both the upper and lower trajectories. There is no need to emphasize that this state does not allow us to say that the particle is in fact in the upper or in the lower trajectory. In our earlier discussion in chapter three we did not explicitly consider the particle's initial wave function; now, however, because this determines the distribution of the possible hidden variables (that is, the possible positions of every individual particle as well as their relative initial velocities), we must explicitly take account of this function in detail. To make the treatment simple, we assume that at the initial instant it is different from zero and assumes constant values on the inside of a tiny spherical region whose equator coincides with the median plane between the magnetic expansions, perpendicular to direction z. I will not go into the details of the analysis given by the pilot wave theory. For our purposes it should be enough to point out the results synthetically pictured in Figure 9.5 below. Owing to the symmetry of the problem, it can be shown that no particle can have a vertical component (positive or negative) of velocity as long as it stays on the median plane. It immediately follows from this that the particles are incapable of passing through this plane. And this is precisely how the theory takes account of the fact that, according to quantum formalism, one-half of the particles end up at U (up)

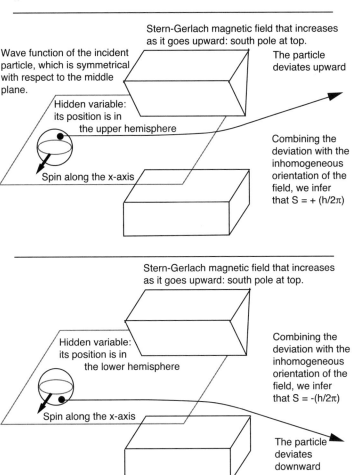

FIGURE 9.5. Description, in terms of the pilot wave theory, of the measurement of the spin components of a particle with spin 1/2, using a Stern-Gerlach magnet. The knowledge of the hidden variables (the initial position of the particle) determines in a simple and direct manner whether the outcome of the measurement will be "the particle has spin down" or "the particle has spin up." In fact, here we are using quantum language. For the theory in question, which is rigidly deterministic, the process requires that "the particle follow an upward trajectory" or "a downward trajectory." These assertions are objective, in the precise sense that there is no need to carry out a measurement in actual fact, that is, to insert an exposure film, to make them true.

and one-half at D (down). The initial hidden variables are the positions of the particles. By the stated hypotheses, half of them will be in the upper hemisphere and half will be in the lower. According to the argument just now presented, particles with the first quality will follow precise trajectories taking them near to U, while those with the second will fall near D. The illustration represents a simple and clear example of how a physical process can be described, taking the de Broglie–Bohm view. It makes it possible to understand how quantum probabilities become epistemic ones: the fact that I am not able to predict where a particle will end up depends on my ignorance of the hidden variables. If I could have known that the particle was in the upper hemisphere to begin with, I would be able to predict with certainty that it would follow an upward trajectory, and vice versa.

We can now discuss the contextuality of the spin variables in this model. Of course, within both classical physics and quantum mechanics, the statement that "the component z of the spin is found to be up (or down)" depends crucially on the orientation of the magnetic field or on the direction along which the field increases. If we reverse the direction of the field at every point, or, alternatively, if we keep the direction of the field but reverse the direction along which the intensity increases, we must conclude that the particles that undergo an upward deviation have been found to have spin down, and vice versa. In fact the deviation depends on the fact, discussed in chapter 3, that there is a tiny magnetic pole associated with spin, and that the north and south poles of this magnet, by being immersed in more or less intense fields, are subject to greater or lesser forces. The operations that lead to the inversion of the field while keeping a fixed intensity, or that preserve the direction but invert its intensity, are illustrated[12] in Figure 9.6, which articulates the situation of Figure 3.6 (above, in chapter 3).

Now let us look at an experiment carried out in the spirit of the pilot wave theory (Figure 9.7). The execution of one of the transformations shown in Figure 9.6 does not change the symmetry of the problem, and thus it is still true that no particle will be able to cross the median plane: the particles lying in the initial upper hemisphere will be deflected upward, the others downward. Since the association between the deviation and the orientation of the spin is inverted with respect to the preceding case, we can conclude (as one does in both classical and quantum mechanics) that the particles that wind up at U have downwards spin while the others have upwards spin. This splendidly illus-

HIDDEN VARIABLES

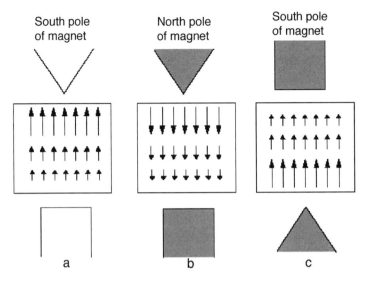

FIGURE 9.6. The various possible experimental setups for measuring the vertical component of spin of a particle by means of a Stern-Gerlach device. The first case is the one considered in Figure 3.6, while the other two correspond to the fact (represented graphically by the shaded parts) that the (north-south) polarization of the magnet has been reversed, and that, apart from this operation, the geometrical disposition is not changed (case b) or inverted (case c). According to quantum mechanics, any one of these operations requires that a particle with spin up in configuration a would move upward, but in the other two configurations would go down.

trates the contextuality of the theory in regard to observable spin: assigning the hidden variable (that is, specifying at the beginning whether the particle belongs to the upper or lower hemisphere) does not determine the spin property possessed (such as "spin is upward, along the z axis") independently of the context. The assertion just now made for a particle with a fixed initial position will be true if the magnet is oriented (or the field varies) in one way, and false with the opposite orientation.

My presentation of the hidden variable theory (and of its postulates and problems) will conclude at this point. We can now move on to the central issue of *nonlocality*.

CHAPTER NINE

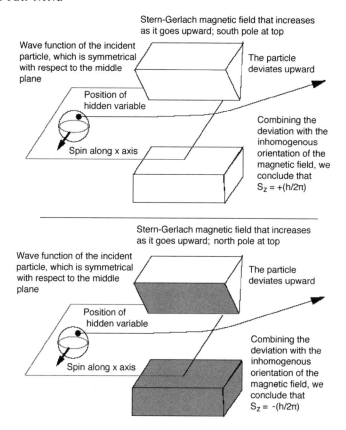

FIGURE 9.7. Contextuality of spin variables in the pilot wave theory. The inversion of the magnetic field does not change the fact that particles with the same initial position finish in the same final position, but in an inverted field, the same final position corresponds to the opposite spin property.

APPENDIX 9A: WHY VON NEUMANN'S REQUIREMENT CANNOT BE SATISFIED

To show how von Neumann's requirement (see section 9.2) is excessive, we can begin with a classical context and consider for the sake of argument the three observables ℓ_x, ℓ_y, and ℓ_z, which are the components (on a three-dimensional perpendicular set of axes) of the angular momentum of a particle with respect to its origin (Figure 9A.1). Their expressions in terms of position and impulse components would be as follows:

HIDDEN VARIABLES

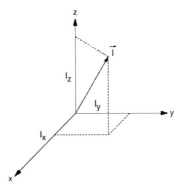

FIGURE 9A.1. The angular momentum of a particle and its decomposition in terms of its components along the axes of an orthogonal triad.

$$\ell_x = yp_z - zp_y \equiv m(yv_x - zv_y),$$
$$\ell_y = zp_x - xp_z \equiv m(zv_x - xv_z),$$
$$\ell_z = xp_y - yp_x \equiv m(xv_y - yv_x). \quad (9A.1)$$

In classical terms, the physical state of the system is completely specified by assigning the position and velocity of the particle. In other words, to know the "state" of the system at time t means knowing the values of the various magnitudes x(t), y(t), z(t), vx(t), vy(t), and vz(t) at the instant of time in question. Consequently, through a simple insertion of these values into the preceding formula, we will know the determinate values for the magnitudes ℓ_x, ℓ_y, and ℓ_z.

Naturally we can also be interested in the value of component ℓn of the angular momentum along an arbitrary direction n, specified by a unit vector and having components nx, ny, and nz with respect to our axes. By the Pythagorean theorem these would have to be related as follows:

$$n_x^2 + n_y^2 + n_z^2 = 1. \quad (9A.2)$$

Of course, the component ℓ_n can be expressed in terms of the components along the x, y, and z axes in the following way:

$$\ell_n = n_x \ell_x + n_y \ell_y + n_z \ell_z. \quad (9A.3)$$

This calls for two observations. Above all, once the state of the system is known, the three components of the angular momentum are defined with an absolute precision. This means that likewise the observable ℓn would have a precise value. Furthermore, this is the linear combination of the values indicated above.

CHAPTER NINE

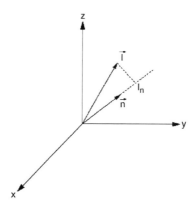

FIGURE 9A.2. The component of angular momentum along the direction n, specified by a vector of unit length, of components n_x, n_y, and n_z.

In other words, if we designate the values of the components along the axes at time t as $\ell_x(t)$, $\ell_y(t)$, and $\ell_z(t)$ (see Figure 9A.2), we will have

$$\ell_n(t) = n_x\ell_x(t) + n_y\ell_y(t) + n_z\ell_z(t). \qquad (9A.4)$$

In conclusion, acccording to classical physics, a physical magnitude that is technically called "a linear combination" of other magnitudes (i.e., that can be expressed as the sum of the said magnitudes, multiplied by appropriate numerical factors) takes on a value that will correspond with certainty to the same linear combination of the values of the accompanying magnitudes.

But now let us move to the quantum world. Within this conceptual framework, the components $\ell_x^{(Q)}$, $\ell_y^{(Q)}$, $\ell_z^{(Q)}$, and $\ell_n^{(Q)}$ will be quantized observables, and each one of them (as explained in section 3.4) can assume only one of the $(2\ell + 1)$ values $m(h/2\pi)$, with $m = (-\ell, -\ell + 1, \ldots, \ell - 1, \ell)$. As always, this assertion means that in a process of measurement of any one of these magnitudes, we can obtain only one of the outcomes now listed. Furthermore, as I emphasized in section 3.4 with reference to angular momentum and spin, the observables in question are incompatible and cannot be measured simultaneously, nor can they all be thought of as possessing definite values. Consequently, even in a case where the state of the physical system is perfectly specified (that is, if we know the state vector associated with the system), the magnitudes in question will not, in general, have precise values. The only information the theory can give us is the probability of obtaining one of the listed outcomes, if a measurement is taken of the relative observable.

We now consider a group of systems, all identically prepared in the state $|\Psi\rangle$, and then subject them all to the same measurement, for example, of $\ell_x^{(Q)}$. We will

HIDDEN VARIABLES

then obtain a series of outcomes, each of which coincides with one of the values m(h/2π), distributed acccording to the precise probabilities assigned by the theory. Multiplying these values by their relative frequency, that is, by the fractions of the system for which this outcome has been obtained, and adding up all the outcomes, we obtain a number that represents the average of the values obtained. We can indicate this average value by $<\ell_x^{(Q)}>_\Psi$, which is what the theory predicts. Using $p_\Psi(m)$ for the probability of obtaining the value m(h/2π) in a measurement of $\ell_x^{(Q)}$, on a system in the state $|\Psi>$, we will have:

$$<\ell_x^{(Q)}>_\Psi = \sum_{m=-1}^{m=1} p_\Psi(m)[m/h2\pi] \qquad (9A.5)$$

In physical terms, the meaning of the quantity here—that is, the average of the magnitude we are interested in—should be clear enough. It represents the element of primary interest in every statistical analysis of random events. For example, it is clear that we can obtain the average height of all Italians by multiplying the values for height times the fraction (or, according to the law of large numbers, times the probability that the height would have this value) of Italians that have that height.

The procedure just carried out for the observable $\ell_x^{(Q)}$ can now be repeated for an ensemble of equally prepared systems (that is, all in the same state $|\Psi>$), but with reference to the other components,[13] obtaining the relative averages $<\ell_y^{(Q)}>_\Psi$, $<\ell_z^{(Q)}>_\Psi$, and $<\ell_n^{(Q)}>_\Psi$. Now a general characteristic of quantum formalism can easily be demonstrated here: for any state $|\Psi>$ these average values are connected by the same relationship that connects their associated operators:

$$<\ell_n^{(Q)}>_\Psi = n_x<\ell_x^{(Q)}>_\Psi + n_y<\ell_y^{(Q)}>_\Psi + n_z<\ell_z^{(Q)}>_\Psi. \qquad (9A.6)$$

The reader will now easily grasp why von Neumann's assumption (9.2) was equivalent to denying *a priori* the very possibility of a hidden variable theory. All we have to do is identify *n* with the direction equidistant from all three axes so that

$$n_x = n_y = n_z = \frac{1}{\sqrt{3}}, \qquad (9A.7)$$

and then

$$\ell_n = \frac{1}{\sqrt{3}}\ell_x + \frac{1}{\sqrt{3}}\ell_y + \frac{1}{\sqrt{3}}\ell_z. \qquad (9A.8)$$

In a theory of hidden variables, once the variables have been assigned, all the observables in question have a precise value among the ones that are permitted. In our case we would have

CHAPTER NINE

$$\ell_x(\lambda) = r(h/2\pi), \quad \ell_y(\lambda) = s(h/2\pi), \quad \ell_z(\lambda) = t(h/2\pi), \qquad (9A.9)$$

with r, s, and t positive or negative integers or zero between $-\ell$ and ℓ. But even the observable ℓ_n is the component of the very same angular momentum along an appropriate direction and is thus subject to the same rules of quantization. This means that once the λ variables have been assigned

$$\ell_n(\lambda) = q(h/2\pi), \qquad (9A.10)$$

with q a negative or positive integer or zero. If von Neumann's stipulation is imposed, these precise values of the variables under consideration should satisfy the same linear relationship as the relative observables, giving us

$$q(h/2\pi) = \frac{1}{\sqrt{3}} r(h/2\pi) + \frac{1}{\sqrt{3}} s(h/2\pi) + \frac{1}{\sqrt{3}} t(h/2\pi) \qquad (9A.11)$$

that is,

$$\sqrt{3} q = r + s + t, \qquad (9A.12)$$

which is manifestly absurd,[14] considering that the number on the left side of the equation is irrational, but the number to the right must be a positive or negative integer.

APPENDIX 9B: WHY SOME VARIABLES ARE UNAVOIDABLY CONTEXTUAL

In this appendix we will illustrate by a simple example how some variables are inevitably contextual. Suppose we have a particle with spin 1. This implies that the particle has an intrinsic angular momentum (like a spinning top), which when squared has the precise value $S^2 = 1(1 + 1)(h/2\pi)^2 = 2(h/2\pi)^2$. Now the relative components in any direction can assume only the values $(h/2\pi)$, 0, $-(h/2\pi)$ (on this, see also the discussion in section 3.7). For convenience, we can measure the angular momentum in units (of $h/2\pi$) so that the square of the spin will be 2 and any of the components can assume only one of the three values $+1, 0,$ and -1. We now imagine three orthogonal directions in space, to which we assign three vectors each of unit length p, q, and r. We use the symbols S_p, S_q, and S_r for the components of spin in the three directions. Since assigning hidden variables unambiguously determines the value of each observable, and since the value has to be one of those permitted by quantum theory, it follows that, once λ has been assigned, each quantity $S_p(\lambda)$, $S_q(\lambda)$, and $S_r(\lambda)$ will have a precise value among the three mentioned above. Obviously, since these values are $+1, 0,$ and -1, the pos-

HIDDEN VARIABLES

sible values for the squares of the components $S_p^2(\lambda)$, $S_q^2(\lambda)$, and $S_r^2(\lambda)$ can be only 0 and 1. As already observed, the square of the spin will certainly have the value 2 (this is the case not only in the theory we are now considering but also in quantum mechanics), and by its own definition, this observable, which we will simply designate as S^2, is the sum of the squares of the relative components along three arbitrary orthogonal directions. Thus the following relationship (nothing other than the Pythagorean theorem) would hold:

$$S^2(\lambda) = S_p^2(\lambda) + S_q^2(\lambda) + S_r^2(\lambda). \qquad (9B.1)$$

Given that $S^2(\lambda) = 2$, what we know about the quantities $S_p^2(\lambda)$, $S_q^2(\lambda)$, and $S_r^2(\lambda)$ implies that for every assigned λ, two of these will have to be equal to 1 and the other will have to be 0. The attention of readers should be called to the fact that the three observables under consideration (in the case of spin 1) are compatible and can be simultaneously measured in one and the same experiment without disturbing each other.

We can now take the decisive step: this will lead us to discuss an interesting problem about the possible ways to color a sphere under certain constraints. In the preceding argument, p, q, and r were any three arbitrary vectors mutually perpedicular to each other. Let us think now of a sphere of unit radius, along with the three vectors each of unit length. These vectors will locate three points on the surface of the sphere P, Q, and R (Figure 9B.1). We decide to color two of the points blue (corresponding to the two directions along which the square of the

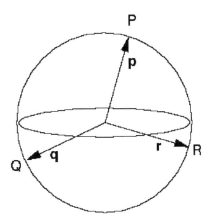

FIGURE 9B.1. Three vectors of unit length p, q, and r, specify three points P, Q, and R on the sphere that will be colored according to the precise rule discussed in the text. The possibility (or impossibility) that all three points of each triad will respect these prescriptions determines the noncontextual or contextual nature of the theory.

CHAPTER NINE

spin component has the value of 1), and the other one red (where the corresponding square is equal to 0).

Here is our fundamental question: can a theory exist that completely specifies the state of the system (by fixing the λ values), and then go on to assign values to all the components of spin, i.e., values that are precise (and only compatible with those permitted by quantum mechanics)? Since the directions we are considering are arbitrary, the problem is immediately transformed into the problem of coloring the sphere: is it possible to color the sphere (using only the colors blue and red) in such a way that each "mutually perpendicular triad" meets the sphere's surface at one red point and two blue points? The answer is "no." Without going into the technicalities or giving a rigorous demonstration (it can be found in the paper by F. de Stefano and myself cited in the bibliography), I will instead present an intuitive and, I trust, rather convincing argument. Let us consider one of these triple arrows, and from among the three points where the triple arrow meets the sphere, take one point to be colored red, according to our prescriptions. We note that there are an infinitude of triple perpendiculars, obtained from this one simply by rotating it at an arbitrary angle around the axis uniting the center of the sphere with the point on the surface. The two vectors perpendicular to the rotating axis describe the sphere's equator. The conclusion should be clear (see Figure 9B.2): taking any point colored red as the pole, the equator associated with this pole will be completely colored blue. We are not able to color the sphere as we wanted, because there are so many "blue" points for each single red one.[15]

What does this impossibility mean? Does it mean that there can be no theory that would allow us to think of all the observables as having definite values? Certainly not. It means, as Bell emphasized, that we have to accept the contextuality of some observables. In fact, with reference to the example just considered, the escape route consists in recognizing that if the global context corresponds to the fact that the observables $S_p^2(\lambda)$, $S_q^2(\lambda)$, are being measured, then, for example, $S_p^2(\lambda) = 1$ is true, whereas if the context corresponded to the fact that the observables $S_p^2(\lambda)$ and $S_t^2(\lambda)$ were being measured, where t represents a direction different from q and from r, then it would be true that $S_p^2(\lambda) = 0$. Contextuality clearly poses some problems that deserve a more profound discussion, but the observation just now made, combined with the consideration of the compatibility or incompatibility of various measurements, enables us to understand their significance. As emphasized before, the components of spin (in the case of spin 1) along perpendicular axes are compatible observables. Therefore, the first experimental situation in which a measuring apparatus measures both $S_p^2(\lambda)$ and $S_q^2(\lambda)$ is perfectly possible. In the same way, $S_p^2(\lambda)$ and $S_t^2(\lambda)$ are compatible observables and thus the second case too, with its own different experimental setup, can likewise be realized in practice. But q and t are *not* two perpendicular vectors, and quan-

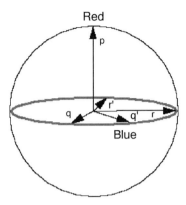

FIGURE 9B.2. Demonstration of the impossibility of coloring the sphere according to the preset rule. If the polar point is colored red, by a rotation of the vertical axis, an infinity of triads is generated (two of which would be p, q, r, and p, q', r', as shown in the figure). It follows that the entire equator would be colored blue. This implies the impossibility of meeting the desired prescription.

tum formalism implies that $S_q^2(\lambda)$ and $S_r^2(\lambda)$ are incompatible observables. Consequently, the first experimental situation is incompatible with the second. We see that, by paying the price of contextuality (that is, by accepting the fact that the specific values depend on the entire experimental context and are not unambiguously determined by the simple assignment of hidden variables) it is possible to keep the requirement that all the observables have precisely defined values and to get out of the contradiction of $S_p^2(\lambda)$ being equal to 1 in the first case and $S_p^2(\lambda)$ being equal to 0 in the second. Therefore, for the same values of hidden variables, the same observable can possess different values in each context. In any case, different values can be correlated only to incompatible contexts.

CHAPTER TEN

Bell's Inequality and Nonlocality

> Bell never wrote down a single, local deterministic theory. Rather, he proved, without ever having to consider any dynamical details, that no such theory can in principle exist. The entire class was killed at a stroke—a classic "no-go" theorem. [Further,] Bell's theorem really depends in no way upon quantum mechanics. It refutes a whole category of (essentially) classical theories without ever mentioning quantum mechanics. Abner Shimony has appropriately given the name "experimental metaphysics" to this type of definitive empirical resolution of what appears to be a metaphysical question.
> —*James T. Cushing*

IN THIS CHAPTER we will discuss what has been described by all the experts in the conceptual and philosophical foundations of quantum mechanics as—to use Nobel laureate Josephson's words—*the most important recent contribution to science:* namely, the derivation by John Stewart Bell of his (now) famous inequality. When Bell died so unexpectedly and prematurely at age 62 in October 1990, many of his colleagues spoke of him as the only man of his generation who could be put at the same level as Bohr and Born for his profound understanding of the conceptual implications of the theory. He was referred to as "the man who proved Einstein wrong" (*New Scientist,* November 24, 1990). Although this assertion is essentially correct, it needs careful articulation to keep it from being taken too far. We will come back to this point and consider it more carefully, in order to allow the reader to reach a correct appreciation of it.

To illustrate the evolution of Bell's thought we can begin with the expressions Bell himself liked to use in commenting on the pilot wave theory:

> Bohm's 1952 papers on quantum mechanics were for me a revelation. The elimination of indeterminism was very striking. But more important, it seemed to me, was the elimination of any need for a vague division of the world into "system" on the one hand, and "apparatus" or "observer" on the other. I have always felt since that people who have not grasped the ideas of those papers ... and unfortunately, they remain the majority ... are handicapped in any discussion of the meaning of quantum mechanics.

A few years later, he would reaffirm his appreciation of Bohm's theory:

> This theory is equivalent experimentally to ordinary, nonrelativistic quantum mechanics—and it is rational, it is clear, and it's exact,[1] and it agrees with experiment and I think it's a scandal that students are not told about it. Why are they not told about it? I have to guess here that there are mainly historical reasons, but one of the reasons is surely that this theory takes almost all the romance out of quantum mechanics. This scheme is a living counter-example to most of the things that we tell the public on the great lessons of twentieth century science: things like the uncertainty principle—that particles do not have velocities as well as positions; things like the necessary role of the observer in modern physics—there just isn't any; things like the necessary appearance of hazard, or pure chance, in modern physics; this theory is deterministic and it accounts for all the quantum phenomena fully. So what's wrong with it?

To answer this question, we need to take a little detour, and trace the development of Bell's thought in the years immediately preceding the derivation of his famous inequality. He observed that Pauli, Rosenfeld, and Heisenberg could find no formal or logical flaw in Bohm's reasoning and had to resort to categorizing it as a "metaphysical"or "ideological" argument. But Bell had no fear of dealing with conceptual theorizing: he was convinced that there was a profound lesson to be learned from the very existence of the theory, even though it needed to be reconsidered. It was precisely thanks to this drive of his to deepen the theory that in 1963, during a sabbatical leave from CERN to the United States, he focused precisely on "what was wrong" in Bohm's theory: its fundamental nonlocality. Events that can occur at a certain point of space have instantaneous consequences, on the particle level, in very distant regions. For example, as we mentioned in the preceding chapter, the trajectory of a single particle in one place can become immediately changed as the effect of an action in another place. Bell was quickly convinced that this was the central problem, and focused on the assertion made at the conclusion of the EPR argument, that the evidence for nonlocality would be apparent only in the framework of the "complete" theory. He asked, "Is nonlocality an accidental element, bound up with the specific pilot wave model, or is it something else of more profound meaning?" With the utmost clarity, he concentrated his attention on Bohm's formulation of the EPR argument, a formulation that made use of spin variables in place of position and impulse variables. He was trying to elaborate a model à la Bohm for this specific case, that is, one that would be deterministic and in agreement with quantum theory, but still *local*. He completely failed. He then asked himself if it could not be demonstrated in general that it was impossible to account for quantum correlations in a local framework, and in this he succeeded, arriving at his famous inequality. To his own surprise, he had furnished proof that, in obeying quantum law, microsystems exhibit unimaginably strange behavior.

Formally speaking, Bell's inequality is extremely simple. Shimony's statement about Bell on the occasion of the 1990 meeting of the American Society for the Philosophy of Science seems particularly appropriate: "Bell's theorem, for which he is most famous, was more a triumph of character than of intellect. The difficult thing about it was the realization of what was understood and what was not understood in the discussion of hidden variables. Bell's honesty about his own understanding provided the impetus for his formulation and proof of the theorem."

Even though the theorem is simple, we are not yet ready to consider it in its most general form. I think the simplest and most effective way to bring the reader to a state of full understanding is to adapt the brilliant idea of Euan Squires (1994), who in his *Mystery of the Quantum World* develops an analogy he calls "a music hall interlude," based on David Mermin's penetrating analyses. I will rehearse this analogy in the following section. The connection between this little story and the important problem we are studying may not be immediately apparent to the reader, but the connection will be carefully explained in the section that follows it, where we will consider an EPR experiment requiring no more knowledge for its comprehension than what we presented in chapter eight concerning correlated photons.

10.1. TELEPATHY OR A CHEAP TRICK?

Let us leave the world of physics for a moment to consider Alice and Bob, who are going to put on an amazing exhibition in a theater or music hall. They are seated on the stage, in two separate cubicles. We are assured that

BELL'S INEQUALITY AND NONLOCALITY

they cannot communicate with one another in any way. At regular intervals, two people from the audience, let's say one from the left section and the other from the right, each give Bob and Alice a card with a 1, a 2 or a 3 printed on it. The choice of numbers appearing on the cards they give to the actors is completely random. At this point, Alice and Bob are asked to write YES or NO by their own choice, on the cards they have been given. The situation is illustrated in Figure 10.1. Now it is clear that Alice does not know the number on Bob's card, nor does Bob know the number on hers. What makes the show interesting is that, although the responses that Alice and Bob write down on their cards are completely unpredictable (independently of the number on the card each actor receives, their answers 50% of the times will be YES, and 50% NO, with a completely random sequence of YESes and NOs), astoundingly, every time they are handed

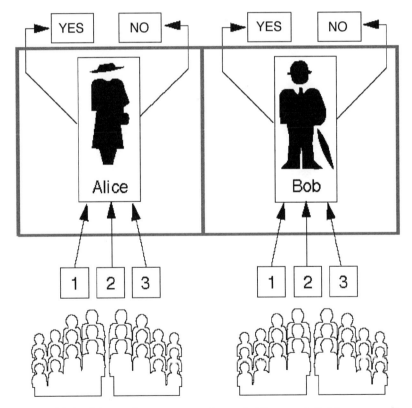

FIGURE 10.1. Schematic representation of Alice and Bob's astounding performance.

229

CHAPTER TEN

cards with the same number on them, they write down the same answer, both writing YES or both writing NO. After they have done this a number of times, Alice and Bob announce they are telepathic, and say they are ready to give repeat performances of the feat. Their proof that they are telepathic is simple: since neither one of them can know what number is on the other one's card, and since, for each different number, the total numbers of YES and NO responses they give are evenly distributed at random, the perfect coincidence of their answers in *all* cases where the numbers they are given are the same can only be explained by the fact that they are telepathically communicating with each other about the numbers on the cards and the responses they each are giving!

Let us say the "Alice and Bob Show," supported by a well-orchestrated publicity campaign, becomes famous. It is presented numerous times, becoming an object of public fascination. And of course a lively debate develops between those who are convinced that Alice and Bob are telepathic and those who refuse to believe it. Let us try to analyze these public reactions. As usual, a certain number of fans will not be very inclined to rational analysis. These people accept the explanation that Alice and Bob *are* telepathic and exalt the pair to stardom. Fortunately, not everyone is so naïve. Among these more skeptical folk there is a first group who have a very simple way to explain the phenomenon and exclude any possibility of telepathy: they assume that before the performance, Alice and Bob have agreed on what answers they will give according to the numbers on the cards they get. For instance, they could decide that on the first set of cards they will respond according to this pattern:

1	2	3
YES	NO	NO

Obviously, to reproduce the statistics of their responses they will have to change their pattern with every set of cards. They could agree to make the following kind of response for the second set:

1	2	3
NO	NO	YES

and they would go on and do this for every set, changing the agreed-on pattern, making sure that the total number of YES answers is equal to the total number of NO answers. Of course, this explanation takes all the mystery out of the performance, but it makes intelligible the perfect coincidence of their answers when they both receive the same number.

Let us take a brief excursion back into the realm of the physics of correlated photons. Suppose we have a source emitting a pair of photons in the "entangled" state we discussed before (see chapter 8), in connection with the EPR experiment:

$$|\Psi> = \frac{1}{\sqrt{2}}|1,V>|2,V> + \frac{1}{\sqrt{2}}|1,H>|2,H>. \tag{10.1}$$

We then suppose that two observers, situated in two separated areas where the photons are subject to measurement, can carry out three different tests of polarization—horizontal, 60°, and 120°. We identify the observers with Alice and Bob and we assume that they carry out the first, second, or third of these tests according to the numbers given (1, 2, or 3, respectively) on the cards. We suppose, further, that they write YES on the card if the photon passes the test and NO if it does not. Since we know that each photon passes or fails the test in a completely random fashion and with a probability of 1/2, and since we also know that, whenever the pair of photons is subjected to the same test (it does not matter which of the three), *both* photons overcome it or *both* fail it, then clearly these photons are reproducing the same performance as Alice and Bob.

We can now consider the EPR argument from a different point of view. These authors (Einstein, Podolski, and Rosen) are like the small group of shrewd spectators who recognized that the simple explanation (and the *only* explanation, if we exclude telepathy) for the perfect coordination of answers consisted in the actors agreeing beforehand on the patterns of answers. These agreements (that is, the tables given above) are the exact analogy of the elements of physical reality according to the analysis by EPR. Even before undergoing some test, photons ought to have certain characteristics, or are in agreement in some way about how to respond to all possible tests. Here, then, is the true meaning of EPR's conclusion that the theory is *incomplete*: quantum mechanics, to continue the analogy, is not in a position to tell us how Alice and Bob have agreed so as to be able to come up with the same answers at the right times. Not only this: in the orthodox interpretation—that is to say, in the hypothesis that the state vector represents the most complete information it is possible for a system to have—the theory asserts that it is illegitimate to think that Bob and Alice *can* agree beforehand: they simply must be telepathic.

Let us return to our stage show now, and consider a final group of spectators, those particularly shrewd ones (we can compare them to Bell)

CHAPTER TEN

who have decided to approach the problem of the Bob and Alice Show in a genuinely scientific spirit. These people have been recording the responses of Alice and Bob not only when they receive cards with equal numbers, but in all possible cases, even when, for example, Alice has received a card with a number 2, and Bob one with a number 3, and so on. These observers have established that in view of the global statistics, and in view of the number of times Alice and Bob agreed and disagreed, it appears that there is agreement in practically *one-half* the cases. This allows these serious students to conclude that Alice and Bob *cannot* have made any agreements. Let us see why.

Since Alice and Bob were given cards at random, the possible number of combinations is limited to nine, appearing each time with an equal probability. We can lay them out as follows, giving the card Alice gets first, and then the one Bob gets:

1,1	1,2	1,3	2,1	2,2	2,3	3,1	3,2	3,3

We then ask ourselves how many times Alice and Bob's answers agree and how many times they disagree, in the case where they conspired to reply as in the first table, that is, with YES, NO, NO, to the numbers 1, 2, 3, respectively. It would look like this:

1,1	1,2	1,3	2,1	2,2	2,3	3,1	3,2	3,3
YES,YES	YES,NO	YES,NO	NO,YES	NO,NO	NO,NO	NO,YES	NO,NO	NO,NO
Agree	Disagree	Disagree	Disagree	Agree	Agree	Disagree	Agree	Agree

As we can see, out of the nine possibilities, five replies agree and four disagree. A few observations can be made.

- The relationship of 5 to 4 of agreements to disagreements has nothing to do with a previous decision made about how to reply to each set of cards. It does not matter if the first or the second pattern is chosen, or any other pattern for that matter, as long as there are two of one answer and one of the other (two YES and one NO, or two NO and one YES).
- Among the possible ways of agreeing before the test there would also be the agreement to give the same three same answers (that is, to say NO or YES no matter what number came up). In this case, all cases where this agreement was made would make for an agreement for all nine possibilities.

232

BELL'S INEQUALITY AND NONLOCALITY

We can conclude then, that if Alice and Bob had arranged beforehand about what answers to give, there would *at least* be a prevalence of five agreements over four disagreements, in contradiction with the fact that in reality there were exactly as many agreements as disagreements. Now, according to the laws of great numbers, given two events that have a 5/9 and 4/9 probability (respectively) of occurring, it would be practically impossible for these events ever to occur an equal number of times in a sufficiently large total. The obvious conclusion is as follows: Alice and Bob cannot have arranged anything beforehand. But the perfect concordance of their responses in *all* the cases where they were given cards with the same number admits of no explanation other than telepathy. In some mysterious way Alice and Bob manage to "communicate with each other" when they are given the same number.

Let us leave our stage show now, and go back to physics.

10.2. Quantum Telepathy

Let us consider again the case mentioned above, the EPR experiment on a pair of photons in an entangled state. We have already seen the outcomes obtained by two observers, whom we can continue to refer to as Alice and Bob, when they measure the polarization plane along three directions at 60°, and agree to associate a YES answer with the passing of a test by a photon, and a NO answer with a failure. So far, they reproduce the first two characteristics of the performance of the two actors in the show, namely:

- For any direction and for either observer, the number of YES and NO answers is distributed at random and with equal frequency for both.
- Whenever the observers carry out a measurement of polarization in the same direction, they obtain concordant results, that is to say, either two YES or two NO answers. There is never a case where one photon passes the test and the other fails it.

Let us go ahead now and analyze the outcomes that quantum mechanics predicts in case Alice and Bob take measurements in different directions. To simplify the matter, let us suppose Alice's measurement takes place an instant before Bob's. We assume that Alice has made her measurement along the horizontal, and has obtained a YES answer, that is,

CHAPTER TEN

her photon has passed the test. By the form of the quantum state, and in accordance with the postulate of "the reduction of the wave packet," the system undergoes an instantaneous transition like the following:

$$|\Psi\rangle = \frac{1}{\sqrt{2}}|1, V\rangle|2, V\rangle + \frac{1}{\sqrt{2}}|1, H\rangle|2, H\rangle \quad \underset{\text{outcome: YES}}{\overset{\text{horizontal measurement}}{\Rightarrow}} \quad |1,H\rangle|2,H\rangle \quad (10.2)$$

But now Bob's photon is horizontally polarized. We ask ourselves, what probability does a photon like this have of passing a 60° or 120° polarization test? The reply quantum theory gives is not at all surprising: it is the probability of the square of the angle's cosine (see table 2A.1), or 1/4.

But we forge ahead with our analysis, supposing that Alice's photon does not pass the test, and that she gets a NO answer. Again, we can ask ourselves, "What is happening according to quantum mechanics?" Instead of the previous formula, we have a reduction to the state |1,V> |2,V>. Bob's photon is now vertically polarized, and since we are interested in their responses that agree with each other, we have to ask what the probability is for a vertically polarized photon not to pass a test of 60° or 120°? As we know, this probability is supplied by the square of the sine of the angle between the two directions (see table 2A.1) which once again has the value of 1/4. The whole situation is illustrated in Figure 10.2 below.

Owing to the symmetry of the problem, the argument now being developed is completely independent of Alice's and Bob's choice of directions, provided these are different. And now we can calculate according to quantum mechanical principles, and considering all possibilities, what the probability is for the answers of Bob and Alice to agree. We proceed step by step.

- In one-third of the nine possible cases, the two receive cards with the same number, decide to carry out the identical measurement, and receive identical answers.
- In the remaining two thirds of the cases they will carry out different measurements, which, as we have already shown, yields a one-fourth probability of getting an agreement.
- We can now evaluate the total probability of answers that agree. It will be equal to the (1/3) probability that the measurements happen in the same direction, multiplied by the (1) probability that in these circumstances the answers provide an agreement, to which is added the (2/3) probability that the measurements happen in different directions, multiplied by the (1/4) probability that in these cases the outcomes will agree. Thus we have

Total probability of agreeing outcomes: 1/3 + 2/3 × 1/4 = 1/3 + 1/6 = 1/2.

BELL'S INEQUALITY AND NONLOCALITY

$$|\Psi\rangle = (1/\sqrt{2})\{|1,V\rangle|2,V\rangle + |1,H\rangle|2,H\rangle\}$$
$$= (1/\sqrt{2})\{|1,60\rangle|2,60\rangle + |1,150\rangle|2,150\rangle\}$$
$$= (1/\sqrt{2})\{|1,120\rangle|2,120\rangle + |1,210\rangle|2,210\rangle\}$$

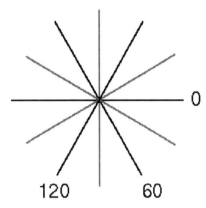

FIGURE 10.2. The exact quantum analogy of Alice and Bob's astounding performance. Suppose we have a pair of photons in the entangled state indicated at the top of the figure, and that two observers carry out measurements for plane polarization along the three directions marked in dark lines. The choice of direction corresponds to one of the numbers on the cards, and the YES or NO answers correspond to the fact that the photon has or has not passed the test. The lightly shaded lines (orthogonal to the dark lines) represent the directions along which the other photon is polarized in case the first one does not pass the test.

In this way, quantum mechanics reproduces the third characteristic of Alice and Bob's demonstration, which is that once all the possible cases are considered, there is an equal probability of having agreeing and disagreeing outcomes.

It should now be clear just how Alice and Bob will be able, in reality, to accomplish this. Each does not know the number on the partner's ticket — but they *do* have a flow of photons in the entangled state and polarization detectors. If they choose the directions along which they carry out their measurements according to the number written on *their cards*, and reply YES or NO according to whether *their* photon overcomes the test or not, they will reply precisely in such a way as to reproduce all the characteristics we have pointed out in the preceding section. And because these outcomes seem to imply telepathy, but Alice and Bob are not telepathic (they simply follow the "instructions" given them by their pho-

CHAPTER TEN

tons) the result is clear: the photons *themselves* must be telepathic, that is, without agreeing (in the language of EPR, without there being elements of reality that preexist the measurement and characterize their behavior in all possible tests), and only responding in each case to a specific test, they always behave in agreement when subjected to the same test.

Some comments are in order. The historical and logical development of the EPR position and that of Bell are as follows:

- The position of EPR: the analysis of the experiment, when limited to the case of equal tests and the recognition of perfect correlations between the outcomes, compels us (if we do not believe in the telepathy of Alice and Bob or photons) to assert that there ought to be previous agreement (objective elements of reality), and therefore the theory, because it can neither admit these elements nor describe them, is incomplete.
- The position of Bell: if we take account of a complete statistical analysis of the experiments of different tests carried out on the pair (of performers or of photons), all prior agreements can be excluded. In other words, by its predictions, quantum mechanics (as we shall soon see verified in practice) is irreconcilable with the hypothesis that elements of reality are objectively possessed by a pair of photons before they are subjected to testing.

Before concluding this section, I would like to turn to a point raised in the preface, namely, the assertion that "Bell proved Einstein wrong." A better way to get at the true meaning of this statement can be found by considering Mermin's presentation of Bell's inequality (1981). Mermin begins by recalling the words that Einstein said to Abraham Pais (1982), which Pais used as the opening sentence of his book, *"Subtle is the Lord."* Pais wrote, "I was accompanying Einstein on a walk from the Institute for Advanced Study to his home, when he suddenly stopped, turned to me, and asked me if I really believed that the moon exists only if I look at it." Mermin then goes on to consider Pauli's observation on this, which we have already quoted: "As Otto Stern said recently, one should no more rack one's brains about the problem of whether something one cannot know anything about exists all the same, than about the ancient question of how many angels are able to sit on the point of a needle. But it seems to me that Einstein's question are ultimately always of this kind." At this point Mermin, embarking on a discussion of Bell's inequality, made the following very appropriate comment: "Einstein and Pauli were both wrong. The questions with which Einstein attacked the quantum theory do have answers, but they are not the answers Einstein expected them to have. We now know that the moon is demonstrably not there when nobody looks."

Of course, this last assertion cannot be taken literally as referring to the moon, since the consideration of such a system would require a preliminary solution of the problem of the quantum description of macroscopic systems, a very delicate problem over which a very lively debate has broken out that we will study in the last part of this book. But if Mermin's statement is understood *as referring to microsystems* and as the assertion that Bob's photon does not have any definite state of polarization or does not possess any objective property, unless someone (e.g., Alice, indirectly) performs an observation, then the statement makes very good sense. Above all, it permits a full appreciation of the enormous conceptual value both of Einstein's critique and of the work by Bell that it stimulated.

In these two sections I have laid out the essential elements of the problem. My analysis has made use, albeit tacitly, of the hypothesis of locality (that is, that what happens on Alice's side does not have any concrete effect on what happens on Bob's). In the next section we will formulate precisely Bell's locality postulate, so that it should become clear under what hypotheses it is valid to conclude, in general, that quantum mechanics cannot be completed.

I would like to conclude this section with a quotation from Einstein that is particularly apt for the example just discussed, and shows how lucidly he intuited (even while refusing to accept) the deepest implications of the theory, long before Bell's own analysis: "It seems hard to sneak a look at God's cards. But that He plays dice and uses 'telepathic' methods (as the present quantum theory requires of Him) is something that I cannot believe for a single moment."

10.3. Bell's Locality Requirement

More than once we have referred to Bell's theorem as a rigorous demonstration that a deterministic and local theory with predictive capability equal to quantum mechanics cannot exist. Clearly, the reader who has followed the twists and turns of the debate thus far will have a sufficiently precise idea about what a "locality requirement" might mean. Nevertheless, since this question decidedly comes to the fore in Bell's inequality, it seems an excellent time to explain precisely just what Bell means by the expression. To grasp this point it should be enough to refer to a specific situation of the EPR type, involving two spatially separated systems, and the outcomes of the possible measurements of appropriate observables. It must be noted that here we are not speaking of the properties objectively possessed by the systems, but rather explicitly and exclusively of measurement outcomes, understood in the sense of the most intransigent supporters of the orthodox interpretation.

We consider a hypothetical theory, and we assume that there is a way to specify completely the state of the system in question, and that such a specification completely assigns the probabilities of the measurement outcomes (note well the word "probabilities"—this means that our argument covers probabilistic hidden variable theories as well). We are interested in measurements that involve both or only one of the constituents. Suppose that one of the constituents is in region A, the other in region B, and we indicate with a and b the observables we are interested in, and with α and β the measurement outcomes. Typically, a and b will indicate, e.g., the two directions along which the polarization states of two photons in an entangled state of the EPR experiment are measured, and α and β will indicate the relative outcomes by which each of them can assume one of the two values: "the photon passes the test" or "the photon does not pass the test." Finally, by λ we indicate all the variables, explicit or hidden, that specify the physical state of the system as exhaustively as formalism permits. Notice that at the present juncture we do not have to say to which conceptual scheme we are making reference: if we are in the realm of standard quantum theory, λ would be identified with the state vector $|\Psi\rangle$; if we are in classical mechanics, λ would be identified with the positions and velocities of all the particles of the system. Clearly, our interest will be directed to "hidden variable" theories on a par with quantum mechanics in their predictive capability. We fix our notation as follows:

- We indicate with $p_\lambda^{AB}(a,b;\alpha,\beta)$ the probability of obtaining the outcomes α and β in an experiment in which, at the two wings A and B of the experimental setup, the observables a, b are measured in the case of a physical system characterized by the value λ of the variables.
- We will also be interested in the case when a measurement is carried out in only one branch of the experiment. The probabilities of obtaining outcome $\alpha(\beta)$ when a measurement is carried out on the system in A (in B) will be indicated by the following symbols: $p_\lambda^A(a, *; \alpha)$ and $p_\lambda^B(*, b; \beta)$, respectively.

Before going any further I would like to emphasize that, even if no assumption of determinism has been made, determinism has not, on the other hand, been *excluded*. In fact, determinism requires that once the state of a system has been completely specified, all the observables will have a precise value. This value can then be incorporated into the framework simply by assuming that all the functions in question (that is to say, the probabilities of obtaining a certain result) can assume only the value 1 (certain outcome) or 0 (impossible outcome).

We are now in a position to formulate precisely Bell's locality requirement, which, analogously with what was done for Einstein's locality (*E.L.*, see section 8.5 above), we can call *B.L.*:

$$B.L.: p_\lambda^{AB}(a,b;\alpha,\beta) = p_\lambda^A(a, *; \alpha) \times p_\lambda^B(*, b; \beta).$$

This relationship asserts that the probability of obtaining a pair of results in a joint measurement at the two extremes of the apparatus is simply the product of the probabilities of obtaining either one of the results, independently of whether or not the other measurement is being carried out.[2]

With these preliminaries settled, we can now move on to the general derivation of Bell's theorem.

10.4. Bell's Inequality

Bell's theorem can be expressed by saying that no theory (deterministic or probabilistic) that respects the *B.L.* requirement can reproduce all the probabilities that quantum mechanics assigns to physical processes, not even when assuming that the preparation of the system does not completely determine the variables, and thus that the experimentally significant probabilities ought to be obtained by means of averaging over values of the hidden variables. The theorem's demonstration consists in the simple and direct derivation of an inequality that ties together the probabilities of joined events in such a way as to be implied in a completely general way by *B.L.*, but to be violated by the corresponding quantum probabilities.

. . . . .

Let us see how this conclusion is arrived at. For the time being we can refer to the case of a system whose state has been completely specified, that is, for which λ has a precise value, and we make explicit reference to the case of polarization measurements on a pair of photons in our entangled state. In this case, the possible outcomes α and β are identified with the fact that the photon overcomes or does not overcome the test of polarization to which it is subjected. Fixing the directions a and b of the plane polarization tests of the photon in A and in B, we will then have four joint probability functions, which we shall label, respectively,

239

CHAPTER TEN

as $p_\lambda^{AB}(a,b;yes,yes)$, $p_\lambda^{AB}(a,b;yes,no)$, $p_\lambda^{AB}(a,b;no,yes)$, $p_\lambda^{AB}(a,b;no,no)$. In terms of these functions we can now define a new function, traditionally indicated as $E_\lambda(a,b)$, defined as the sum of the probabilities of obtaining concordant answers minus the sum of the probabilities of obtaining discordant ones.[3]

$$E_\lambda(a,b) = p_\lambda^{AB}(a,b;yes,yes) - p_\lambda^{AB}(a,b;yes,no) - p_\lambda^{AB}(a,b;no,yes) + p_\lambda^{AB}(a,b;no,no). \tag{10.3}$$

It is now rather simple, with recourse to the locality requirement B.L. to derive an inequality that involves four functions of the type $E_\lambda(a,b)$. The explicit computation, which requires only elementary algebraic operations, is developed in an appendix at the end of this chapter. Here we need to consider only its form:

$$|E_\lambda(a,b) - E_\lambda(a,d)| + |E_\lambda(c,b) + E_\lambda(c,d)| < 2. \tag{10.4}$$

Of course, the quantities that appear in this inequality are not immediately significant in the case of a hidden variable theory because it is not possible to prepare a system for which the value of the hidden variables is known with certainty (or, equivalently, can be controlled). As already explained at length, in order for this to be comparable with the quantum theory whose statistical predictions we want to reproduce, we need to average the quantities being tested over the distribution of the hidden variables of the ensemble of systems we are considering. We suppose that the physical procedure corresponding to the preparation of a quantum mechanical state (in our case of the entangled state of two photons) allows for a certain distribution of the hidden variables. The physically measurable quantities that should reproduce the corresponding quantum expressions are the averages of the functions $E_\lambda(m,n)$, taking account that in our group of systems, for example, N_i particles over a total of N are characterized by the value λ_i, of the hidden variables:

$$E(m,n) = \sum_i \frac{N_i}{N} E_{\lambda_i}(m,n). \tag{10.5}$$

At this point it is trivial to show (see Appendix 10B) that these new quantities satisfy exactly the same relationship as was derived for the single elements of the ensemble, that is to say,

$$|E(a,b) - E(a,d)| + |E(c,b) + E(c,d)| \leq 2. \tag{10.6}$$

This is Bell's famous inequality, stated in its most general form (which in fact was worked out by John F. Clauser, Michael A. Horne, Abner Shimony, and Richard A. Holt, [1969] and contains some advantages over the original), valid for four arbitrary directions a, b, c, and d.

BELL'S INEQUALITY AND NONLOCALITY

We can now have the satisfaction of calculating, according to quantum theory, the quantities that appear in the equation in the case of the photons. The reader will be easily be persuaded that

$$p_\lambda^{AB}(a,b; yes, yes) = p_\lambda^{AB}(a,b; no, no) = \frac{1}{2}\cos^2(a \hat{\bullet} b),$$

$$p_\lambda^{AB}(a,b; yes, no) = p_\lambda^{AB}(a,b; no, yes) = \frac{1}{2}\sin^2(a \hat{\bullet} b), \qquad (10.7)$$

where $(a \hat{\bullet} b)$ indicates the angle between the directions a and b. These expressions imply something that can be easily verified by anyone with a little trigonometry:

$$E(a,b) = \cos[2(a \hat{\bullet} b)]. \qquad (10.8)$$

We now choose the following values (in degrees) for the four directions that appear in the expression to the left of Bell's inequality:

$$a = 0°, b = 22.5°, c = 45°, d = 67.5°. \qquad (10.9)$$

When the quantum probabilities are considered as true, the term to the left of equation (10.6) assumes the value

$$[\cos(45) - \cos(135)] + [\cos(-45) + \cos(45)] = \left|\left(\frac{1}{\sqrt{2}} + \frac{1}{\sqrt{2}}\right)\right| + \left|\left(\frac{1}{\sqrt{2}} + \frac{1}{\sqrt{2}}\right)\right| = 2\sqrt{2}$$
$$= 2.828, \qquad (10.10)$$

which is appreciably greater than 2. Therefore, for the choices made for the angles and for the state of the system, Bell's inequality is greatly violated.

. . . . .

Before concluding this section it will be a good idea to recapitulate the major ideas.

- With reference to the case of the EPR experiment with two photons in the entangled state, we considered a completely general theory (probabilistic or deterministic) in which the state of a system is characterized *in the most complete way possible* by the assignation of the λ variables.
- For this theory, it has been assumed that the assignment of the λ variables *univocally* (i.e., nonambiguously) *determines the probabilities of the outcomes* of single or correlated measurements at the two wings of the apparatus and that *these probabilities satisfy the condition of B.L.*

CHAPTER TEN

- Subsequently an inequality has been derived for a precise combination of expressions that yields the difference between the probabilities of concordant and discordant outcomes for any four arbitrary directions of polarization, for every precise value of the hidden variables.
- We have observed that the same inequality holds true for the averages of the same expressions over an ensemble, in which the hidden variables are distributed completely at random.
- Finally, it has been shown that the inequality thus obtained is violated by the corresponding expressions of quantum mechanics.

The conclusion ought to be obvious: we have furnished a proof that there does not exist any conceivable theory by which the assignment of the λ variables determines the probabilities in question in such a way as to satisfy *B.L.*, and that can yield the quantum values as the result of an operation of averaging over such variables, no matter how these are thought to be distributed. In particular, if it is assumed that the probabilities we are seeking can assume only the values +1 (certain result) and 0 (impossible result), or in other words, if we consider the possibility of a deterministic completion of quantum mechanics, the theorem now derived shows in full generality that it is in fact impossible. It is important to notice that by the very way we have argued, that is to say, without having to compromise in any way on the conceptual structure, or on the form or nature (deterministic or probabilistic) of the theory in question, or on the distribution of variables that have to correspond to the preparation of the quantum state under consideration, the lesson to be drawn from Bell's inequality is that it is not the special theory or interpretation adopted but nature herself and the peculiar behavior of microsystems that imply nonlocality (that is to say, the violation of *B.L.*) at the level of individual systems.

. . . . .

An observation should be made, even though it may seem trivial: since quantum theory violates Bell's inequality by asserting that single systems are completely characterized by the wave function, then this theory itself would have to be nonlocal. To be convinced of this it should suffice to recall that, as we know, the probability of obtaining the result YES or NO for any test of polarization on a single photon is equal to 1/2, while, because the sole outcomes possible relative to measurements of correlation along the same (randomly chosen) direction are that both the photons pass the test or both of them fail it, and these two outcomes present themselves with the same probabilities, we have[4]

$$p_\Psi^{AB}(a,a;YES,YES) = p_\Psi^{AB}(a,a;NO,NO) = 1/2,$$
$$P_\Psi^{AB}(a,a;YES,NO) = p_\Psi^{AB}(a,a;NO,YES) = 0, \tag{10.11}$$

while, as just now observed:

$$p_\Psi^A(a,*;YES) = p_\Psi^A(a,*;NO) = p_\Psi^B(b,*;YES) = p_\Psi^B(b,*;NO) = 1/2. \tag{10.12}$$

The comparison of these relations with the preceding shows that quantum mechanics violates the *B.L.* condition.

. . . . .

10.5. An Example of "Experimental Metaphysics"

As I have repeatedly emphasized, one of the great benefits of Bell's work consists in the fact that he made it possible to give an experimental answer to a question that would appear at first sight to be a metaphysical one. In fact, the crux of the debate between Bohr and Einstein about the EPR experiment can be summed up by saying that Einstein (owing to what his opponents called a metaphysical prejudice) held it unacceptable that any serious theoretical schema might imply that properties (elements of reality) should be recognized as existing, and at the same time that the schema did not allow one to describe nor even to think how they could be possessed by the systems under investigation. It goes without saying that the one possibility for avoiding this conclusion lay in renouncing any claim to locality, and this was a step neither Einstein nor his most tenacious opponents were willing to make. The orthodox position on this dilemma could be reduced to the simple observation that the question was irrelevant since it could admit of no verification in practice. This is the real crux of the issue, so effectively illustrated by Mermin (1981), and set forth in section 2 of the present chapter. As the reader should understand by now, Bell showed that the question *admits* of a nonambiguous response that can be deduced from the outcomes of precise experiments. If measurements are carried out concerning the correlations between the polarization states (in different, arbitrarily chosen directions) of two photons, and if these measurements confirm the predictions of the theory, then we must conclude that it is contradictory to think that the properties brought into the open by the measuring process existed before the measurement itself without being ready to make the "great refusal," that is, to renounce any claim for the locality of natural phenomena.

In brief, the problem of whether it is possible to hypothesize that remotely separated and noninteracting systems have objective properties that cannot be influenced instantaneously at a distance becomes the problem of verifying whether the quantum correlations are verified in fact in the experiments we can carry out in our laboratories. Now a series of experiments of this kind were carried out between 1972 and 1982, and all but two experiments (whose results were seriously suspect) confirmed the quantum predictions and showed, we might say, that "Einstein was wrong," or, equivalently, that "microsystems *are* telepathic." I will not take time to analyze this experiment or its technical details, but shall limit myself to the observation that some of them involve photons, while others involve particles with mass.[5] I would, however, like to focus on one subtle point that allows some who do not feel ready to abandon the requirement of locality B.L. to assert that such experiments are not yet conclusive. Let us investigate the reasoning behind this objection.

A typical experiment intended to verify Bell's inequality envisages a photon source in an entangled state (and thus with strict correlations of polarization) and the positioning on the two wings of the apparatus of two devices that carry out measurements of polarization. The simplest version of an experiment like this requires (1) a collimation of the emitted photons in such a way that the only pairs selected are those that would propagate in the direction where the counters are found, (2) a suitably accurate registration of the instant of revelation with respect to a temporal interval that is the average of two successive processes of emission, by guaranteeing that the two photons which arrive will be the two partners of a definite entangled pair, and finally, (3) a counting of "concordant" and "discordant" outcomes at the two counters for various orientations. At this time it is worthwhile to recall and reinforce the point that Bell's entire argument is based on the hypothesis of locality. Let us analyze the experiment in question from this perspective: It is clear that this hypothesis presupposes—to put it in the language of our story—that Alice and Bob have no way to communicate with each other and thereby transmit information about the numbers on their cards. Einstein thought this eventuality was guaranteed by imagining that the spatial separation of the two counters could be such that there could not be enough time for a signal to be propagated, in such a way that any physical action originating from the counter that was first set off could influence the other system (the other photon) or the other counter before it too went off in turn.

Now, if we inspect this experimental situation closely, we may encounter a fundamental objection that, though it would require physical effects that have never been put into evidence and are physically meaningless, would allow refutation of the conclusions of the experiment. In fact, if in carrying it out it happens (as it actually does in practice) that the orientation of the polarizing lenses on each branch should be chosen well before the registration at the counters, one could hypothesize (I repeat, this would be to imagine a rather fantastic eventuality) that the act itself of arranging and orienting the lens in A (B) in order to measure the polarization of the plane along *a* (*b*) could in some way influence not only the process of emission of the photon pairs (compelling them to assume precise polarizations in the considered directions), but also the efficient functioning of the counter placed on the other branch of the experiment. And then the experiment itself would be inconclusive.

In 1982 Alain Aspect and his collaborators J. Dalibard and G. Roger took up this challenge, and, using the latest technological innovations, set up one of the most refined experiments ever carried out. The basic idea of Aspect's experiment could be expressed by saying that the decision concerning the orientation of the polarizer along A is made "at the last minute," that is to say, in such a way that not even a signal propagated at the speed of light could depart from A and reach the photon or the polarizer of B (in its turn oriented at the last minute) before the second test is carried out. A schematic representation of the Aspect-Dalibard-Roger experiment is shown in Figure 10.3. There is no need to consider the various parts of the experiment in detail. The crucial constituents are the two switches SA and SB, which can permit the photons in A and B to travel toward the polarizers Pa and Pb, or can deviate them toward polarizers Pc and Pd which make measurements along the four directions shown. The switches are activated or deactivated every 10 nanoseconds (i.e., every

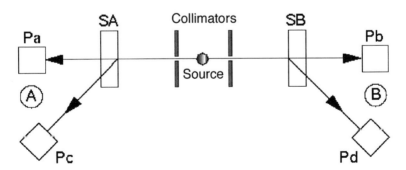

FIGURE 10.3. The experiment of Aspect and collaborators: this represents the most convincing experimental proof of quantum nonlocality.

CHAPTER TEN

hundred-millionth of a second). The distance between the two counters is such (about 13 meters) that a light signal would take about 40 nanoseconds to be propagated from A to B. The situation could be described as follows: the card brought to Alice must be handed to her, and she must write her answer on it, so fast that even if she used a radio which we were not able to discover and to screen out, she would not have time to communicate the number on her card to Bob before Bob himself would be given his card and give his answer. If this were the setup, it would not be necessary to make sure that Bob and Alice did not have hidden cell phones or some other means of communication, since the signal transmitted from A could never reach B in time to instruct it how to respond.

At this stage all we have to do is report the outcome of Aspect's experiment: the quantum predictions were so fully confirmed that the discrepancies between the outcomes and those implied by any hidden variable theory could not be imputed to systematic errors or random statistical fluctuations (we should recall that the difference between quantum predictions and those of any deterministic local theory is remarkable: in fact, the value of the first member of the inequality is equal to $2\sqrt{2} \cong 2.828$ for the standard theory and less than 2 for any conceivable hidden variable theory). Translated into concrete terms, this fact simply means that the experiment excludes the possibility of a local completion of the theory. Einstein was truly in error because neither he nor any of his opponents had contemplated the possibility that natural processes were fundamentally nonlocal.

10.6. THE LAST BASTION OF THE SUPPORTERS OF LOCAL MICROREALISM

Personally, I take the experiment of Aspect and his collaborators as conclusive: photons really are telepathic, or to use more scientific terminology, they cannot be considered as possessing any local characteristics that determine whether or not they will pass the test before the test is carried out. Nevertheless, they still react the same way for the same test. For those who find this unacceptable (and it is worth mentioning now that this does not imply any direct violation of relativity, a subject to be treated in the next chapter), a last escape route is still left open, in connection with the low level (not of the precision but) of the efficiency of registration by the measurement apparatus. In this section I shall attempt to clarify the technical aspects of this problem. If someone is still particularly stubborn and refuses to admit what is being maintained (i.e., that no logical necessity requires microscopic processes to fit within the conceptual categories we have elaborated on the basis of our macroscopic experience), this one avenue of escape is still possible, which can be closed off only thanks to some remarkable advances in technology. Accordingly, there are those who cling to this idea and claim that Aspect's experiment is not decisive. Again, let us inspect their reasoning.

At the two wings of the apparatus are photon counters that produce different signals as soon as the photon passes or does not pass the test to which it is being submitted. Now, as anyone might expect, no counter will be perfectly efficient. In the specific case at hand, the efficiency in fact is rather low. What does this mean in practice? Simply that in an appreciable number of cases one or both of the counters do not register any photon at all. Clearly such eventualities are simply set aside, so that the evaluation of the correlations is based only on the total of instances when a measurement has been carried out and has given an outcome in both wings of the apparatus.

How could this obvious, natural, and trivial fact, common to all processes of measurement, invalidate Aspect's conclusion that Bell's inequality is being violated? The explanation (as baroque as it may appear to me) would make use once again of our charming example of Alice and Bob. The fact that one counter does not register an event is likened to the case where one of the two actors (or both of them) may write nothing at all on the card, making that particular test be rejected as worthless. But—and this is the crucial fact—it can be imagined that they agreed beforehand to make this decision not to write anything down. Following Mermin again, we see how it would happen that Alice and Bob are in a position to carry out their extraordinary experiment, respecting all its features. Above all, they decide to limit their agreements to the case when two replies of two different kinds are given (that is to say, two YES and one NO, or two NO and one YES, varying of course from case to case). This strategy is obvious: since the choice to give three equal responses independently of the numbers that appear on their tickets yields two agreeing responses, and since on the other hand the datum that must be reduced is just the 5 to 4 proportion of agreeing replies to disagreeing ones, it is fitting to avoid this kind of choice. But Alice and Bob further articulate their agreement by deciding that one of them (a different one each time) will not reply to one of the two questions which anticipate the an-

CHAPTER TEN

swer. To illustrate, they could make the following choice: Alice will behave according to the following pattern:

1	2	3
YES	NO	NO

and Bob will adopt the following pattern:

1	2	3
YES	#	NO

or even this one:

1	2	3
YES	NO	#

The symbol # means simply that he will write nothing on the card that bears the indicated number. As before, this kind of agreement is changed from time to time with respect to the answer to be given to different numbers, but what they decide is that one of the two gives one reply of one kind and two replies of another, and the other person gives the same reply for the first kind and the same for the second, but writes nothing at all for the remaining case. When we analyze the outcome, the reader will easily be convinced that the facts we are going to list do not depend in any way on making special agreements, so long as the given rules are respected. For example, one could refer to the first of the two cases mentioned above, meaning the one represented in the table

1	2	3
YES	NO	NO

for Alice, and this one

1	2	3
YES	#	NO

for Bob. The results would then be as follows:

- In all the cases where the actors received the same card and both replied, their answers would agree.
- The complete statistics of the outcomes of all possible tests if we eliminate, as we stipulated, the cases where one reply was missing, would be as follows:

1,1	1,3	2,1	2,3	3,1	3,3
YES, YES	YES, NO	NO, YES	NO, NO	NO, YES	NO, NO

As we can see, keeping the perfect agreement of replies in the case of the same questions, there is an equal distribution of agreements and disagreements when all possible tests are considered (i.e., even in those cases when the cards carry different numbers). In this instance Bob and Alice will successfully carry out their astounding experiment while respecting the fact (noticed by the more attentive members of the audience) that the total statistics bring a perfect equilibrium between agreeing and disagreeing replies.

How then do we read the escape route just described in physical terms? The escape simply asserts that before the photons have been subjected to any test, they agree not only on the response to give but also when not to give any response at all. It could be summed up by saying that every photon of every pair, in addition to carrying a note in its pocket instructing it how to react to any given test, also carries a token that it has agreed to show, or not to show, to the counter, that says, "please in this case do not reveal whether or not I passed the test." It seems to me superfluous to emphasize that everything that we know about the functioning of the counters contradicts such an assumption. The fact that in some cases a counter fails to reveal a particle or a photon that has just passed through it is universally understood to be a purely random event—not intrinsically random, but resulting from the complexity of the apparatus and the many uncontrollable elements that govern the registration process. Here, the position is completely overturned in an extremely peculiar way: the nonfunctioning of a counter would not be a random event but would be a precise and perfectly determined event, governed by the impinging photon itself, and what is more, photons, counters, and the entire experimental setup would conspire with the utmost precision in carrying out a colossal deception. The natural processes would, in their foundations, be perfectly deterministic and the correlations between the two outcomes at the branches of the experiment would be perfectly settled before the measurements could be made. But an incredible conspiracy of events and the occurrence of processes that have no explanation or rationale in any known theory would bring it about that the photons and/or the counters would be such as to make us believe that a quantum world exists instead of the classical one! And these agreements would have to be subtly controlled so as to vary with the kind of correlations we decide we want to study, and to vary also with different levels of efficiency in the counters we use in our experiments.

I trust that the reader by now has located the focal point of the argument. It has been shown that "in principle" there could be secret agreements between Alice and Bob, just so long as the counters are not perfectly efficient. But this escape seems so incredible that it is hard to believe how anyone could take it seriously. As Mermin has also observed, many among the most convinced supporters of the necessity of elaborating a completely local and deterministic theory are inspired by the positions taken by the great Einstein (which, if the analysis of the foregoing chapters is considered, they appear to be reading in a certain one-sided way). But these same researchers have forgotten a brilliant statement of that great scientist: "subtle is the Lord, but not malicious." And how else could it be understood but as perversely malicious that, all the while nature is really governed only by clear and simple local and deterministic laws, there would nevertheless exist incredible[6] and incomprehensible physical processes that systematically conspire to make us believe that nature's laws are fundamentally stochastic and nonlocal?

The escape route just discussed is well matched with another one that is advanced from time to time under the name of "the preestablished harmony hypothesis." It should be clear to every attentive reader that an essential ingredient of the analyses both of EPR and of Bell is represented by the hypothesis that the observer can choose by his own free will which measurement to carry out, that is to say, how to orient the polarizing filter. If the freedom to do this is denied, and replaced by the idea of a rigid and absolute determinism governing all the processes of the universe, another escape route is opened up. Say, for example, that at the moment of the big bang, the entire future was written into the initial conditions; in particular, that it was written that Aspect would undertake his experiment on that particular day, orienting the polarizers in such and such a way, so that the photons would have to react the way they reacted. Clearly, if such a position is assumed, the verification of the violation of Bell's inequality would prove nothing at all. But, from our perspective, both in the case of the point of view presented earlier and in this one now being discussed, all motivation to pursue scientific investigation would disappear. If nature wants to deceive us on purpose, or if it is already implicit in the nature of things from the beginning of time that I would carry out certain experiments and elaborate certain theories, if it were already written from time immemorial that Schrödinger would come up with his equation and Einstein would be disturbed by the orthodox interpretation, that Bell would derive his theorem and Aspect would obtain the results he obtained, what sense would there be (even if we could not stop ourselves in this case) in doing scientific research at all, in striving to grasp the secrets of nature, or of God?

10.7. A New and Conceptually Very Significant Version of Bell's Inequality

Let us briefly reconsider the argument that leads to Bell's inequality. It can be expressed as follows: any local theory—that is, any theory that (a) assumes that both photons have definite properties even before being subjected to various polarization tests (that is to say, that the photons "know" beforehand how they are supposed to react to each of the three measurements in question) and (b) guarantees the perfect agreement of outcomes in the case of identical tests—such a theory cannot reproduce the statistical distribution foreseen by quantum mechanics concerning the outcomes of correlation measurements, even when different tests are considered for the two photon "partners." The logical structure is of the following type: let a group of experiments be considered for which the theory makes *certain predictions* (perfect correlations for identical measurements), then any local theory that accounts for these outcomes cannot reproduce the *probabilistic predictions* of quantum theory relative to other experiments. The refutation of locality is thus possible only through recourse to a large number of repeated correlation experiments involving polarization tests along different directions, and with a statistical analysis of the various outcomes.

Recently, Daniel M. Greenberger, Michael Horne, and Anton Zeilinger have demonstrated (with reference to a much more complicated physical situation than one of the experiments we were discussing before) how any local theory equivalent to quantum mechanics can be excluded on the basis of a single experiment, without having to take into account the

CHAPTER TEN

statistics of many events. The logical structure is this: consider a group of experiments for which the theory makes certain predictions; then any local theory whatsoever that reproduces such predictions cannot also reproduce the *certain predictions* of quantum mechanics relative to another experiment. The detailed exposition of this brilliant argument is technically more complex than what has gone before, making it necessary to put off until the end of this section a treatment of the specific physical system and the actual measurements involved. For now, I will present the argument in a simple, easily visualized form, imagining a performance similar to the one with Alice and Bob in section 10.1. The situation is the same, except that now we have *three* performers (Alice, Bob, and Charlie) who cannot communicate with each other, and instead of having three cards for each performer, there are only three cards in all, bearing the numbers 1, 2, and 2, and these cards are shuffled before being handed to the contestants (see Figure 10.4).

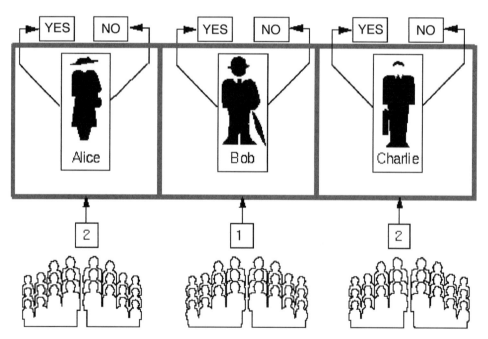

FIGURE 10.4. The performance, analogous to the one in Figure 10.1, which excludes any local theory equivalent to quantum mechanics, but without recourse to the statistical analysis of a great many experiments.

BELL'S INEQUALITY AND NONLOCALITY

What happens in this new performance can be described as follows:

- After looking at his or her card, each performer writes a YES or NO on the card, in an apparently random fashion, with an equal distribution of YESes and NOs, no matter what number the card has on it.
- Looking at the three replies that Alice, Bob and Charlie give every time, it is established that they *always* contain an odd number of NOs: that is, either two of them write YES and one writes NO or all three write NO on their cards.

Of course, the explanation for what has happened up to this point is obvious, as long as we take recourse to the idea of "preexisting agreements." In fact, in the present case it is not necessary to assume that the performers agree on how to respond depending on the number they find on their card, since here, unlike the performance of Section 10.1, there are no specific correlations to reproduce. It will be enough here for the performers to decide among themselves (changing their agreement from time to time) who alone among them is going to put the NO, or that all three of them are going to write NO.

But now let us consider the astounding part of the experiment. It is arranged that, without Alice, Bob, or Charlie knowing it, the audience puts a number 1 on all three cards. What happens now can only be explained by telepathy: the number of NOs that appear on the performers' cards is suddenly even: there will be either three YESes or one YES and two NOs.

How can this happen? Clearly, the interpretation we used before, that the performers agreed on how to reply without any reference to what numbers were on their cards, will not do, because if that were the case, they would continue to respond with an odd number of NOs. But we can contemplate the possibility that there is some strategy they could agree on before the experiment that would control the way they respond to the numbers on the cards, in such a way as to succeed in giving the above reported answers in case they receive three cards with number 1 on them. The long and tedious analysis that now follows shows explicitly that such a strategy would be impossible.

. . . . . .

To reach this conclusion, we need to consider all the possible ways that Alice, Charlie, and Bob have of agreeing among themselves (that is, how they decide to reply when they receive cards with certain numbers on them), in such a way as to guarantee the outcomes of the first part of the experiment, that is, in the case of

CHAPTER TEN

1, 2, and 2 being written on the cards, that there will always be an odd number of NOs. We list in the three columns the possible replies of the contestants in response to the numbers they receive on their cards. As will be seen clearly later, it is possible to choose randomly two of Alice's answers and one of Bob's. After this is settled, then all the remaining responses are unambiguously determined by the rule that the number of NOs must be odd.

To show the reader how this must be proved, we begin by considering a specific case, that is, one where it has been decided that Alice will answer YES if her card bears the number 1 or 2, and Bob will respond YES if his card has the number 1.

Alice		Bob		Charlie	
1	2	1	2	1	2
YES	YES	YES			

We now argue as follows. We suppose that Alice receives the card with the number 1. Now, Bob and Charlie will both receive cards with 2 on them. Since we have to end up with an odd number of NOs, the only possible choices are that Bob writes YES and Charlie NO, or vice versa:

Alice		Bob		Charlie	
1	2	1	2	1	2
YES	YES	YES	YES		NO
YES	YES	YES	NO		YES

However, the second of the above possibilities can be rejected by the following argument: If Bob had received the card with the number 1, then both Alice and Charlie would receive cards with 2s, and they would all reply YES so that there would be three YESes and no NO. For the choice we are considering then, relative to the first three answers, we would have to have the following:

Alice		Bob		Charlie	
1	2	1	2	1	2
YES	YES	YES	YES		NO

It should be easy to see that even the last response not yet ascertained (that is, Charlie's reply to the number 1) is fixed unambiguously. In fact, if it happens that

254

BELL'S INEQUALITY AND NONLOCALITY

Charlie gets the number 1, since his partners would both reply YES, he would have to reply NO to keep to the rule.

Concluding this first analysis, once the first three replies are determined in the way shown, all the others are determined as shown in the following table:

Alice		Bob		Charlie	
1	2	1	2	1	2
YES	YES	YES	YES	NO	NO

We can now change the three replies, considering, for example, the case

Alice		Bob		Charlie	
1	2	1	2	1	2
YES	YES	NO			

and we now reason in the same way to determine how we ought to fill up the empty blanks in the table. We will not repeat the argument in detail but we go on immediately to list, in the following table, the *eight* possible modes of agreement among the performers that are the *only ones* (among sixty-four possibilities) that assure us that in this experiment of one card with the number 1 and two with the number 2, there will *certainly* be a response that contains an odd number of NOs. From the table one can deduce what would happen in the case in which all three contestants receive a card with the number 1.

Alice		Bob		Charlie	
1	2	1	2	1	2
YES	YES	YES	YES	NO	NO
YES	YES	NO	NO	YES	YES
YES	NO	YES	NO	NO	YES
YES	NO	NO	YES	YES	NO
NO	YES	YES	NO	YES	NO
NO	YES	NO	YES	NO	YES
NO	NO	YES	YES	YES	YES
NO	NO	NO	NO	NO	NO

255

CHAPTER TEN

In the case we are considering, the responses corresponding to the agreements represented in the various rows would be, respectively,

YES,YES,NO; YES,NO,YES; YES,YES,NO; YES,NO,YES; NO,YES,YES; NO,NO,NO; NO,YES, YES; NO,NO,NO.

As we see, these all contain an odd number of NOs and they contradict the certain outcome of the experiment in question, that is to say, that there would be an even number of NOs. Therefore, if Alice, Bob, and Charlie reply as foreseen (with certainty) by the quantum theory in all the experiments of the type 1,2,2, then executing a single experiment—such as the 1,1,1 type—shows that they cannot be in agreement before the measurements have been carried out: the only explanation for the certain appearance in this last case of an even number of NOs, would result from telepathic communication that the cards they are getting suddenly have the same number on them.

. . . . .

The conclusion that can be drawn from the analysis just completed, at least from the experimental point of view, is that the execution of a single experiment (that is, one that foresees three tests of the first type) and the relative outcome (an even number of NO answers) all by itself allows us to *exclude* the existence of any local theory equivalent to the quantum theory, that is, one that reproduces the certain results of the other experiments.

To conclude the section we only need to mention the physical process involved and the types of measurements that correspond to the little story we told. Suppose we have a photon source (Figure 10.5) that emits three photons in an entangled state, at once, and in three different directions:

$$|\Psi\rangle = \frac{1}{\sqrt{2}}\big[|1,V\rangle|2,V\rangle|3,V\rangle + |1,H\rangle|2,H\rangle|3,H\rangle\big], \quad (10.13)$$

that is to say, the source produces the superposition of the state of three photons with a vertical polarization and the one of three photons with a horizontal polarization. At positions A, B, and C are located our "performers" who can choose whether to carry out a measurement of plane polarization at 45° or one of circular polarization.[7] The theory predicts that when one measurement of plane polarization at 45° is carried out, and two measurements are carried out aimed at determining if the circular polarization is to the left, in every case (that is to say, no matter in what branch the single measurement for plane polarization is carried out) the outcomes are always that *two* or *no* photons pass the test. If Alice,

BELL'S INEQUALITY AND NONLOCALITY

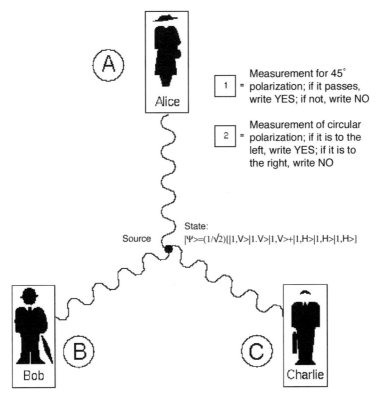

FIGURE 10.5. This is how Alice, Bob, and Charlie can do their astounding performance with three photons in an entangled state.

Bob, and Charlie decide to carry out a test of 45° polarization if their card has the number 1 and a test for circular polarization if their card has a 2, and to respond YES or NO according to whether their photon passes the test or not, then we see that their answers always contain an odd number of NOs. But the theory also says what happens if they carry out three tests of plane polarization at 45°.

In fact, as the reader can verify for himself by using the well-known relations between the states of horizontal and vertical polarization, and the polarizations at 45° and at 135°, the state just considered can also be expressed as follows:

$$|\Psi\rangle = \frac{1}{2\sqrt{2}}\left[|1,45\rangle|2,45\rangle|3,45\rangle + |1,45\rangle|2,135\rangle|3,135\rangle \right. \quad (10.14)$$
$$\left. + |1,135\rangle|2,45\rangle|3,135\rangle + |1,135\rangle|2,135\rangle|3,45\rangle\right].$$

257

CHAPTER TEN

It is immediately apparent[8] that either all three photons will pass the test (the first term of the superposition) or (all the remaining terms) only one of the three will pass it. These are the "three YESes" or the "one YES and two NOs" written in response to the cards.

I must apologize to the reader for the pedantic nature of the analysis. But I want to emphasize what seems to me extremely important: to make clear how the quantum theory permits the preparation of a state in such a way that only a single test on it (together with confirmation that the outcome of the test is in fact the one predicted by the theory) is sufficient to exclude any possible deterministic and local completion of the formalism.

APPENDIX 10A: PRELIMINARIES TO BELL'S INEQUALITY

Expressing the probability functions that appear in the definitions of $E_\lambda(a, b)$ using the *B.L.* condition, one obtains

$$E_\lambda(a, b) = [p_\lambda^A (a, *; \text{YES}) - p_\lambda^A (a, *; \text{NO})] \times [p_\lambda^A (b, *; \text{YES}) - p_\lambda^A (b, *; \text{NO})],$$

from which immediately follows

$$E_\lambda(a, b) - E_\lambda(a, d) = [p_\lambda^A (a, *; \text{YES}) - p_\lambda^A (a, *; \text{NO}) \times \{[p_\lambda^A (b, *; \text{YES}) - p_\lambda^A (b, *; \text{NO})] - [p_\lambda^A (d, *; \text{YES}) - p_\lambda^A (d, *; \text{NO})]\}.$$

Observe that the absolute value of the product of any two numbers is equal to the product of their absolute values and recall that since the photon in A either passes or does not pass the test, we have

$$p_\lambda^A (a, *; \text{YES}) + p_\lambda^A (a, *; \text{NO}) = 1.$$

Making use of this, we see that the first factor of the equation becomes

$$p_\lambda^A (a, *; \text{YES}) + p_\lambda^A (a, *; \text{NO}) = 1 - 2 p_\lambda^A (a, *; \text{NO}).$$

Since $p_\lambda^A (a, *; \text{NO})$ lies with certainty between 1 and 0 (being a probability) the expression just now written is contained within 1 and −1 and thus its absolute value is less than or equal to 1; whence it follows that

$$|E_\lambda(a, b) - E_\lambda(a, d)| \leq |\{[p_\lambda^A (b, *; \text{YES}) - p_\lambda^A (b, *; \text{NO})] - [p_\lambda^A (d, *; \text{YES}) - p_\lambda^A (d, *; \text{NO})]\}|.$$

BELL'S INEQUALITY AND NONLOCALITY

And by following precisely the same line of reasoning it is trivial to show that

$$|E_\lambda(c, b) + E_\lambda(c, d)| \leq |\{[p_\lambda^A (b, *; \text{YES}) - p_\lambda^A (b, *; \text{NO})] \\ + [p_\lambda^A (d, *; \text{YES}) - p_\lambda^A (d, *; \text{NO})]\}|.$$

Summing up the two last equations, then, we have

$$|E_\lambda (a, b) - E_\lambda (a, d)| + |E_\lambda (c, b) + E_\lambda (c, d)| \leq |r - s| + |r + s|,$$

where we have defined

$$r = [p_\lambda^A (b, *; \text{YES}) - p_\lambda^A (b, *; \text{NO})], \; s = [p_\lambda^A (d, *; \text{YES}) - p_\lambda^A (d, *; \text{NO})].$$

We note now that for the expression $|r - s| + |r + s|$ the four following cases are possible, with the values given:

r − s	r + s	r	s	|r − s| + |r + s|
positive	positive	positive	?	2r
positive	negative	?	negative	−2s
negative	positive	?	positive	2s
negative	negative	negative	?	−2r

Since, as pointed out earlier, the absolute values of r and s both turn out to be less than or equal to 1, and since, as the table shows, the numbers of the last column are always positive, we have

$$|E_\lambda (a, b) - E_\lambda (a, d)| + |E_\lambda (c, b) + |E_\lambda (c, d)| \leq 2,$$

which is equation (10.4).

APPENDIX 10B: BELL'S INEQUALITY

If we consider the average of the values of $E_\lambda(a, b)$ over an ensemble in which we have Ni particles characterized by the value i of the hidden variables over a total of N particles, the average $E(a,b)$ of this quantity over the ensemble will be

$$E(a,b) = \sum_i \frac{N_i}{N} E_\lambda (a, b).$$

CHAPTER TEN

Considering, now, the expression analogous to the one in Appendix 10A for the averages, we will obtain

$$|E_\lambda(a, b) - E_\lambda(a, d)| + |E_\lambda(c, b) + E_\lambda(c, d)|$$

$$= \left|\sum_i \frac{N_i}{N}[E_\lambda(a, b) - E_\lambda(a, d)]\right| + \left|\sum_i \frac{N_i}{N}[E_\lambda(c, b) - E_\lambda(c, d)]\right|.$$

It can be noted that, for any sum whatsoever, its absolute value is less than or equal to the sum of the absolute values of the terms (something that is obvious because if not all the terms have the same sign, cancellations occur). From this it follows that by taking the sums and the fractions outside the sign of the modulus the right hand side of the last equation increases:

$$|E_\lambda(a, b) - E_\lambda(a, d)| + |E_\lambda(c, b) + E_\lambda(c, d)|$$

$$\leq \sum_i \frac{N_i}{N}|[E_\lambda(a, b) - E_\lambda(a, d)]| + \sum_i \frac{N_i}{N}|[E_\lambda(c, b) - E_\lambda(c, d)]|$$

$$= \sum_i \frac{N_i}{N}\{|[E_\lambda(a, b) - E_\lambda(a, d)]| + |[E_\lambda(c, b) + E_\lambda(c, d)]|\}.$$

Because the expression in parentheses has already been shown (this is the conclusion of Appendix 10A) to be less than 2 for any λ_i whatsoever, and because by definition $\sum_i \frac{N_i}{N} = 1$,

there follows the inequality we are interested in:

$$|E_\lambda(a, b) - E_\lambda(a, d)| + |E_\lambda(c, b) + |E_\lambda(c, d)| \leq 2.$$

CHAPTER ELEVEN

Nonlocality and Superluminal Signals

> If, however, [the two points] have space-like separation, then a causal connection between them would be an action at a distance. And even if one abstains from using a causal locution in this situation . . . the correlation . . . constitutes a kind of non-locality (which perhaps is sui generis and appropriately named "passion at a distance"). Consequently, even though there is evidently some tension between relativity theory and . . . the experimental results, there is nevertheless a kind of "peaceful coexistence" between quantum mechanics and relativity theory.
> —*Abner Shimony*

THE FACT FULLY ILLUSTRATED in the preceding chapter should certainly have struck the reader profoundly: i.e., that correlations among measurement outcomes in regions spatially far distant from one another—correlations by themselves, and not in connection with a specific interpretation of the theory—require the nonlocality of natural processes. We have been led to conclude that, in a definite act of measurement, what happens (or the action that the observer chooses to perform) at a precise moment and in a certain place can to some extent instantaneously influence another system removed to any arbitrary distance from the first system. In a most natural way, this raises a question of great conceptual, scientific, and epistemological importance: do these nonlocal characteristics of physical processes create an incurable conflict between quantum mechanics and that other great pillar of today's science—namely, the theory of relativity—since this theory is based precisely on the assumption that no physical effect can be propagated faster than the speed of light?

This crucial problem has been amply discussed in recent scientific literature. It is interesting to note that, although there was widespread conviction that quantum formalism did not bring about a direct violation of

CHAPTER ELEVEN

the relativistic postulates, the rigorous demonstration of this fact only took place in the decade of the 1980s—and that was long after the inescapable nonlocality of natural processes had been identified.

It seems a good time now to direct the attention of the reader to a curious fact of human behavior: the problem we are considering (a highly technical subject that seems suited only for the initiated) has enjoyed a remarkable diffusion among people who are normally very little interested in scientific questions. This should not be surprising. One of the frequent themes, for example, of current science fiction is the possibility of traveling at speeds faster than light, or at least of being able to transmit superluminal messages. This latter ability is indispensable for a science fiction hero who has to take part in events occurring in distant parts of the cosmos—so distant that it would require thousands or millions of years to travel from one place to another even if you went at the speed of light!

This lively interest on the part of a huge public has had an effect on those responsible for diffusing information: the journalists. Whether they work for daily newspapers or for popular science magazines, journalists are constantly on the lookout for some "scoop" to get the public's attention, and have consequently placed a remarkable, but misplaced emphasis on any research whatsoever that contemplates or suggests the possibility of "traveling" or "communicating" at speeds faster than the speed of light.

But leaving aside further reflections on this quite understandable tendency of the media, let us return to our theme. Above all, I would like to direct the reader's attention once again to the most characteristic features of quantum nonlocality, the features that put in relief the nonclassical nature of physical reality at a much more radical level than would follow from its conflict with the requirements of relativity. The analysis of the following section reviews some of the points we have already considered, but with a view to bringing into the open a few characteristics that might not have been fully understood before, even by a very attentive reader. After these preliminaries I will go on to show that despite its peculiarities, quantum nonlocality (as Shimony points out) does not in fact give rise to an incurable conflict with the theory of relativity.

11.1. THE PECULIAR CHARACTERISTICS OF NONLOCALITY

There are at least three aspects of quantum nonlocality that are absolutely surprising and reveal the wide divergence of quantum theory from all classical theories. To illustrate these, we begin with a typical phenomenon of classical physics: gravitational attraction. Newtonian physics assumes that all bodies endowed with mass exert an attraction on other bodies with a force proportional to the product of their masses and inversely proportional to the square of the distance between the two bodies. Although he was very concerned over this problem, Newton did not advance any hypothesis explaining how this action between massive bodies could take place, especially when it came to the specific modes governing the propagation of the effects. According to his original conception, if I move a body here, all the bodies in the entire universe, even the most distant galaxies, immediately feel the effect of my action.[1] At first sight, quantum non-locality may not seem any more surprising than this "Newtonian nonlocality." But such is not the case, as we shall soon see.

Above all else, quantum nonlocality is characterized by the fact that its "effects" do not vary with distance. While the displacement of a mass, even in the nonrelativistic view of Newton, produces ever smaller effects the more distant the region (since the force decreases with the square of the distance), in an EPR situation the measurement carried out at one extreme of the apparatus (or better, on one of the two constituents of an entangled state) has precisely the same effect on the other constituent, independently of whether this other constituent is located a few meters away (as in Aspect's experiment) or at the far side of the galaxy. If, for example, the photon to the right is subjected to a test of vertical polarization and passes the test, the photon on the left, wherever it may be located (even a billion light years away), instantly acquires the "property" of being polarized in the vertical plane (and that is a property it could not have had prior to the act of measurement).

Secondly, the nonlocal connections between the members of an entangled system are peculiarly selective. If I move a body with mass in this room, all the masses of the universe are affected by the change. In particular, even from a relativistic perspective, all the masses situated in the same region feel the effect around the same time. But if I have at my disposal a great number of paired photons or electrons in an entangled state, and the members of the various pairs are separated by the same distances and are in the same regions, my action intended to identify the polarization state or the spin component of this specific photon or electron "influences" only and exclusively its partner, that is, the one that was generated with it and remains entangled with it. Other photons or electrons which are in the same remote region or in other places, but are not entangled with the constituent subjected to the measurement process, feel no effect of my action: for them it is as if nothing happened.

Finally, the "effect" (with all the precautions required for this expression, which the attentive reader will certainly understand by now, but the profound meaning of which he will understand still better in what is to follow) on the partner, the "change" induced in its wave function, the fact that through this change a property emerges that could not have been attributed to it beforehand, is absolutely instantaneous. Even Newton who, as already mentioned, did not want to make pronouncements on the specific nature of gravitational effects (*hypotheses non fingo*, or "I do not invent hypotheses"), leaves room for supposing that those effects would have to be in some way mediated through "messengers" (particles for example) that were undetectable. And thus even he foresaw the typical position of modern classical (as opposed to quantum) physics, i.e., that we must assume that a certain time has to intervene between the action and the effects of the action in the remote region. On the contrary, quantum nonlocality brings about decisively faster-than-light, or superluminal, "effects," which Aspect and others have demonstrated in the laboratory.

The preceding observations show the appropriateness of Shimony's suggestion that we call these special nonlocal effects "passion at a distance" rather than "action at a distance." We can reinforce this point by showing how, in fact, it is impossible to make use of this surprising characteristic of natural processes to send messages at velocities faster than the speed of light.

11.2. The Impossibility of Communicating at Superluminal Speeds

Despite its nonlocal aspects, quantum formalism does not permit us to use the phenomenon of the reduction (or collapse) of the wave packet to send superluminal signals, a fact we may illustrate with reference to an EPR experiment of the type discussed in chapter eight. We take a pair of photons in an entangled state (as in equation [8.7], and whose characteristic properties are synthesized in Figure 8.3). We already know that before the observer at A (our friend Alice) carries out a measurement of polarization along any direction of her choosing, the probabilities of the two possible outcomes, i.e., that the photon passes through the filter or is absorbed, are both equal to 1/2. But we also know that, at the precise moment when Alice has carried out her measurement, it is legitimate to assert that the observer at B (Bob) who is a little further removed from the source than Alice (so that his possible measurement would take place after Alice's), would obtain exactly the same result if he carried out the same test. We now suppose that Alice and Bob have a common source that sends them a pair of photons at regular intervals in the same entangled state, and we admit that they agree beforehand about the common direction they will choose for their measurements, and we ask ourselves: is it possible for Bob to know whether Alice has carried out a measurement or not? The answer is clear: it is absolutely true that Alice's and Bob's results present a perfect correlation, that is to say, they are always identical, but it is also true that Alice's results are genuinely distributed at random and that the two possible outcomes (the photon passes or does not pass the test) occur with equal probability. Then we put ourselves in Bob's shoes. What are the outcomes of his measurements? Clearly, they are identical with Alice's and likewise distributed completely at random among the two possible alternatives. What would happen if Alice decided not to carry out a measurement? Since we know that the system is in the state (8.7), the theory assigns a 1/2 probability to the two possible outcomes of a measurement of polarization in any arbitrary direction, so that, even in this circumstance, Bob would then have a random succession of positive and negative outcomes.

This is a completely general truth, not dependent on the particular physical system we are investigating, nor on its state, nor the type of measurement we have chosen to carry out. The outcomes of the measurements of each experimenter are exactly distributed in the same way whether or not the partner carries out any kind of test on the system. Only a posteriori, by communicating (say, by a radio message, and thus at the speed of light) the succession of results, could Alice and Bob become aware of the peculiarity, and see the nonlocal nature of the process they are studying: the two series, while completely random, are absolutely identical. The reason why it is impossible for them to use this process to communicate with each other superluminally should be obvious. It finds its origin in the fact that the nonlocality of the process is completely beyond their control: in fact, Alice can certainly decide by her own free choice to carry out or not carry out a measurement, as well as what direction to choose for the test, but in no way can she control the outcome of the test. Clearly, if she could influence the outcome by, for example, changing in favor of one or the other possible alternatives the fraction of times where there is a certain result (say, by managing to induce a 60%–40% distribution of successful and unsuccessful outcomes), then Bob would of course discover that his outcomes (which are perfectly correlated to Alice's) would present the same disequilibrium, and would know that Alice had chosen to carry out the measurements. But Alice cannot in any way control the genuine stochasticity of her outcomes.

The same problem appears if we assume the point of view of a hidden variable theory instead of quantum mechanics. The fact that such theories are predictively equivalent to quantum theory is already enough to assure us that what we have just stated applies as well to this new method of looking at physical phenomena. But it is necessary to point out nevertheless that there is a notable conceptual difference between the two cases, owing to the fact, mentioned several times before, that while quantum probabilities in the orthodox interpretation are genuinely nonepistemic, the theories with hidden variables require that these probabilities have an epistemic character, that is to say, that they can be attributed exclusively to our ignorance of the values of the variables themselves. As we shall see in section 4 of the present chapter, this fact implies that if we could ever have access to the hidden variables, it would in fact be possible to send signals at speeds greater than the speed of light. But it is important to remember that in a theoretical scheme of this kind it is logically inevitable to assume that we would not in any way be able to control the hidden variables; if the contrary were true, it would be easy (by properly selecting, on the basis of the values of these variables, the members of an ensemble quantally described by a precise state vector) to prepare ensembles that would violate the predictions of quantum mechanics. In other words, a hidden variable theory that permitted us to control these variables would result in a disagreement (a disagreement demonstrable in a laboratory) with quantum mechanics which would then be falsified. The conclusion is obvious: since the experimenter does not have access to the hidden variables, it is not possible for him to use the nonlocal aspects of physical processes. The reasons for this impossibility are very different in the two perspectives. According to quantum mechanics, nonlocality cannot be made use of for the purposes we are considering because of the genuinely stochastic nature of the measurement outcomes; in any hidden variable theory the outcomes are determined and not random, but depend on variables that must be assumed inaccessible on principle.

As I have already indicated, the conclusion about the impossibility of communicating faster than the speed of light has been shown to be universally valid. But a rather peculiar fact about this matter will not escape the attention of someone with an interest in the history of scientific progress. When the founding fathers of the theory introduced the postulate of the reduction of the wave packet in a process of measurement, the peculiar aspects of entanglement were not yet fully understood, and for a long time it was clear that their analysis was concerned with a nonrelativistic theory. The postulate appeared to imply instantaneous effects at a distance (in the precise sense we have discussed in the foregoing chapters), which would have resulted in a conflict with the undeniable requirements of relativity theory. On the other hand, and surprisingly, the reduction of the wave packet, a genuinely nonrelativistic process, combines in such a peculiar fashion with the stochastic aspects of the formalism as not to allow any instantaneous effects or signals at a distance.

In chapter eight we mentioned that Sir Karl Popper proposed an experiment that, in his opinion, would allow us—provided quantum predictions were valid—to make use of the phenomenon of the reduction of the wave packet to produce instantaneous physical effects at a distance. If such an argument were shown to be correct, anyone can see how enormously important it would be. In fact, Popper's thesis is that, if the experiment confirmed the quantum predictions, it would imply a violation of the theory of relativity, while, in order for the experiment to be compatible with the relativistic requirement of the impossibility of superluminal signals, it would have to be in conflict with quantum mechanics. In either case, the result would be Nobel prize material, by revolutionizing all our ideas about natural phenomena. We shall see that such is *not* the case, and that the argument contains a fundamental flaw deriving from an improper use of quantum formalism. An analysis of the proposed argument, and the identification of the error that invalidates it, will permit us to illustrate further the strange aspects of the theory and will help to deepen our understanding of the phenomenon of the reduction (or collapse) of the wave packet.

11.3. A CURIOUS PROPOSAL FOR COMMUNICATING FASTER THAN THE SPEED OF LIGHT

In his collection of writings entitled *Quantum Theory and the Schism in Physics*, Popper proposed a variation of the EPR *Gedankenexperimente*. He presented an argument which, if correct, would imply a violation of quantum predictions or of relativity requirements. Let us follow his argument step by step.

CHAPTER ELEVEN

1. A source Σ emits pairs of particles *in opposite directions* (this is a precise assumption made by Popper), which are propagated toward two screens R and L that are equipped with adjustable slits. Behind these screens are found two semicircular guides, along which are arranged particle counters at regular intervals, such as, for example, Geiger counters (Fig. 11.1a). We identify as x the direction that connects the source with the Geiger counter in the very center of the guide, and as y the direction perpendicular to that. In the first phase of the experiment, the slits of the two screens R and L are large enough to make it legitimate to ignore diffraction effects. The particles emitted from Σ are therefore propagated in a rectilinear fashion, activating those counters from which one can "see" the source (that is, the straight line connecting them with the source is not interrupted by the screen). In Figure 11.1b, these counters are shaded.

2. At this point, Popper refers to an analysis that he had made a few pages before, and investigates what happens at one of the extremities of the apparatus when the size of the slit in the screen is changed. He begins with a pertinent observation: since restricting the slit (for example, at screen R) amounts to an ever more precise determination of the position

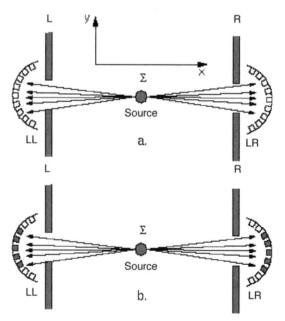

FIGURE 11.1. Schematic view of Popper's proposed experiment.

268

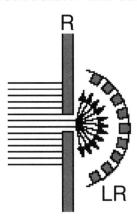

FIGURE 11.2. The effect of narrowing the right-hand slit on the impulse of the particles that go through it. The straight lines to the left of the screen correspond in reality to the lines emerging from the source in Figure 11.1.

along direction y of the particles that pass through the slit, as a consequence of the uncertainty principle there is a diffraction effect (an uncertainty of the y component of the momentum) that will increase the angular dispersion of the particles after the slit. (It would help in this connection to review Figure 1.8 and the discussion of the uncertainty principle in section 3.7). At the limit, as the slit is made narrower, the diffraction figure will necessarily become so large that the particles will be able to propagate themselves with a high probability in any one of the directions indicated to the right of the screen (Figure 11.2).

It thus happens that, while in the case of wide slits, only the darkened counters in the illustration (11.1b) can register particles, for relatively narrower slits, even the nonshaded counters of Figure 11.1b will be activated. So far, the argument is perfectly correct.

3. We now consider the concluding part of Popper's argument. First of all, by basing his example precisely on the EPR proposal, he assumes that the quantum state of the particle pairs under examination is such as to imply a perfect correlation of their positions: that is, when the particle to the right is found to be at a certain point by a measurement, then its partner will be exactly localized at a point symmetrically related to the source. On the basis of this hypothesis Popper now proceeds to develop his reasoning as follows: thanks to the perfect correlation between the particle partners, a measurement that reduces the indeterminacy in position of a particle at R compels (by the process of reducing the wave

CHAPTER ELEVEN

packet) an exactly equal reduction of the indeterminacy of the other particle's position. In other words, as an effect of narrowing the slit in R, the particle at L will become localized, and, as an inevitable consequence of the fact that its position and its momentum in the direction of the y axis must satisfy the uncertainty principle even if the slit at L has in fact remained unchanged, this particle will now exhibit an indeterminacy of possible directions of propagation that corresponds to the indeterminacy of the particle on the right. The conclusion will now be obvious: when the slit in R is narrowed, there will also be an appreciable probability at L (without the slit at L needing to be changed in any way) that the particle there is going to activate the counters (unshaded in Figure 11.1b) that could not have been activated unless there had been a "localization" of particles on the other side.

It is clear that since R and L can be arbitrarily distant from each other, and since pairs of entangled photons keep arriving, one after another, at the two regions under study, it will happen that almost immediately one of the nonshaded counters of L will reveal a particle. But now the observer at L (let's say, Bob) will be able to conclude that someone (say, Alice) at R has taken care to make the slit narrower there: Alice and Bob could thus easily arrange to make use of this process to communicate at superluminal speeds.

The refutation of Popper's argument requires a correct understanding *both* of the meaning of the existence of correlations in position of particle pairs, *and* of the various modalities of wave packet reduction. Let us begin with an assertion that will be much clearer a little later on, when we discuss the case of particle pairs that are "almost perfectly correlated" in their positions. When we consider a pair of particles, the more precise the correlations in position, the more indeterminate are the momentum values of the two partners, whether or not a process of measurement is carried out. In the extreme case Einstein considered (which had quite different ends in view from what Popper proposed) the complete correlation in position implies that each of the two particles presents a complete scattering, or dispersion of the momentum values (even though the pair of momenta are strictly correlated). In other words, in Popper's hypothetical case, even without narrowing the screen to the right, the particle that passes the screen has a completely indeterminate momentum, and can thus very surely activate one of the counters of side LR. The same goes for the particle to the left. Therefore, the conclusion described at (1) is erroneous, if, as the author assumes, the state of the composite system is really such as to imply perfect correlations, for such a state, *all* the counters

270

NONLOCALITY AND SUPERLUMINAL SIGNALS

have an appreciable probability of being activated. But this is still not the end of the story. In fact, although it is technically rather difficult, it is possible to demonstrate that not even in the realistic case of imperfect correlations will there be any change in the outcomes of the measurement to the left if the slit is narrowed on the right. In what follows, I shall develop, for the interested reader, the fine points of the argument leading to this conclusion.

. . . . .

As already emphasized, we are interested in analyzing the case where the correlation between the particle pairs is only approximate. This can happen when the quantum state of two particles is the superposition of various states (for simplicity's sake we can limit ourselves to three only), each of which corresponds to the fact that the particle pair is found in regions symmetrically placed with respect to the source. Consider, for example, three states Ψ_{AR}, Ψ_{BR}, Ψ_{CR}, for the particle to the right, corresponding to three normalized wave functions, differing from zero only in regions A, B, and C (Figure 11.3), respectively, and which assume a practically constant value in the same regions. Let us also consider three states Ψ_{A*L}, Ψ_{B*L}, Ψ_{C*L}, for the particle to the left, symmetrical with the previous states with

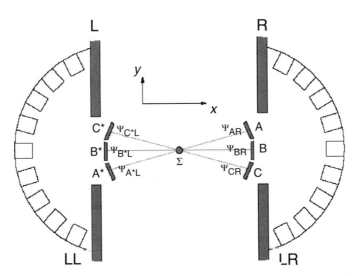

FIGURE 11.3. Analogue of the situation imagined by Popper, in the case where nonprecise correlations are assumed between the positions of the particle pairs.

CHAPTER ELEVEN

respect to the source. These premises being laid down, we assume that the composite system is described by the state

$$\Psi = \frac{1}{\sqrt{3}}[\Psi_{AR}\Psi_{A^*L} + \Psi_{BR}\Psi_{B^*L} + \Psi_{CR}\Psi_{C^*L}]. \qquad (11.1)$$

We then ask ourselves what would be the implications of such a state regarding the positions of the particles. The attentive reader knows very well that the particle to the right has an equal probability of being found in each of the three regions, and that within each region there is an equal probability of being found anywhere in that region. If we limit ourselves to making a measurement of position intended to determine if the particle is in A or B or C (that is, if we put a screen on its path whose opening corresponds exactly to one of the three regions), then the wave function (11.1) is reduced to the term of the sum corresponding to that outcome. Suppose we now consider precisely a measurement of this type that corresponds to verifying if the particle is in region B (obviously this can be realized by narrowing slit D as indicated in Figure 11.4) and by supposing as well that the particle passes the test. As we know, after the measurement the state becomes

$$\Psi_{\text{after}} = \Psi_{BR}\Psi_{B^*L}. \qquad (11.2)$$

If we consider the particles that have overcome the test, equation (11.2) means that their partners to the left will certainly be found in region B* if they are subjected to a process of measurement of position. The reader will have observed that

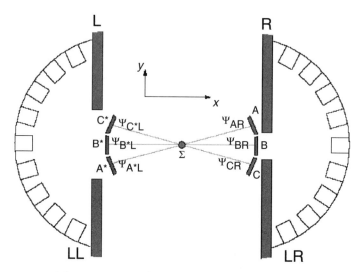

FIGURE 11.4. The situation of Figure 11.3, when the slit to the right is narrowed so as to select only particles in region B.

NONLOCALITY AND SUPERLUMINAL SIGNALS

there are correlations between the positions of the particle pairs but that, in this concrete example, these are accurate only to the extent allowed by the transverse extension of each of the states under consideration (that is, if a particle in a measurement is found in A [B,C], its partner will with certainty be in A* [B*, C*]).

But this is not the end of it. The particle to the right can fail the test, and then the state of the system after the measurement will be

$$\Psi = \frac{1}{\sqrt{2}}[\Psi_{AR}\Psi_{A^*L} + \Psi_{CR}\Psi_{C^*L}] \qquad (11.3)$$

which guarantees that the particle to the left has an equal probability of being found in A* or C*, if subjected to a process of measurement.

At this point we must ask ourselves about the momentum distribution of the particles. Let us consider any one of the three states to the right, for example Ψ_{BR}. Since this corresponds to the fact that the particle is localized in direction y with precision Δ (this refers to the extension, in this direction, of the region where the wave function is different from zero), it will be characterized by a corresponding dispersion of momentum compatible with the uncertainty principle. To put ourselves in an initial situation similar to the one considered by Popper, we suppose that the dispersion of momentum is remarkably small, in such a way that the possible directions of particles emerging from region B differ rather little from each other. The same reasoning holds for the state under consideration and for all the other states that enter into our problem. A global view of the situation is presented in Figure 11.5: As we see, it is

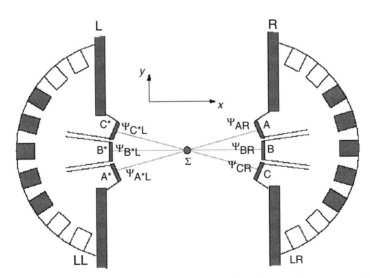

FIGURE 11.5. The situation of the two preceding figures, but now indicating the impulse scattering associated with the states of every particle.

precisely only the shaded counters in the illustration that can be activated by the incident particles.

Two important observations need to be made. Above all, if we want to increase the precision of the correlations, we must consider wave functions with smaller extensions in the transverse direction. But this would also bring about, for each of these functions, a greater dispersion of momentum and thereby the possibility of activating the nonshaded counters, even in the absence of any measurement. Popper's case, as already emphasized, is precisely the extreme case: perfect correlations of position imply arbitrary distributions of momentum (even if these too are correlated) and thus, contrary to his initial assertions, for the type of state he considered even when the screens are in the positions of Fig. 11.1, all the counters have an equal probability of being activated.

Let us pose ourselves the question, whether or not an argument à la Popper can be developed even in case of approximate correlations. Now, because narrowing the aperture to the right from the value Δ to a much smaller value Δ introduces a dispersion of momentum in this region, it is quite conceivable that an analogous dispersion could be produced to the left, and thereby bring about Popper's hypothetical effect. But this idea can nevertheless be shown to be mistaken, by resorting to the very postulates of quantum formalism. Let us suppose we reduce the aperture to a mere slit (Figure 11.6) corresponding to the central zone BM of re-

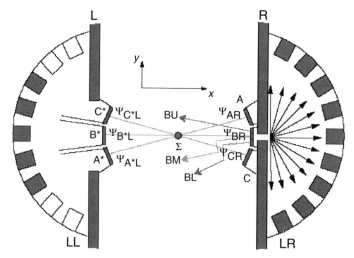

FIGURE 11.6. Narrowing the slit to the right provokes a scatter of the directions of propagation of the particles going through. However, as discussed in detail in the text, this has no influence whatsoever on the situation to the left. Once again, it is not possible to use the reduction of the wave packet to send signals faster than the speed of light.

NONLOCALITY AND SUPERLUMINAL SIGNALS

gion B. Undoubtedly this process induces an indeterminacy of momentum that will cause the nonshaded counters in LR to be activated. However—and this is the crucial point—what happens to the wave function on the left? To answer the question, it will be convenient to rewrite the wave function (11.1) in such a way as to make evident (remembering the considerations presented above in section 6.4) the terms that correspond to the outcome that interests us (the particle is in BM) and to the opposite outcome (the particle is outside MC):

$$\Psi = \frac{1}{\sqrt{3}}[\Psi_{AR}\Psi_{A*L} + \frac{1}{\sqrt{3}}(\Psi_{BUR}\Psi_{B*L} + \Psi_{BMR}\Psi_{B*L} + \Psi_{BLR}\Psi_{B*L}) + \Psi_{CR}\Psi_{C*L}]. \qquad (11.4)$$

In this equation we have indicated as BU, BM, and BL the three subregions (upper, middle, and lower) into which the region B is divided by the central part (BM) of the opening; each subregion is of amplitude $\delta \ll \Delta$, and we have indicated the normalized states that correspond to the subregions as Ψ_{BUR}, Ψ_{BMR}, and Ψ_{BLR}, respectively. Now, what does the postulate of the reduction of the wave packet tell us in the case when the particle to the right passes the test? It tells us that from the state (11.3) we have to extract the term that corresponds to the fact that the particle to the right is in BM. In other words, after passing through the opening to the right, the wave function for those particles that pass the test is

$$\Psi_{after} = \Psi_{BMR}\Psi_{B*L}. \qquad (11.5)$$

. . . . .

In conclusion, the correct application of the theory leads us to assert that the particles to the left whose partners have passed the opening of width δ to the right are now described by the function Ψ_{B*L}, the extension of which corresponds to region B*. While the particles to the right that pass the test will activate the nonshaded counters, the momentum dispersion of their partners to the left does not change. If we then take account as well of the particles to the right that did not pass the test (those, of course, whose partners are not intercepted or eliminated), it can be concluded that absolutely nothing changes on the left in regard to the activation of the counters.

At this point I cannot help quoting Popper's text:

> If the Copenhagen interpretation is correct, then any increase in the precision of our mere knowledge about the position of the particles going to the right should increase their momentum dispersion; and this prediction should be testable. If, as I am inclined to predict, the test decides against the Copenhagen

CHAPTER ELEVEN

interpretation . . . it would mean that Heisenberg's claim is undermined that his formulae are applicable to all kinds of indirect measurements (a claim that the adherents of the Copenhagen interpretation—such as von Neumann—would undoubtedly regard as part of quantum mechanics). What would be the position if our experiment (against my personal expectation) supported the Copenhagen interpretation—that is, if the particles at the left whose position has been indirectly measured show an increased scatter? This could be interpreted as indicative of an action at a distance; and if so it would mean that we have to give up Einstein's interpretation of special relativity and return to Lorentz's interpretation and with it to Newton's absolute space and time.

While the first assertion is correct—that is, that the particles to the right show an actual increase in "scatter" or momentum dispersion—the other two statements are completely erroneous. The Copenhagen interpretation *implies* that the measurement to the right does not change the momentum distribution to the left. Therefore, what Popper claims to be his expectation would certainly be satisfied . . . but alas! *This is precisely the clear and unambiguous prediction of the theory in its orthodox interpretation!* It would be difficult to imagine a more radical misunderstanding of the effects of nonlocality, or a more erroneous use of the formalism.

11.4. WHAT IF WE COULD CONTROL THE HIDDEN VARIABLES?

It would appear useful now to illustrate the fact (mentioned in section 11.2 above) that, whereas in a genuine quantum context the formal elements cannot in any way be identified that would permit us to have any control over nonlocality, in hidden variable theories, by contrast, it would be possible, if we could have access to the variables themselves, to communicate faster than the speed of light and produce instantaneous physical effects at a distance. Let us consider a situation à la EPR, with entangled states of a system of two particles with spin 1/2, and analyze the process first from the standard quantum mechanical point of view, and then from the perspective of de Broglie and Bohm. In the course of this analysis we will show how contextuality—a typical characteristic of hidden variables theories which we already explored in section 9.7—plays a crucial role for this kind of process. The reader will be able to understand why we were compelled in chapter 9 to analyze (with perhaps

NONLOCALITY AND SUPERLUMINAL SIGNALS

an excessive pedantry) this key aspect of the theories in question, and should be able to realize something that is probably already suspected: that contextuality itself presents nonlocal characteristics.

We can set forth precisely the physical process in question. There is a source of particle pairs with spin 1/2, propagating in opposite directions toward two observers. As regards the wave functions that describe the spatial movement of the particles, we will assume only that they are identical (except for the direction of propagation), symmetrically disposed with respect to the source, and sufficiently localized, i.e., to the extent that we are able to speak of the instants when they reach the two observers. These latter, as before, are positioned at slightly different distances from the source (Alice, the observer on the right, will be reached by her particle a moment before its partner particle reaches Bob on the left side of the source). To make the argument as simple as possible without losing any of its rigor, it will be convenient to assume that the spatial wave functions are symmetrical with respect to a horizontal plane that passes through the source. Such specifications having been made, let us pay attention to the spin component of the system's state vector. The wave function of the spin is the exact analogue of the entangled state of two photons (see section 8.7: this was the theme of chapter 8 and of most of the subsequent chapters). As will soon become very clear, everything we are concerned with will be merely the translation, into the language of spin variables, of the analogous situations of light polarization that we have previously analyzed in such detail.[2]

Our notation can be specified as follows: we indicate the spin states of particle 1 (by convention, this is the particle to the right) as $|1, z\ up\rangle$ and $|1, z\ down\rangle$, which corresponds to the fact that this particle has spin "up" or "down" with respect to the vertical axis z. Analogously, $|2, z\ up\rangle$ and $|2, z\ down\rangle$ are the corresponding spin states for particle 2. In accordance with our premises, we now assume that the spin state of the two particles is entangled, and takes the following form in our notation:

$$|\Psi\rangle = \frac{1}{\sqrt{2}}\big[|1, z\ up\rangle |2, z\ down\rangle - |1, z\ down\rangle |2, z\ up\rangle\big]. \qquad (11.6)$$

This state, in fact, is the exact analogue of state (8.7) in the case of photons. The reader may perhaps be surprised by the change of the | sign in (8.7) for the − sign here. This is necessary because in the case of spin the state (11.6) is really the isotropic state, that is, the one which has the same form no matter to what direction reference is made. In other words, just as in the case of the photons in state (8.7) it is not important to specify the

CHAPTER ELEVEN

(orthogonal) planes of the two members of the composite system, insofar as the explicit form of the state is the same for any choice of orientations (recalling equation [8.10], so fundamental for the understanding of EPR). Also, in the above equation it is possible to substitute for the index z any index n signifying any arbitrarily chosen orientation, so that the state in question could also be indicated in the visually most significant way as

$$|\Psi\rangle = \frac{1}{\sqrt{2}}\left[|1,\Uparrow\rangle|2,\Downarrow\rangle - |1,\Downarrow\rangle|2,\Uparrow\rangle\right]. \qquad (11.7)$$

The form (11.7) puts clearly into view that what really counts is that in one of the two superposed states the two particles have spins oriented in opposite directions along an arbitrarily chosen direction, and that in the other state the orientations are reversed. This is precisely the fact that implies the strict analogy with the case of the photons—an analogy that requires only a slight qualification. As is clear in the equation just offered above, this analogy guarantees that in a measurement for the spin of one of the particles (say, of particle 1) along some arbitrary direction, there are equal probabilities of finding spin "up" or "down" (it is enough for the present purpose to think of the arrows as referring to an arbitrarily chosen direction); the analogy also means that, in the case at hand, if the particle's spin is discovered as "up" in the measurement, then the process of the reduction of the wave packet, which (as we know very well) will bring the state $|1,\Uparrow\rangle|2,\Downarrow\rangle$, guarantees that if a measurement of spin for the same direction is carried out on the other particle, the certain result there will be "down." Once again, the outcomes of identical measurements are perfectly correlated, the only difference being that in the case of photon pairs, both are always found to have the same polarization state when subjected to the same measurement, while in the present case, the outcomes of identical spin measurement are opposites. In practice, since the measurement of the spin component is carried out by sending the particle into a region with an inhomogeneous magnetic field, and since the passage across this field brings about an "up" or "down" deviation of the particle's trajectory, the state (11.7) implies that when each of the two particles is made to pass through a Stern-Gerlach apparatus with an orientation identical to the other, if one particle is deviated in one direction, the other will certainly undergo the opposite deviation.

With these qualifications stated, we can enter into the actual experiment. For our own purposes, we consider two possibilities: both Alice and Bob make measurements in the direction z. Bob chooses an orientation for his apparatus and keeps it fixed in all instances. Alice, on the other side,

NONLOCALITY AND SUPERLUMINAL SIGNALS

by her free choice *either* orients her magnets in the same direction as Bob or reverses the direction of the magnetic field (or the direction along which it increases—the reader may wish to consult Figure 9.6).

Let us analyze these two alternatives from within the perspective of the standard quantum scheme. Suppose the two magnets are oriented the same way. What will happen? In 50% of the cases, Alice's particle will be deviated upward, the reduction of the wave packet will bring about a state $|1,\Uparrow\rangle|2,\Downarrow\rangle$, and thus Bob's particle will be deviated downward. With the same probability, that is to say, in 1/2 of the cases, Alice's particle will be deviated down and Bob's will go up. The situation is schematically illustrated in Figure 11.7, in which the wave functions in space of the two particles are shown as tiny spheres before they reach the region of the apparatus, to underline the assumption that will be relevant for future analyses, i.e., that the wave functions are symmetrical with respect to the horizontal plane.

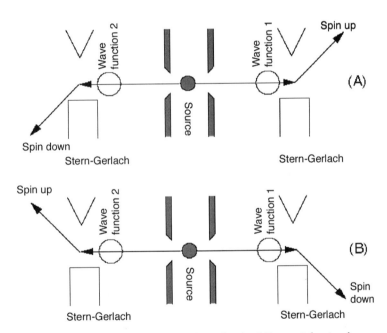

FIGURE 11.7. EPR experiment on pairs of spin 1/2 particles in the entangled state (11.7). Spin measurements are carried out by making the particles pass through regions between two Stern-Gerlach magnets with vertical orientation. The outcomes are perfectly anticorrelated, which is to say, if one particle is found to have its spin pointing upward, the other one will be found to have its spin pointing downward. The two only possible outcomes (A) and (B) occur with equal probability.

279

CHAPTER ELEVEN

We proceed now to discuss the case when the orientation of the magnetic field on the right is reversed (Fig. 11.8). We already know that when the field is reversed, the association between the orientation of spin and the deviation of the trajectory is turned upside down. This means that when Alice obtains an outcome that corresponds to the first term of the state (11.7)—that is, she finds that particle 1 has "upward" spin—now the particle itself will be deviated downward. But the reduction of the wave packet will nevertheless bring about the state $|1,\Uparrow\rangle|2,\Downarrow\rangle$, and since Bob has not reversed his own magnetic field and particle 2 in this state has spin downward, this particle will be deviated downward. I apologize for my pedantic manner: the reader should clearly understand that if by equal orientations of the fields the particles are with certainty deviated in opposing directions, they ought to become deviated into the same direction when the fields are oriented in the opposite way. It follows from this, as illustrated in Figure 11.8 that in case Alice obtains the outcome "particle 1 is being deflected *upward*," the same will happen for Bob's particle.

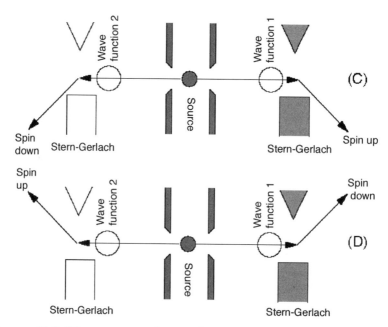

FIGURE 11.8. The same experiment of Figure 11.7, in the case where the magnetic fields of the two Stern-Gerlach apparatuses are oppositely oriented. It is assumed that Bob is keeping the orientation of his magnet fixed.

We now proceed to consider the same physical process from the perspective of the pilot wave theory. Everything is the same except for one fundamental fact: despite the extension of the wave function (indicated by the little spheres), in every individual case both particles will possess a well-defined position, even if this position is unknown (in fact, the positions are the hidden variables of the theory). Of course, these positions will be distributed in such a way as to respect quantum probabilities—that is to say, they will be governed by the square of the modulus of the wave function. If, as we have supposed, the wave functions have a practically constant value on the inside of the circles in the figure, then, in an ensemble of identically prepared systems, the particles will be distributed in such a way as to be found at any one of the points within the circle.

At this point, let us admit a rather hazardous hypothesis. As we well know, we have no way of managing the hidden variables—that is to say, we cannot know, for example, if in a specific case particle 1 is actually found in the upper or lower half of the sphere. But we suppose there exists a little witch: "Bohm's witch," let us say, on the analogy of "Maxwell's demon." As is well known, the hypothetical Maxwell's demon of thermodynamics is in a position to know something impossible for a human being to know: the position and velocity of every single molecule in a gas. The demon is supposed to be able to operate (either by himself or through the mediation of a servant) a little window on a wall dividing a container of gas into two compartments. The demon watches the particles on the left and right that are moving about in the area close to the window. If a particle on the left looks as if it is about to come into the area, he closes the window against it, making it bounce back into the left-hand side. On the other hand, if he sees a particle coming from the right he suddenly opens the shutter to allow the particle to pass into the left-hand chamber. In this fashion the demon can accomplish something impossible in the real world: the particles that began as being equally distributed in the two halves of the container end up by being all on the left. The process leads to a violation of the second law of thermodynamics. I have presented this one example of the demon's various possible exploits because it is the simplest and most intuitively comprehensible.

Let us turn then to our problem. What peculiar gifts belong to "Bohm's witch"? Well, she would simply be able to see the hidden variables of the system. This means that, unlike the experimenter who is only able to work with the wave function and not the actual positions of the particles of the system in every individual case, the witch is able to know, system by system, exactly where the particle—let us say the one on the right—

CHAPTER ELEVEN

actually is. At this point we should recall that in the de Broglie–Bohm theory the spin variables are contextual. In particular, in the case under investigation (see the discussion in section 9.7), as a consequence of the symmetry of the problem, at the first measurement, that is the measurement on the right, the particle can never pass along the horizontal plane independently of Alice's choice of field orientation: if a particle is in the upper part of the sphere, it will be deviated upward, if it is in the lower part, it will go downward. But the theory, we know very well, perfectly reproduces the predictions of quantum mechanics about the outcomes of spin measurements. This fact implies especially that if the particle to the right is found "to have upward spin," then certainly[3] the particle to the left "will have to have downward spin."

It is now easy to see how Bohm's witch is able to act (or how she makes suggestions to Alice to act) in order to produce superluminal effects in Bob's region (Figure 11.9). The witch is waiting in the region where the particles are about to reach Alice, and "observes" their hidden variables, that is, their positions. We suppose that at the beginning Alice has her Stern-Gerlach magnet oriented like Bob's, pointing upward, and we suppose that the witch sees that the arriving particle is in the upper hemisphere of the little sphere (Figure 11.9a). This time, the witch says nothing to Alice. Alice lets the magnet stay where it is, and the particle is deviated upward (this happens because of its position, and not by the orientation of the magnet), and Alice concludes that she has discovered "upward spin" by her measurement. Since Bob, as required by quantum mechanics, ought to obtain the opposite outcome, that is, "downward spin," and since his magnetic field is oriented like Alice's, his particle will surely be deviated downward.

Now let us suppose that the witch sees particle 1 of the next pair arrive, and notices that this one is in the lower half of the little sphere that represents its wave function (Figure 11.9b). This time the witch tells Alice to reverse her magnetic field. Because of the contextuality of the spin variable, the particle is going to move downward. But the downward deflection with *reversed magnetic field* means Alice will end up with the outcome "upward spin." It follows from this that Bob's measurement must yield "downward spin," and it being granted that Bob has not changed the direction of his magnet, his particle will in fact be deflected downward.

This is how we get our stupendous result: by choosing to change the setting of her magnetic field according to the witch's instructions, that is, by leaving it unchanged every time the incoming particle is located in the

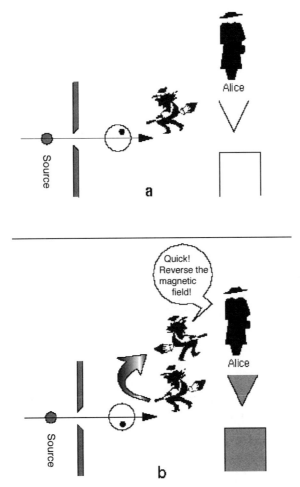

FIGURE 11.9. The surreal Bohm's witch who has access to the hidden variables—meaning, she knows the position of each particle in every single process—will be able to take advantage of the contextuality of the spin variables to suggest to Alice how to orient her own magnetic field in such a way as always to obtain the outcome "upward spin" in her measurement. Since the perfect anticorrelation of spin measurements as predicted by quantum mechanics is always correctly reproduced by the theory, and since Bob does not change the orientation of his apparatus, by always getting the outcome "downward spin," he will see all his particles go down.

CHAPTER ELEVEN

upper half of the wave function, and reversing it every time the particle is in the lower half, Alice is able to produce the surprising physical effect of having Bob's particles always go down. For, if Alice had not inverted her magnetic field in the second instance, Bob's particle would have gone upward.[4]

This example provides a clear and explicit proof that, in this model, if the hidden variables could be known—that is, the position of each individual particle—it would be easy not only to communicate at speeds faster than the speed of light (Bob, very far away, seeing that suddenly none of his particles is going up, knows that Alice is sending him a message); it would also mean bringing about concrete physical effects at a distance. In this example, Alice is causing an instantaneous deflection downward, by the reversal of her magnetic field, of a particle that would otherwise be going up.

Nevertheless, as I have frequently remarked, the hypothesis that one could have access to the hidden variables (which are assumed to exist and to have definite values) must be excluded a priori if we want to have a theory that is not in conflict with the experimental results of quantum mechanics. The performance we have just described is possible for Bohm's hypothetical witch but not for human beings. For, just as the second law of thermodynamics makes it impossible for perpetual motion to exist, i.e., it asserts that human beings cannot violate the second law of thermodynamics, so the nonlocality of physical processes cannot be used to transmit superluminal signals or produce instantaneous distant physical effects.

11.5. WHAT IF WE COULD CLONE INDIVIDUAL PHOTONS?

To continue our entertaining speculations on our theme, we can conclude this chapter by investigating another recent proposal for using nonlocality to send signals faster than the speed of light. Of course, this too will be shown to be impracticable, but the idea is instructive all the same. Among other things the proposal is of very topical interest today, when the process of cloning has become the object of so much attention along with the spectacular successes of bioengineering. So let us suppose, with the inventor of this idea, that it is possible to construct a machine that could clone any photon, and produce a great number of them, each one identical with the original (Figure 11.10). The process of cloning would have to respect not only properties such as color (i.e., fre-

284

NONLOCALITY AND SUPERLUMINAL SIGNALS

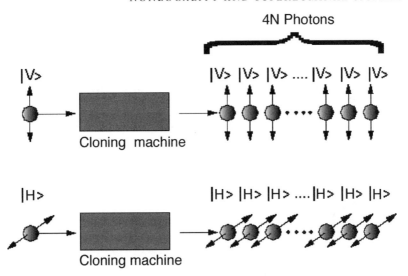

FIGURE 11.10. The hypothetical apparatus that plays an important role in the argument of this section. It has the ability to produce an enormous number of identical copies of (i.e., it clones) the photon that enters it. In particular, every photon that goes out of it has the same polarization as the one that enters. Here we consider the case of vertical and horizontal incident polarization, but for the argument to be valid it is necessary for the apparatus to function the same way for all other polarizations as well.

quency), but also the state of polarization of each individual photon. We then imagine we have at our disposal an instrument that could be triggered by a photon of a certain polarization, and then produce, and emit, an enormous number (let us say 4N, with N as a very large number) of photons absolutely identical (including polarization) with the one it started with. We can summarize the cloning process by a formula expressing what happens according to whether the original photon has vertical or horizontal polarization:

$$|V\rangle \Rightarrow |4N, V\rangle, \quad |H\rangle \Rightarrow |4N, H\rangle \qquad (11.8)$$

and we assume that the same should hold for any polarization state whatsoever.

Taking up again the example of a pair of photons in an entangled state (8.7) that we used to discuss the EPR paradox, we can analyze the precisely specified experiment illustrated in Figure 11.11. The initial state involves two photons propagating in opposite directions and exhibiting the correlations characteristic of their state. While the two photons, as usual,

CHAPTER ELEVEN

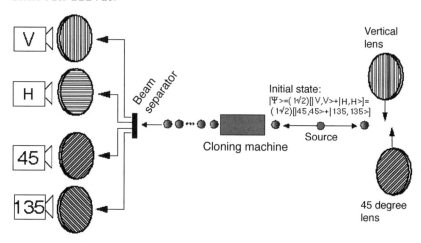

FIGURE 11.11. The experimental apparatus that supposedly permits superluminal signals. To the right, Alice can choose to carry out either a vertical measurement or a measurement at 45°. As discussed in the text, the two alternatives require that, at the left, *either* one of the two upper counters *or* one of the two lower counters *not* register any photon.

are very far removed from one another, the observer to the right (Alice) decides to execute a measurement of plane polarization and can freely choose between a vertical polarization and a 45° polarization.

We begin with the first event. Alice makes a photon pass through a filter with a vertical polarization plane. Let us say it passes the test. We know (and this is necessary for guaranteeing perfect correlation between the outcomes of identical measurements on photon pairs prepared in this way) that the photon to the left will be polarized vertically immediately after the measurement (it does not matter how far away this happens). At this point the photon enters the cloning apparatus that multiplies photons, keeping the same polarization. There will then be 4N photons coming out the other end of the machine precisely identical in polarization with the first one. It is then a very easy matter to use a simple optical device (semitransparent mirrors) to divide up the emergent light beam into four beams, each containing N vertically polarized photons. These four beams are then introduced into four devices (designated as V, H, 45, 135), which are nothing other than filters polarized along the four planes (vertical, horizontal, 45°, 135°), with counters placed behind each filter. What will happen? Everyone knows the right answer: we know that a vertically polarized photon passes with certainty through a filter with vertical polarization and is certainly absorbed by a filter with horizontal polarization, while it has a 1/2 probability of passing a filter with 45° or 135° polarization. At the end we will have all the photons of the first beam passing the test, none of the second beam, and one-half each of the third and fourth beams passing their tests. We can list in order the quantities registered by the four counters as N, 0, N/2, N/2.

Of course, we cannot ignore the fact that in the first measurement (to the right), we could have had the opposite result, that is, the photon could have been absorbed. This would imply that immediately after the test, its partner would have been horizontally polarized. This one too could have been cloned and the resultant beam of cloned photons separated into four beams. We would then have four beams of photons to the left, all horizontally polarized. Taking account of the rules governing the passage of photons of this type through the filters they encounter, we can immediately see that the resulting records of the four counters will be 0, N, N/2, N/2. The two scenarios we have now described will occur with equal probability, and at random, as the experiment is repeated.

Now let us suppose Alice decides to measure the polarization along the 45° plane and that her photon passes the test. Now, the photon to the left (and by cloning all the photons of the four beams to the left) will be polarized at 45°, and as such will pass with certainty a test of polarization in this direction, and will fail with certainty a test in the orthogonal direction (i.e., one at 135°), and will have a 1/2 probability of passing tests of vertical and horizontal polarization. The corresponding foursome of quantities registered by the counters will be N/2, N/2, N, 0. Analogously, in the equally probable case that the photon to the right does not pass the test of 45° polarization, the photon to the left will be polarized at 135°, and consequently, through cloning, separation, and registration, the foursome will be N/2, N/2, 0, N. The situation is presented in Table 11.1.

Now comes the interesting part. In the first two foursomes, there is a zero in the first or second place, whereas in the second two, there is a zero in the third or fourth place. If Alice chooses to carry out a measurement of vertical polarization, she cannot in any way control which of the first two possibilities will occur (for Bob) on the left. Analogously, if Alice chooses to carry out a measurement for 45° polarization, she cannot in any way control which of the two possibilities of the third and fourth foursomes will be realized on the left. However, if we assume that Alice can freely choose to carry out a measurement of vertical or 45° polarization, her choice will determine whether Bob will have the first two or the second two foursomes. Simply by observing which counter does not register any photons, Bob can know what kind of measurement Alice chose

TABLE 11.1.
The table shows the counts of the four counters to the left, depending on the type of measurement carried out by Alice, and the genuinely random outcomes of her measurement.

Alice's choice	Outcome of measurement	Situation on the left	Series of outcomes			
vertical polarization	the photon passes the test	four beams of vertically polarized photons	N	0	N/2	N/2
vertical polarization	the photon does not pass	four beams of horizontally polarized photons	0	N	N/2	N/2
45° polarization	the photon passes the test	four beams of photons with 45° polarization	N/2	N/2	N	0
45° polarization	the photon does not pass	four beams of photons with 135° polarization	N/2	N/2	0	N

to make: if she chose a measurement for vertical polarization, Bob's first or second counter will register zero photons; in case she measured for 45°, Bob's third or fourth counter will register zero. Since, at the moment of being measured by Alice, the photons can be any arbitrary distance apart, Bob will be able to know, without any additional communication on Alice's part, whether she decided in each case to carry out a measurement of vertical polarization or 45° polarization. Imagine, then, a source fixed somewhere in space that generates, for example, one pair of photons per second, and an agreement set up whereby the two observers agree to use Morse code (if Alice chooses vertical polarization, it means a dot, if she chooses 45° polarization, it equals a dash), then Alice could send Bob a message faster than the speed of light.

Now, what is wrong with this argument? If there is nothing wrong with it, we would have to conclude that it is possible to violate the laws on which the theory of relativity is founded. But it can easily be shown that, even though nothing in principle may prevent us (the technical difficulties involved in the cloning of photons may be insuperable, but that does not really matter since this is obviously a *Gedankenexperiment*) from making an apparatus for cloning photons of two assigned orthogonal polarization planes (such as vertical and horizontal), quantum mechanics itself, that is, the hypothesis that even the cloning machine is a system that must obey the laws of the theory, implies that it would not be able to clone photons at 45°. I do not want to be misunderstood: nature does not in some way preselect vertical or horizontal polarization; rather, every pair is equal to every other. To put it simply, any conceivable apparatus that obeys quantum laws can clone photons with two arbitrarily chosen orthogonal polarizations, but not photons with other polarization states. Consequently, it would be possible to create an apparatus that clones photons polarized at 45° or at 135°, but now this apparatus would not be able to clone ones polarized vertically or horizontally. However, it should be obvious that for the experiment just described to work as it should and allow for superluminal communication, it is necessary to have a single cloning machine that can clone all four types of photons needed for the process (that is, photons with V, H, 45°, and 135° polarization). At this point it would seem worthwhile to take a few lines of text to explain why such a machine cannot exist if we suppose that it, like all physical systems, must obey the laws of quantum mechanics.

. . . . .

CHAPTER ELEVEN

Let us pretend we have a system whose input is a photon polarized vertically or horizontally, and whose output is 4N identically polarized photons. Using the language of the theory, the properties of the machine in question would be expressed by equation (11.8), where the arrow describes the evolution of the state from the beginning to the end of the cloning process. Suppose we send a photon into it with 45° polarization, and we ask ourselves what we have for an output? To answer this question we simply express the state $|45\rangle$ as a linear combination of the states $|V\rangle$ and $|H\rangle$, and observe that the evolution of the system, being governed by quantum formalism, implies that the linear combination of the two states evolves in the same combination of their evolutes. As we know (recalling equations [2.8] and [3.2]),

$$|45\rangle = \frac{1}{\sqrt{2}}[|V\rangle + |H\rangle]. \tag{11.9}$$

Taking account of the fact, as asserted by (11.8), that the two terms $|V\rangle$ and $|H\rangle$ of the sum evolve, respectively, into the two states $|4N,V\rangle$ and $|4N,H\rangle$, and that, as just emphasized, the evolution preserves the linear combination, it follows that if the cloning machine is set up to clone with the state (11.9), the outcome will be the following:

$$|45\rangle \Rightarrow \frac{1}{\sqrt{2}}[|4N,V\rangle + |4N,H\rangle]. \tag{11.10}$$

Our unavoidable requirement that the machine must clone photons with any arbitrary state of polarization would require that the state at the exit would be one that contains 4N photons polarized at 45°. Explicitly, this state would be

$$|4N,45\rangle = |45\rangle|45\rangle\cdots|45\rangle|45\rangle$$
$$= \left\{\frac{1}{\sqrt{2}}[|V\rangle + |H\rangle]\right\}\left\{\frac{1}{\sqrt{2}}[|V\rangle + |H\rangle]\right\}\cdots\left\{\frac{1}{\sqrt{2}}[|V\rangle + |H\rangle]\right\}\left\{\frac{1}{\sqrt{2}}[|V\rangle + |H\rangle]\right\}, \tag{11.11}$$

where the number of factors to the right is precisely equal to 4N. If we work out the products we see immediately that the state $|4N, 45\rangle$ contains two terms, which derive from the product of all the first terms and of all the last terms of the various factors of (11.11), which are really the two states that appear to the right of (11.10). But these two states have an extremely small coefficient $(1\sqrt{2})^{4N}$ and, moreover, to the right of (11.11) appear also all the states with an arbitrary distribution of photons in two groups, one containing K photons with one polarization and the other containing (4N – K) photons with the other polarization (here, K takes the values 0,1,2, ..., 4N). The state (11.11) has nothing to do with state (11.10); that is, our apparatus cannot clone photons at 45°. I will not delay further to show, although it would not be difficult to do so, how even the form (11.11) of the state is such as not to allow Bob to know if Alice chose a vertical measurement or one at 45°.

. . . . .

In conclusion, the experiment considered in this section is impossible not because the argument on which it is based is logically inconsistent, but because nature would not permit us to build a machine that could clone photons of *any* polarization, because *all* physical processes obey quantum laws. It is very strange, again, that the very structure of the theory does not allow us to realize the science-fiction fantasies of faster-than-light signals it appears to suggest.

I think it would be fitting to close this chapter with a conceptually relevant observation. We have been led to recognize, with Shimony, that the peaceful coexistence of the two great pillars of modern science—quantum mechanics and the theory of relativity—a peace that seems seriously threatened by the nonlocal character of the former, is in fact endangered neither by the consideration of quite ingenious experiments, nor by the adoption of nonorthodox points of view toward the formalism (such as hidden variable theories), nor even by imagining a bizarre adaptation of biogenetics for the cloning of light quanta. Nonlocality, an actual fact proven by experimental evidence that leaves no room for reasonable doubt, presents characteristics of such a peculiar kind as to *keep it from being used* to produce instantaneous effects between distant physical systems.

CHAPTER TWELVE

Quantum Cryptography

> It may be roundly asserted that human
> ingenuity cannot concoct a cipher which
> human ingenuity cannot resolve.
> —*Edgar Allan Poe*

It may be disappointing to learn that we cannot communicate at speeds faster than light; however, there is good news, too, for the truly quantum nature of natural processes does in fact make possible another dream: the construction of secretly coded messages that cannot be decoded by any spy, in such a way as to contradict Poe's assertion given above. The present chapter will be dedicated to illustrating how such a stupendous feat can be accomplished. The basic idea of the procedure, as well as the ideal cryptographic mechanism will be rather easy for the reader to understand at this stage, now that he or she has a grasp of the phenomenology of photon experiments that lies at the foundation of quantum cryptography. Before we undertake to explain how the genuine stochasticity of natural processes enables us to reach this goal, it will be convenient to introduce the reader briefly to the problem of message encryption. We will also dedicate a few pages to explaining the binary system that will be used in what follows.

Not all readers are aware, perhaps, of Poe's interest in cryptography or of his remarkable abilities as a decoder. Poe gave a major thrust to literary cryptography with his story about a secret message, *The Gold Bug*. As a writer, Poe combined a sense of fantasy with a logical rigor, and this brought him under the spell of cryptography. In 1839 he wrote in *Alexander's Weekly Messenger* that "In spite of the anathemas of the over wise we regard a good enigma as a good thing. Their solution affords one of the best possible exercises of the analytical faculties, besides calling into play many other powers." His work stirred the curiosity of readers and started a process that led to Poe's reputation (mostly undeserved) of being one of the greatest cryptographers that ever lived. Accordingly, the statement reported at the head of this chapter might be considered a kind of exhibitionism coming from the sense of superiority he had acquired about his reputed talents. Even so, it must be admitted that Poe's statement has received some astounding confirmations in recent years. The best illustration I think can be found in the first chapter of the splendid book by David Kahn, *The Codebreakers: The Story of Secret Writing*. In that chapter the author describes how in the early 1940s the Americans succeeded in breaking the ingenious system of cryptography in use by the Japanese. A brief digression on this interesting story will pave the way to my theme.

Kahn's chapter is entitled "One Day of Magic," playing on a double meaning of the word *magic*. Literally the word denotes an extraordinary event, something done "like magic," but at the same time "Magic " was the code word for the American operation dedicated to breaking their enemy's very difficult code. The "day" in question was December 7, 1941, the day the Japanese sent an encoded message to their Ambassador in the United States, instructing him to convey what was in essence a declaration of war to the American Embassy, although it had the form of a suspension of the existing peace treaties. It was no coincidence that, simultaneously with the transmission of these instructions, the attack on Pearl Harbor was set in motion.

Let us temporarily sidestep the dramatic events of those days to consider the background of the exhausting rivalry between the Americans and Japanese in elaborating, and breaking, ever more sophisticated codes. Beginning in 1920 the Americans had intercepted and deciphered almost all the Japanese codes, but in 1934 the Japanese navy acquired a new coding machine from the Germans, called "Enigma," that had been further modified to become the most complex and secret system of cryptography in use at the time. To their disappointment the Americans discovered that the messages they recognized to be of major importance (thanks to their knowledge of the message destinations and other easily identifiable elements) had been codified in a manner radically different from anything before, and so they had to put a maximum of effort into breaking the new system. The system the Japanese were using was given the name "Purple," and Magic, as I said, was the name for American operation to break Purple. Before entering into the details of this titanic struggle, it is interesting to jump ahead in time to consider a significant event that highlights how extremely sophisticated Purple was as a code. On April 28, 1941, almost a year after the Americans had succeeded in breaking it and had been systematically deciphering all the messages written in it, the German consul in Washington, Hans Thomsen, informed his Minister of the Exterior that he had learned from an absolutely trustworthy source that the Americans

293

were able to decipher messages written in Purple. On May 29, the German government informed the Japanese, in anxiety that their own plans would be discovered. The Americans deciphered this message, and were anxiously expecting the Japanese to change their cryptographic system completely. But, to their surprise, very important messages kept coming in the old code. In October of that same year a crisis in the Japanese government occurred, followed by a series of changes. One of the first acts of the new administration was to summon those responsible for secret codes and to make explicit interrogations about their security. The head of the Japanese Secret Service stated without hesitation that the system in place was absolutely secure, considering its extreme complexity.

That in fact it was extremely refined is proved by the huge effort the Americans invested in operation Magic. The hero of the operation was William Friedman, an assistant to the official in charge of the Communications Department. Friedman, universally considered to be the greatest cryptanalyst of all time, was aware of the enormous difficulty of the undertaking as soon as he set out to work on Purple. For months upon months not the tiniest crack could be found. The solution was finally found in August 1940, after almost 20 months of the most intense analysis ever applied to the cracking of a code. As Kahn quite fittingly shows in his book, the solution came at a terrible price. The continuous pressure on his mind as it was engaged in this gigantic brainteaser, which he could not forget during meals or sleep, giving him insomnia or sudden heart palpitations at night, together with the pressure to succeed and the awareness that a failure would bring serious consequences to the entire nation; the desperation lasting for endless weeks when no solution was in sight, and then the constant frustrations that followed the rare moments of hope; the mental shocks, tension, fatigue, and the need for absolute secrecy—all this was pounding in Friedman's brain. After the solution was found in August, he had a nervous collapse in December, and had to spend three and a half months recovering in Walter Reed Hospital, before he was allowed to return to work with a reduced schedule.

But Friedman's really prodigious performance was what led to the "Day of Magic," December 7, 1941. On that day, at 1:28 A.M., the Japanese sent out a message in code to their ambassador in Washington that would be intercepted by the American navy at Bainbridge Island, near Seattle. The message was decoded and handed to the SIS (Signal Intelligence Service), the section of the army responsible for communi-

QUANTUM CRYPTOGRAPHY

cations. The translation from Japanese was done by 7 A.M. The Japanese language expert Lt. Alwin D. Kramer, who at the time had been engaged in translating a whole series of messages, immediately grasped the enormous importance of this one, containing the order to notify the American Secretary of State at a precise (and very unusual) time about the dispatches he had received affirming that the Japanese government believed it was impossible to reach an accord and that it intended to interrupt all negotiations immediately. Kramer grasped the significance of the fact that the Japanese wanted to communicate a break of the treaty at a precise time, and he went immediately into action. The message was sent to the highest levels of government. I do not plan to dwell on the details of this affair nor explain why (despite the suspicions in government circles) the American fleet at anchor in Pearl Harbor was not notified in time of the danger. But I would like to recall one of the greatest passages from Kahn's pages: "This moment, with Kramer running through the empty streets of Washington bearing his crucial intercept, an hour before sleepy code clerks at the Japanese embassy had even deciphered it and an hour before the Japanese planes roared off the carrier flight decks on their treacherous mission, is perhaps the finest hour in the history of cryptology."

But I must now leave this dramatic story that has taken us so far from our immediate purposes. It was useful as a way of illustrating Poe's dictum: the violation of Purple certainly shows that the human mind can reconstruct any processes dreamed up by another human mind, no matter how contorted and complex.

Such considerations offer us the occasion to anticipate the "trick" that lies behind quantum cryptography. For, whereas a human mind can almost miraculously penetrate the mysteries of another mind, no human intelligence can grasp the secrets of what we are studying: microsystems. In fact, as we have already seen, the titanic powers of scientists during the first half of the twentieth century not only permitted us to grasp unsuspected aspects of reality, they also brought us to the conclusion that it is absolutely impossible to predict the precise outcome of a quantum process when it can have diverse outcomes. As we shall see, in quantum cryptography the key role is played, once again, by the genuine randomness of natural processes. This characteristic aspect of microscopic processes constitutes the fundamental element in them that permits the construction of inviolable codes, allowing us to refute Poe's assertion quite precisely, that human ingenuity can always decode what human ingenuity puts into code.

CHAPTER TWELVE

12.1. Cryptography: Terminology and Basic Procedures

Before we go any further, it would be wise to clarify a few things about the secret transmission of information, and to set out a precise terminology for the various elements and processes that come into play. The problem we are concerned with, of course, is transmitting a message to a legitimate addressee in such a way that only the addressee can understand it. The text to be transmitted, which is known as the *plaintext*, is subjected to a process of *encryption* which transforms the text into a coded text, or *cryptogram*. An essential element of the enciphering process to which we will soon direct our attention is the *key*, the knowledge of which allows someone to transform the cryptogram back into the plaintext. The sender of the message must send his message to the addressee by some transmission channel, such as telephone, radio, or FAX. Of course the sender will be concerned that the channel is secure, but normally, especially when it is a question of long or frequent messages, security for the channel of communication, called the *ordinary channel*, can end up being very expensive, difficult, or just impracticable. To get to the root of the problem, then, one must assume that the message will be *intercepted* by a spy who will try to *cryptanalyze* it. This term indicates the operation of working one's way from the cryptogram back to the plaintext without the use of a key. The legitimate addressee must also move from the cryptogram back to the text, but this process is called *deciphering* in order to emphasize the difference between this process and what the spy attempts. The addressee has the key for decoding the message: his operation is very well defined, and in some cases it can be done very simply (all the more easily today, with the help of computers).

Obviously, should the spy know the enciphering method (we will be looking at some simple examples) and be in possession of the key, he will be able to decipher the message exactly as if he were the legitimate addressee. For reasons we cannot now analyze in any detail, it is also a good idea to assume (and this is the unanimous assumption of everyone involved in modern cryptography) that the type of code being used will itself be accessible to the spy[1] (of course it will not actually be known, but the practice of cryptanalysis has shown that it is relatively simple for an experienced spy—especially one properly equipped, that is, with a series of messages written in the same code—to figure out the system being used). Nevertheless, the idea behind the modern theory of encoded transmission is that the real locus of strength in a code is the key: whoever knows the key can translate the message; whoever does not have the key must encounter almost insuperable difficulties in trying to decode it. Naturally, it is necessary for both the sender and the addressee to have the key, and that means to have access as well to an absolutely secure channel of communication for sending the key.

Now, almost anyone who has followed me so far in this discussion of the correct terminology of coding may react with an objection, which although perfectly understandable, is not justified: if the sender and the addressee have a secure channel of communication for the key, why not use that channel for the message and dispense with coding altogether? The answer is easy: a large amount of information will be passing by the ordinary channel, whereas the key is relatively brief, and the channel need be used only for a very short period for transmitting a key. It is much easier to guarantee the security of a channel used for this purpose than for channels used for longer periods, or for a "flood" of information. The reader will immediately understand that there is an extreme but effective method for guaranteeing the secrecy of the secured channel: the sender meets the addressee and gives him the key that will then be used to decipher a relatively large mass of enciphered messages. Figure 12.1 shows the essential elements of the process we are describing in schematic form.

I can conclude this section with a final point: *cryptography* is the constructive part of the process of transmitting secret messages, and this is what the sender does; the destructive part, called *cryptanalysis,* is what the spy does, as he attempts to retrace the steps from the cryptogram back to the plaintext. Obviously, there would be no sense in discussing the problem of cryptography without inquiring into the methods of cryptanalysis, since the primary purpose of the former phase is precisely to elude the attempts of the latter to undo it. Both disciplines together (in reality they are two sides of the same coin) are referred to as *cryptology.*

12.2. Simple Codes: A Few Examples

The simplest codes are codes of substitution. The idea is very simple: a permutation is made of the letters of the alphabet (e.g., the English alphabet with its 26 letters) and a different letter is substituted for each letter of the plaintext:

A B C D E F G H I J K L M N O P Q R S T U V W X Y Z

N D J R Z Q K C B H M S P X A G I F O W T E L U Y V (12.1)

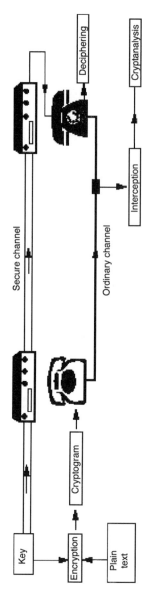

FIGURE 12.1. Schematic view of a secret transmission system between the sender and the receiver. It is expressly taken into account that a spy might intercept the cryptogram along the ordinary transmission channel and will attempt to decipher it. In the upper part of the figure is shown the secure channel used for transmitting the key.

A would be replaced by N, B by D, C by J, and so on. The two lines of letters given above would constitute the key. The sender uses the key by going down, the receiver uses it by going up. The simplest example of a code of this type of historical interest was the so-called code of Caesar, by which every letter is substituted by the letter that appears three places later in the alphabet (when you reach the end of the alphabet, you start again at the beginning). Thus A becomes D, B becomes E, . . .Y becomes B, Z becomes C (of course Caesar did not use the letters K, W, or Y of the English alphabet).

Two observations of importance can be made here. First of all, it must be realized that the number of possible keys is enormous: the permutations of the 26 letters of the alphabet are 4×10^{26} — that is almost a billion billion billion! When it comes to deciphering, if you chose an exhaustive approach by trying out all the possibilities one after another, you would be undertaking a rather difficult enterprise indeed. But codes like this are actually decipherable in a very simple way, thanks to a fundamental characteristic of every language, namely, the differing relative frequencies of each letter. In Italian, for example, the most frequent letters are A,E,I,O, in that order, and appear with about a 10% frequency, while B has a 1% frequency and Q only 0.6%. For example, if the permutation given in (12.1) were chosen, you would have a message where I would hardly ever appear while N, Z, B, and A appeared quite frequently. More refined statistical methods, such as those taking account of diphthongs (in Italian Q is always followed by U), will in a short time succeed in cryptanalyzing a code based on simple substitution.

But now we can try something very clever: let us take, for example, 50 different permutations of the alphabet, and number them 1 through 50; then we write a series of numbers from 1 to 50 in any order. Now, the instructions are as follows: if the first number of the sequence is 5, the first letter will be substituted by using the fifth permutation; if the second number is 41, the second letter will be substituted using the letter that corresponds to it in the forty-first alphabet, and so on. It should be clear to the reader that in this way the frequency with which the various letters of the message occur is disguised: all the letters appear with an almost equal frequency, and that will be too much for a simple statistical approach. But it can be shown that even refined methods like this can be easily broken through with sophisticated cryptanalysis techniques, especially when powerful calculators can be used to explore a large number of possibilities in a short time. For reasons of space I cannot illustrate the most refined techniques of this kind elaborated by cryptographers, nor

CHAPTER TWELVE

describe even in general terms the incredible feats of cryptanalysts in deciphering secret messages.

For our purposes it will be convenient to concentrate on the analysis of a cryptographic procedure that can be proven to be indecipherable or inviolable, but for reasons that I will soon explain, is very difficult to realize within the limits of traditional cryptography. But before we take this up, we need to devote a short section to *binary numeration*.

12.3. THE BINARY SYSTEM OF COUNTING

Any counting system uses a certain collection of various symbols to represent whole numbers, and these symbols define the so-called base of the system. For example, we use the base ten, since we use ten different symbols, that is, the numbers from 0 to 9. To count collections that contain more than nine items we give a special meaning to the positions of the digits. According to this system, the last digit has its own value, but the one preceding it must be thought of as being multiplied by the first power of the base (in our case, that would be 10^1, or 10), and then the digit preceding this one is thought of as multiplied by the second power (10^2, or 100) and so on. To illustrate this, consider the number 3,753:

$$3{,}753 = 3 \times 10^3 + 7 \times 10^2 + 5 \times 10^1 + 3 \times 10^0, \qquad (12.2)$$

where the zeroth power of any number is 1 ($10^0 = 1$).

Other counting systems have been adopted. For example, if base 6 is used (the only digits would then be 1, 2, 3, 4, 5, and 0), instead of 6 one would write 10, instead of 7 one would write 11, and so on. As Jorge Luis Borges observed in *Otras Inquisitiones:*

> Theoretically, the systems of counting are limitless. The most complex one (used by God and the Angels) would have an infinite number of symbols, a different one for every whole number; the simplest one would only require two: zero for zero, then 1 for 1, then 10 for 2, 11 for 3, 100 for 4, 101 for 5, 110 for 6, 111 for 7, 1000 for 8, and so on. . . . . [T]his was Leibniz's invention, who was intrigued (apparently) by the curious hexagrams of the I Ching.

The binary system Borges mentions here is the one used in computers, for reasons that will soon be made clear. One could say, in Borges' style, that these machines represent the "opposite of the angelic minds." At any rate, the reason why the binary system is so appropriate for calculations made by mechanical or electronic instruments comes from the fact that

QUANTUM CRYPTOGRAPHY

the simplest way to make an elementary constituent or "chip" for storing information is to create a system that can find itself in two different states only. For example, a simple switch can be closed (0) or open (1). The chips in our computers can allow the input current to pass or not pass. Of course, chips of this kind are much easier to make than ones that can have several different states. It is also clear that the gain in one direction is compensated for by a loss in another: as we have seen, to write the number 8, instead of using one chip with 8 or more different positions, we need 4 binary chips (8 = 1000, or the first one open and the other three closed). But the balance of advantage over disadvantage from a practical perspective is decisively in favor of the binary system.

Lets take a brief look at how we go from a binary number to its corresponding number in decimal notation, and vice versa. We have the number 1001101. As explained above, this would be understood in the following manner:

$$1001101 = 1 \times 2^6 + 0 \times 2^5 + 0 \times 2^4 + 1 \times 2^3 + 1 \times 2^2 + 0 \times 2^1 + 1 \times 2^0$$
$$= 64 + 0 + 0 + 8 + 4 + 0 + 1 = 77. \quad (12.3)$$

Analogously, if we were to write the number 37 in binary notation, we would have to find the greatest power of two that is less than 37, which in this case would be $2 \times 2 \times 2 \times 2 \times 2$, or $2^5 = 32$. Subtracting this number from 37 gives us 5, and we then repeat the operation, finding $2^2 = 4$ as the greatest power of 2 that is less than five, and we then have a 1 left over. That means that

$$1 \times 2^5 + 0 \times 2^4 + 0 \times 2^3 + 1 \times 2^2 + 0 \times 2^1 + 1 \times 2^0 = 32 + 4 + 1 = 37. \quad (12.4)$$

Therefore, 37 in binary notation is 100101.

I will not linger any more to discuss this type of counting but will only show how to do addition: since (in decimal notation) $0 + 0 = 0$, $0 + 1 = 1 + 0 = 1$, $1 + 1 = 2$, the corresponding operations in the binary system are $0 + 0 = 0$, $0 + 1 = 1 + 0 = 1$, $1 + 1 = 10$. Thus, when we add two numbers and find we have two 1s in the same column, we write a 0 and carry over a 1, exactly as with decimal numbers when the sum of the numbers is more than 9. But it can also help—and here is the last bit of information relevant to our purposes—to adopt the principle of *addition without carrying*, which we indicate as $\oplus$. In this instance we simply add the two numbers in a column, using the rules

$$0 \oplus 0 = 0;\ 0 \oplus 1 = 1 \oplus 0 = 1;\ 1 \oplus 1 = 0. \quad (12.5)$$

CHAPTER TWELVE

The advantage of this convention comes from the fact that subtraction coincides with addition, which is helpful for simplifying the procedure of encoding and decoding. The assertion just made can be verified with reference, for example, to the following elementary case:

$$\begin{array}{cc} 1001101+ & 1100110+ \\ \underline{0101011=} & \underline{0101011=} \\ 1100110 & 1001101 \end{array} \qquad (12.6)$$

12.4. The ASCII Code

For the reasons just given, and because increasingly massive data transmissions, even when nonsecretive, were being made electronically, it made sense to transmit even plaintexts as numbers, using a binary notation for the alphabet. Instead of simply associating every letter of the alphabet with a number (between 1 and 26) that would identify its position in the alphabetic order, the letter would be associated with a number in binary notation, a "string" or sequence of 1's and 0's. Various codes are available for this purpose, but one of the most well known and frequently used is the ASCII code (American Standard Code for Information Interchange) (table 12.1).

It is worthwhile to emphasize that this code has nothing to do with the problem of cryptography, but was developed simply because calculators operate most efficiently in the binary system. Quite analogous to this is the Morse code, which provides every letter of the alphabet with a corresponding expression made up of a combination of two different symbols (dot and dash), a method that has proven very useful for certain types of communication.

Notice how every sign of the ASCII table corresponds to a binary number of 8 digits. A message would then be a shorter or longer string of digits 0 and 1 (but always consisting of a multiple of 8) that would have to be divided into groups of eight to be decoded. Consider the following string:

0100100100100000011011000110111101110110011001010010000001111001011011110111 0101

which, once divided into a series of eight-digit strings, becomes

01001001	00100000	01101100	01101111	01110110	01100101
I	*(space)*	*l*	*o*	*v*	*e*
00100000	01111101	01101111	01110101		
(space)	*y*	*o*	*u*		

QUANTUM CRYPTOGRAPHY

TABLE 12.1.
The short form of the ASCII code. The abbreviations SP, CP, and NL stand for "space," "carriage return," and "new line" respectively.

A	01000001	i	01101001	+	00101011
B	01000010	j	01101010	–	00101101
C	01000011	k	01101011	=	00111101
D	01000100	l	01101100	(	00101000
E	01000101	m	01101101	)	00101001
F	01000110	n	01101110	"	00100010
G	01000111	o	01101111	/	00100000
H	01001000	p	01110000	SP	00100000
I	01001001	q	01110001	CR	00001101
J	01001010	r	01110010	NL	00001010
K	01001011	s	01110011		
L	01001100	t	01110100		
M	01001101	u	01110101		
N	01001110	v	01110110		
O	01001111	w	01110111		
P	01010000	x	01111000		
Q	01010001	y	01111101		
R	01010010	z	01111010		
S	01010011				
T	01010100	0	00110000		
U	01010101	1	00110001		
V	01010110	2	00110010		
W	01010111	3	00110011		
X	01011000	4	00110100		
Y	01011001	5	00110101		
Z	01011010	6	00110111		
		7	00110111		
a	01100001	8	00111000		
b	01100010	9	00111001		
c	01100011	.	00101110		
d	01100100	,	00101100		
e	01100101	:	00111010		
f	01100110	;	00111011		
g	01100111	!	00100001		
h	01101000	?	00111111		

303

CHAPTER TWELVE

We now proceed to present a theorem of information theory of great conceptual relevance. This is the theorem that will suggest how a perfect code can be realized by using the genuine stochasticity of quantum processes.

12.5. The Search for Perfection

I would like to convince the reader of an extremely significant fact, namely, that if two persons who wish to communicate secretly are both (and they alone) in possession of the same random sequence of numbers (which will represent the key in this system), then they are in a position to send inviolable messages. To avoid encumbering the text now with long strings of 1's and 0's, let us use the very simple code that numbers each letter of the alphabet according to its natural position in the alphabet (every letter will uniformly have two digits, so that one through nine will be written 01, 02, 03, etc.). We will also ignore the distinction between upper and lower case. The letter a will be represented by 01, b by 02, and so on, until we come to z as 26. The message above, "I love you," would be 0912152205251521. Now let us suppose that the sender or the receiver has been given a random series of numbers (each of two digits and less than 74) that is longer than the message to be sent, for example, 06, 34, 72, 12, 03, 10, 34, 66, 71, 09, ... Alice, the sender, will take the first five numbers of the series and add them to the numbers of her plaintext, and the cryptogram is the result:

09 + 06 = 15; 12 + 34 = 46; 15 + 72 = 87; 22 + 12 = 34; 05 + 03 = 08, etc.

The message to be sent along the ordinary channel would be 1546873408354987. The legitimate addressee of the message (Bob) has only to use the key at his disposal, that is, the random succession of numbers 06, 34, 72, 12, 03, 10, 34, 66, 71, 09, and then subtract them from the numbers sent in the coded message, giving 09, 12, 15, etc.: "I love you."

The crucial element in this process is the fact that it is easy to prove (as the reader will readily intuit) that knowledge of the coded message gives no information at all about the plaintext. To express this fact in precise terms, I will refer to a theorem developed by Claude Shannon, the father of information theory. Let us suppose that the plaintext is constituted by the sequence $(m_1, m_2, m_3, \ldots, m_n)$ and we indicate the sequence of the

QUANTUM CRYPTOGRAPHY

first n random numbers as $(k_1, k_2, k_3, \ldots, k_n)$ which is the key that both Alice and Bob have, and finally the sequence of numbers $(c_1, c_2, c_3, \ldots, c_n)$ which is the cryptogram itself. The theorem states that knowledge of the cryptogram does not permit us to attribute a greater probability to the sequence of n numbers constituting the message than to any other sequence of n numbers. Stated more simply, it is just as difficult to figure out the message from the cryptogram as it is to figure it out without having the cryptogram to work with: all the labors of the spy to intercept the message will be in vain.

Before concluding, some important observations need to be made. First, in order for the theorem to be valid, the sequence of numbers $(k_1, k_2, k_3, \ldots, k_n)$ must be genuinely random. This gives rise to an extremely relevant problem to which we will turn in the chapter on quantum computers, namely, the generation of truly random sequences of numbers. The second observation is that the numbers of the random sequence have to be at least as great as the numbers of the message, which poses a difficult problem to traditional cryptography, since, in order for the key to be effectively transmitted, we need to guarantee the security of a channel large enough to deliver the message itself. And so the objection above becomes pertinent: Why not use the same channel to communicate the message too? This difficulty is made even more serious with the observation that (as the reader will easily see) after a number has been used once, it is necessary to discard it: if the same sequence is used for more than one message, the cryptogram could be subjected to the same statistical analysis discussed above (the relative frequency of certain numbers pointing to more frequently used letters, etc.) which are capable of undoing the most refined cryptographical techniques.

It is interesting to know that it is likely that during the Cold War the secrecy of the line of communication between Washington and Moscow was based on a code of this kind. The two legitimate users of the code would meet, and would throw a coin a certain number of times, obtaining a string of numbers in base 2 (0 for heads, 1 for tails) that would be kept on file; larger or smaller portions of the string would be used as needed for the messages, depending on how long they were, and would be discarded after being used.

We are now ready to show how easy it would be to use the peculiar properties of microsystems to realize a perfect cryptographic system, or rather, a system that could not be violated without the legitimate users becoming aware of the fact that an attempt to violate it has been performed.

CHAPTER TWELVE

12.6. Using Photons for an Unbreakable Cryptographic System

To show how a system of coding that uses precise quantum effects could be put into operation, we consider the familiar source that regularly produces—let us say once per second—a pair of photons that propagate themselves in opposite directions toward the sender (Alice) and the receiver (Bob), and which will allow them to share a random sequence of binary numbers (this will be the key to decipher the message). The state of the pair of photons will be supposed to be that of equation (8.10) which we will rewrite here in a form that will be useful for our present purpose:

$$|\Psi\rangle = \frac{1}{\sqrt{2}}\left[|1,V\rangle|2,V\rangle + |1,H\rangle|2,H\rangle\right] \\ = \frac{1}{\sqrt{2}}\left[|1,45\rangle|2,45\rangle + |1,135\rangle|2,135\rangle\right]. \quad (12.7)$$

Alice and Bob have agreed beforehand to carry out plane polarization measurements along two chosen directions, namely, at the vertical and at 45°, but they have made no agreement whatsoever about the *succession* of the measurements they will carry out: each of them will choose completely randomly (by freely choosing or by flipping a coin) along which direction each individual measurement will be taken. We note that the choice of the two directions to be used can be made public (and we shall see how many other steps of the process also assume non-secret communication). Alice and Bob have also agreed to write the number 0 whenever their photon does not pass the test and 1 every time it does. Finally, for each individual measurement, they take note of the kind of test (i.e., vertical or at 45°) they decided to apply. The situation is illustrated in Figure 12.2 where the two filters depicted at either side next to Alice and Bob are there to show that in each single case one of the two filters is interposed between the arriving photon and the detector.

At the end of the process (that is, after having carried out a series of many measurements) Alice and Bob will each have a table such as shown in Figure 12.3, which specifies, for each single incident photon, what test it was subjected to and what the outcome of that test was. It will be helpful to point out some fundamental characteristics of the process:

QUANTUM CRYPTOGRAPHY

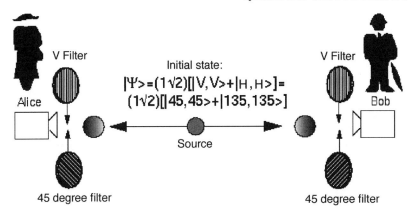

FIGURE 12.2. The experimental setup that Alice and Bob would use to get their random but identical succession of zeros and ones.

- The outcomes of each measurement are genuinely random and are distributed between the two possibilities with an equal probability
- Whenever the measurements were carried out for the same direction the two outcomes inevitably coincide.

At this stage, Alice and Bob make public the choices of direction they made for each measurement. Of course, an eavesdropper trying to get the key for their cryptographic messages is going to know, but only now for the first time, the pattern of their choices concerning the polarization direction of their measurements in each case.

But now a very important operation is going to take place: Alice and Bob eliminate from their tables the cases where they have chosen different directions (the ones shadowed in Figure 12.3b) and keep all the others in their original order. Of course, on the average, they will have chosen the same directions half the time; when the operation is complete, then, each of them will have at their disposal the table shown in Figure 12.3c, that is, they will have a string of 0's and 1's, of about one-half the length of the strings of 12.3a and 12.3b. Now their strings will have two very important characteristics: first, and above all, because they refer to the results of measurements along the same direction, they are perfectly identical (since, as we know, for measurements of this kind it is always true that both photons pass the test or that both of them are absorbed). Secondly, the distribution of 0's and 1's on the string is absolutely random.

At this point the reader will be able to conclude that the problem we posed has been resolved: Alice and Bob possess two random but identical

307

CHAPTER TWELVE

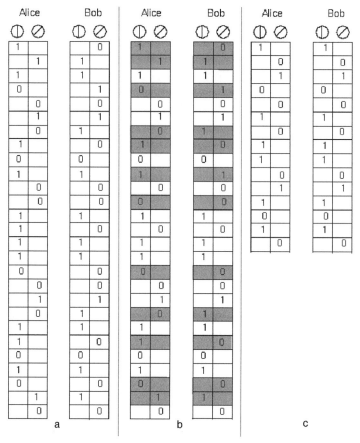

FIGURE 12.3. Alice and Bob's preliminary operations for obtaining a genuinely random succession. In (a) are shown the types of measurements they have carried out and the outcomes they have got (the symbol above the first half column shows a vertical measurement, the other shows a 45° one). In (b), after Alice and Bob have publicly declared the directions chosen for each individual measurement, the shaded cases are the ones where they have performed measurements different from each other. These are then eliminated, leaving column (c) which has the important qualities (random but identical outcomes) described in the text.

series of numbers of the binary system. But a crucial question remains to be asked: Wouldn't a spy also be able to get the same string? To guarantee security from such an eventuality (which would completely destroy their plans), Alice and Bob would have to exclude any would-be interceptors from having the same information about the succession of num-

QUANTUM CRYPTOGRAPHY

bers, and would likewise have to keep anyone from fixing the source (for example, by substituting, without their knowing it, a source that emitted three photons so that the results of the first two could be deduced from the third). For this purpose Alice and Bob now publicly announce the results they have obtained, for example, in the second, the fourth, the sixth measurements, and so on, as in Figure 12.3c, which we repeat in Figure 12.4.

This operation has no other purpose than that of verifying that Alice and Bob have identical results. But obviously, since these results are public information, they are no longer useful and must be eliminated. In Figure 12.4b we emphasize with shading the outcomes that were announced publicly, and in Figure 12.4c we have given the new string, composed of a subsuccession of the original but keeping all its characteristics, namely, the randomness and identical nature of the two strings.

The purpose of this last operation can be inspected more closely. The conceptually relevant point derives from the fact that the theory guarantees (and it is a theorem that admits a simple, rigorous demonstration)

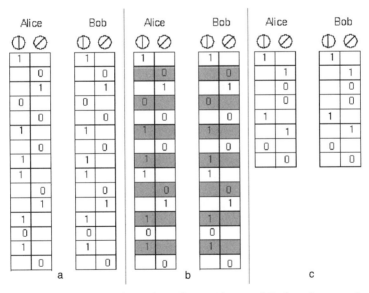

FIGURE 12.4. The procedure that allows Alice and Bob to know whether or not a spy has interfered with their process of agreeing on a random key. It requires making public and then eliminating a certain number of outcomes (the shaded ones) and thereby reducing the actual string at their disposal to (c).

309

CHAPTER TWELVE

that each interference with the original process, carried out with the intention of learning the series of outcomes or of fixing the source, brings inevitably a certain probability of destroying the perfect correlations of the outcomes obtained by Alice and Bob.

It would take too long to provide a rigorous demonstration of the theorem in the general case, but it is possible—and in fact easy, relying on the familiarity that the reader now enjoys with the mechanisms in question—to make the theorem plausible, taking up the case of the spy trying to learn the outcomes that Bob, say, would obtain, by placing one or the other polarizing filter in the line of the light beam. Of course, we would have to limit our study of the effect of the intrusion to the cases where Alice and Bob carried out the same measurements, since the outcomes of the differing measurements were discarded. We know that if one photon passes the spy's test, immediately afterward the two photons of the pair will both be polarized in the direction chosen by the spy. There are two possibilities. Either for that individual instance both Alice and Bob choose the same direction as the spy, and their results will agree. Alternatively, Alice and Bob choose the other direction.

We suppose that the spy has carried out a measurement of vertical polarization and that Alice and Bob choose to carry out a measurement of 45° polarization. We ask ourselves: what is the probability that they will have the same results? Of course, this is equal to 1/2 because there are four equally probable outcomes: both photons pass, or both are absorbed, or the one on the right passes and the one on the left is absorbed, or vice versa. The argument does not depend in any way on the direction chosen between the two on which they agreed or on the result obtained by the spy.

Resuming our analysis, then, in one-half the cases the action of the spy has no effect and in the remaining half the results of Alice and Bob's measurements will be identical. This implies that, despite the intervention of the spy, the probability that Alice and Bob will obtain equal outcomes in a specific measurement is equal to 3/4. But if Alice and Bob announce the results, let us say, of N measurements (e.g., the even ones of their sequence) the probability that there will be no discordant outcome would be the Nth power of 3/4. So then, if N = 100, it would be $(3/4)^{100} \cong 10^{-13}$. This means that the event of 100 measurement outcomes being in agreement even in the presence of the spy's interference can be verified once in every 10,000 billion cases. Even if one takes into consideration possible attempts to control or fix the source, the probability of their 100 results being equal would be just as negligible. The conclusion should be

QUANTUM CRYPTOGRAPHY

obvious: the final operation of Alice and Bob guarantees beyond any reasonable doubt that no spy would be up to the task.

At this point it should be clear how the procedure works. Alice and Bob possess identical random sequences (the strings of Figure 12.4c) of 0's and 1's and they are certain that nobody else knows the sequences. Let us say the string looks like this:

10100101101101011001011010001011101100011101000111011001101010001100110100011001...

and that Alice wants to tell Bob that she loves him, without anyone being able to decipher the message. This is what happens:

1. Taking the string of 80 digits (0's and 1's) that meant "I love you" in ASCII (as at the end of section 12.4 above):

01001001001000000110110001101111011011001100101001000000111110101101111101110101

2. And taking the first 80 of the random series of digits above:

10100101101101011001011010001011101100011101000111011001101010001100110100011001

3. The two are added together (without any carrying over of the sums, of course):

01001001001000000110110001101111011011001100101001000000111110101101111101110101
10100101101101011001011010001011101100011101000111011001101010001100110100011001
11101100100101011111010110010011000111101010011111001110101011010001001101100

4. The last line is the cryptogram to be transmitted over a public channel. Bob then adds the numbers of the random series to the bottom line, without carrying over any sums, and winds up with the original message: I love you.

We can conclude this chapter with a few points of special emphasis. The attentive reader will have understood by now that it is precisely the unavoidably random nature of microscopic processes that permits, in principle, the practically secure transmission of secret information. Of course, the analysis showed that Alice and Bob can find out if anyone is spying on them but that they cannot prevent someone from effectively disturbing the channel they are using to obtain their identical and random strings. So there is still the problem of securing this channel, but there is also a noteworthy difference with respect to the "classical" method of transmitting the key. Whereas in the classical case, if the secure channel of communication is violated the sender and receiver cannot know about it, in the quantum case, they can securely identify any interception!

311

One final observation. The possibility of using systems of this kind in practice depends in a crucial way upon the possibility of maintaining perfect quantum correlations between two distant photons. This is not always easy, since, for example, the collision of a photon with an atom can change its polarization. It would not be easy to construct the ingenious machine we have described. However, it should be interesting to know that this is nevertheless no mere speculation, or just amusement for the reader. Not very long ago an experiment carried out on Lake Geneva using light guides showed that it is possible to keep the desired coherence between a photon pair for distances on the order of about 80 km. Apparently some banking institutions are thinking about installing networks like these that would connect, for example, their branches and their customers within a limited area, such as a city. The potential importance of such a network becomes obvious if we think, for example, of the problem of identifying a client with a certain code name or of transmitting an investment order that must be kept secret.

The example of inviolable cryptography analyzed in this chapter not only makes use of the genuinely random nature of measurements on microsystems; it also depends on the perfect correlations that are characteristic of an entangled state. It should be mentioned that there are also variants of the procedure just described that do not have recourse to entangled states but otherwise retain all the relevant aspects (i.e., the generation of a purely random key and the certainty of being able to identify any attempt at interception).

CHAPTER THIRTEEN

Quantum Computers

> At any rate, it seems that the laws of physics present no barrier to reducing the size of computers until bits are the size of atoms, and quantum behavior holds dominant sway.
> —*Richard Phillips Feynman*

IN THIS CHAPTER we will explore another field where quantum mechanics is in principle capable of bringing a degree of qualitative improvement unsuspected until recent years. I thought it necessary to add the qualification *in principle* because the problem we are going to examine is remarkably different from the subject of the past chapter. In the case of quantum cryptography, the principles involved are extremely simple (the outcomes of measurements on entangled states are random but perfectly correlated); the only technical difficulty arises in the need to keep the coherence between two microconstituents of the quantum system (such as the pair of photons that reach Alice and Bob) that play so important a role in the process. As we pointed out at the end of the last chapter, the modern technology of light guides has already allowed us to verify the feasibility of an instrument for providing the sender and the receiver with a random succession of numbers that forms the key for their system of communication. In the case we consider in this chapter, on the other hand, the practical feasibility of the necessary devices is yet to be tried, even if recent progress in the field of quantum optics and in other areas allows us to get a fleeting glimpse of the possibility of realizing this ambitious project within a few years. Therefore, the principal objective of these pages will be that of making plausible how the utilization of genuine quantum systems for the storage, transmission, and manipulation of information (the basic operations of computers) will make possible so great a degree of progress as to embody a veritable revolution in the field of information technology.

The idea of making a quantum computer was born in a rather natural way in the early 1980s. The challenge in those days was mainly the race toward the miniaturization of electronic components. The engineers were

CHAPTER THIRTEEN

trying to load as many elementary constituents as possible (logical gates) into the silicon chips that were the building blocks of integrated circuits. At that time they were aware that this process would inevitably lead them to constituents small enough to hold a handful of atoms. Now clearly this posed a problem. As everyone knows, when these levels are reached, the systems involved do not function as if they were governed by the laws of classical physics. At the microscopic level, matter obeys quantum laws that need to be taken seriously. What implications would this have? Making use of ingredients governed by the new laws would certainly represent a challenge, requiring a complete overhaul of "elementary circuits": would not the very idea clear the way to sudden and unsuspected developments in information technology?

Richard P. Feynman, the brilliant American scientist from the California Institute of Technology, was among the first to enter this unexplored world in 1984, and he succeeded in giving some precise indications about the way quantum systems could be used to carry out calculations. He brought into the open a problem of great conceptual relevance, that is, how the hypothetical quantum calculator would constitute an excellent simulator of quantum theory itself; in other words, in the course of its normal operation the computer would have to carry out some quantum experiments.

The important research which David Deutsch has been conducting for over ten years in this field should also be mentioned: it has brought to light how a quantum mechanical computer would be able to accomplish what no traditional computer could accomplish.

13.1. QUANTUM INFORMATION THEORY

Information theory is the mathematical theory that deals with the transmission, storage, and treatment of information. The most important contributions to this field came from Claude Shannon (with regard to the first and second aspects), while the problem of the elaboration of data was a very important object of study around 1935, especially through the investigations of the great mathematician Alan Mathison Turing. His analysis of the "machine" named after him (the ideal calculator) signified a revolution in the field of formal logic.[1]

Information is fundamentally physical in nature insofar as every elaboration of data requires a physical support. This assertion may appear obvious and of little importance, but it has significant implications when we contemplate the possibility of using supports that do not obey the laws of classical physics (the laws upon which until recent times all studies on calculators have been based). But recent years have seen such an explosion of theoretical research in this direction as to mark the birth of a really new discipline: a true quantum theory of information. Of course, we cannot fully enter into this fascinating but exquisitely technical field: I will limit myself to the analysis of a few aspects, and hint at some of the implications.

13.2. CLASSICAL BITS AND QUANTUM BITS

Information, then, is always physical and is quantified and incorporated in physical constituents. The fundamental element is the bit, which represents the unit of information equivalent to the choice of one of two possible alternatives. We will use the same terms as well to indicate the device in which information is stored. In fact, the expression "bit" is an abbreviation of *binary digit*, since, as explained in the previous chapter, the binary system is the most convenient one for electronic machines.

We begin with an analysis of the classical situation: information theory is in line with the intuitions of the common man. The greatest number of different messages that can be transmitted using an object that can be found in one of a possible total of N distinguishable states is equal to N. As we noticed before, the case that interests us is that of a system that can be found in only two states (with switch open or closed) which can be represented by means of two numbers, 0 and 1, of the binary system. Using an object like this it is not possible at the classical level to send more than two different messages. Alice will transmit a bit of information corresponding to the state (0) to communicate to Bob that they can meet tomorrow, or one corresponding to the state (1) to communicate to him that they cannot. Bob executes his measurement and, according to the outcome, recognizes Alice's reply without ambiguity.

If we consider a quantum mechanical system that can be found in two mutually exclusive and incompatible states (the most natural example is the familiar photon with vertical or horizontal polarization) it would seem only natural to think that by using such a system we could, just as before, transmit only two possible messages.[2] However, this is not the case, as I want to show in this chapter. To get at this problem we begin with a single bit (Figure 13.1). For obvious reasons, we will indicate its two possible states as (0) and (1) in the classical case and as $|0\rangle$ and $|1\rangle$ in the quantum case. In the figure the two states are represented by two

CHAPTER THIRTEEN

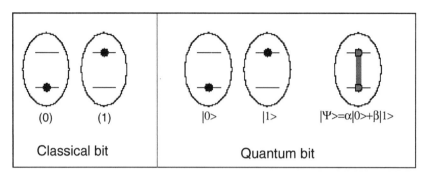

FIGURE 13.1. Schematic representation of the elementary device, or bit, used for information storage. While classically there are only the two possibilities (0) and (1), in quantum terms, if we consider two possible states for the system (|0> and |1>), the system can also be in any arbitrary superposition of the states. This greater wealth of possibilities opens up new ways of handling information.

short lines, and we can think of one of these as being occupied and the other as empty. In the quantum case, the situation is entirely analogous but we must remember that, thanks to the principle of superposition, if a system can be in one state $|0\rangle$ or the other $|1\rangle$, then it can also be in any one of their (normalized) linear combinations:

$$|Y\rangle = \alpha |0\rangle + \beta |1\rangle \qquad (13.1)$$

with $|\alpha|^2 + |\beta|^2 = 1$. The generic state (13.1) has been represented in the figure in such a way as not to make it definite which of the two states is occupied and which is empty (it is of course unnecessary to emphasize once again that in a state of this kind we are not in fact permitted to think that in fact one of the states is occupied or empty). For our purposes it will be advantageous to consider a particular case of (13.1), that is, the one in which two states enter in exactly the same way:

$$|\Psi\rangle = \frac{1}{\sqrt{2}}\left[|0\rangle + |1\rangle\right]. \qquad (13.2)$$

13.3. How to Realize and Prepare a Quantum Bit

The practical way to implement the program of having a quantum bit with two states, and to be able to prepare it in the state that is of interest to us for future applications, is to make use of what are called "quantum dots",

QUANTUM COMPUTERS

that is, quantum points, an expression used to indicate a single ion trapped in a "cage" of atoms (a possibility opened up by recent technology). When the single system that is in the state $|0\rangle$ of lowest energy is irradiated by a pulse of laser light at the right frequency for a precise interval of time, it will undergo a transition to the closest excited state $|1\rangle$. The process is particularly useful at the practical level because it can be "undone" at will. In fact, by repeating the operation just mentioned on a system in state $|1\rangle$ a spontaneous emission of electromagnetic radiation is induced that causes it to return[3] to the state $|0\rangle$. The two states are perfectly distinguishable by a measurement process and they play the roles of states $|0\rangle$ and $|1\rangle$ of the preceding formulas. The quantum dot is thus a quantum bit. But it presents a richer gamut of possibilities. In fact, if we illuminate the ion with the same light but for exactly half the period of time required to raise it from $|0\rangle \rightarrow |1\rangle$, the final state will be exactly the superposition (13.2). At the end of the chapter we will examine the practical difficulties encountered in the management of such a process; for now we will be content to know that, in principle, the goal we have set for ourselves is attainable.

13.4. Storage of Numbers by Means of Quantum Bits

In order to understand the advantages that quantum bits offer in comparison to their classical brothers, we begin by considering an extremely simple problem, that of storing a number in binary notation in a certain collection of bits. For simplicity's sake, we suppose we are interested in a number that requires six digits. To represent one of these, for example, 57 = 111001, we will need six bits. In the classical case, we would use all six bits and would put them into the states (1)(1)(1)(0)(0)(1), respectively. This is how we would store the information we want.

But now suppose we have 6 quantum bits to work with and we prepare all six in the state (13.2). What state do we get? We proceed to carry out the calculations and write down the states of the various bits in order. Evidently, we will have a linear combination $|\Pi\rangle$ of 64 states, obtained by taking different terms of each of the factors. I will not give the entire result but will indicate it with a few terms only:

$$|\Pi\rangle = \frac{1}{8}\{|000000\rangle + |000001\rangle + |000010\rangle + \cdots + |100000\rangle + |100001\rangle$$
$$+ \cdots + |111101\rangle + |111111\rangle\}$$

(13.3)

CHAPTER THIRTEEN

The reader will have already grasped the surprising fact that the state |Π⟩ of the system of 6 quantum bits "contains" not only the number 57 (which appears in the sequence as |111001⟩), but all the numbers from |000000⟩ to |111111⟩, that is, from 0 to 63. Of course, if we wanted to store precisely the number 57 we would have to subject the state (13.3) to the right manipulations (or, better, we could have avoided going to states of the type [13.2] from the beginning). Here, I wanted to show how the same number of bits that are needed to store a number of six digits at the classical level allows us, in a certain way, to have at our disposal the entire range of numbers with a maximum of six digits in binary notation. As we shall see, the state (13.3) will also play a very important role in the process of generating random numbers.

13.5. Quantum Signals: How to Compress Two Bits into One

It would almost be unnecessary to point out that one of the most effective uses of the quantum features of nature is once again based on taking advantage of the entanglement of a pair of microconstituents. In order to understand this we begin as usual with the state (8.10) of two entangled photons, which we will rewrite (in keeping with the notation of the present chapter) by substituting |0⟩ and |1⟩ for the |V⟩ and |H⟩ of that equation. For reasons of simplicity we will abandon the indices 1 and 2, and will indicate, first, the state of the photon nearest to Alice and second, the one nearest Bob. For our system, then, the situation we will have can be expressed as follows:

$$|\Psi\rangle = \frac{1}{\sqrt{2}}\big[|0,0\rangle + |1,1\rangle\big]. \tag{13.4}$$

We now proceed as follows. Alice keeps her photon stored, while the other photon reaches Bob. He can now subject it to one of the following four operations:

1. Allow it to keep its polarization state unchanged.
2. Rotate the polarization plane 90°, formally expressed as follows:

    $|0\rangle \rightarrow |1\rangle, |1\rangle \rightarrow |0\rangle.$

3. Leave the polarization state unchanged if it is |0⟩, but if it is |1⟩, introduce a delay of one-half wavelength, which can be expressed as

QUANTUM COMPUTERS

$$|0\rangle \to |0\rangle, |1\rangle \to -|1\rangle.$$

4. Rotate the polarization plane 90 if it is 0, but if it is 1, rotate the plane by the same amount but also introduce a delay of one-half wavelength, giving

$$|0\rangle \to |1\rangle, |1\rangle \to -|0\rangle.$$

It is important to emphasize that the operations now described are extremely easy to implement with simple apparatuses (sheets of crystal) well known from use in classical optics.

Once he has carried out one of these operations at his free choice, Bob sends his photon to Alice. Resorting to appropriate symbols, here is a table of the four possibilities,[4] in order, of the two photons Alice has at her disposal at the end of the experiment:

1. $|\Psi^+\rangle = \frac{1}{\sqrt{2}}[|0,0\rangle + |1,1\rangle]$ 2. $|\Phi^+\rangle = \frac{1}{\sqrt{2}}[|0,1\rangle + |1,0\rangle]$

3. $|\Psi^-\rangle = \frac{1}{\sqrt{2}}[|0,0\rangle - |1,1\rangle]$ 4. $|\Phi^-\rangle = \frac{1}{\sqrt{2}}[|0,1\rangle + |1,0\rangle]$

The fact of great importance here is that the four states are mutually exclusive and incompatible. This means that it is possible to construct an apparatus that replies in a precise way to the question: in which of the four states here listed is Alice's pair of photons? Clearly, Bob and Alice can be in agreement beforehand about the fact that Bob will carry out one of the four operations with respect to what he wants, among the four distinct messages, to send to Alice. And Alice, by a process of measurement, identifies without any ambiguity the message sent to her.

And this, then, is the compression of two bits into one: Alice and Bob, by exchanging a system (the polarized photon that is propagated toward Bob is being manipulated and sent back), which can be in one of only two mutually exclusive and incompatible states (this is the exact analogue of the classical bit), are able to send two bits of information, namely, four distinct messages.

13.6. Generating Series of Random Numbers

In the preceding chapter we have already analyzed the problem of the generation of truly random sets of numbers. This is a problem of general interest as well for classical information theory, which goes far beyond the uses of cryptographic applications. Let us outline in brief how

CHAPTER THIRTEEN

in practice and by making use of a classical "calculator" one series can realize successions of numbers such that we might call them *pseudorandom*.

The most widespread method makes use of what have been called *linear feedback shift registers*. The register is a collection of N bits or memory cells that at a certain instant represent an N-binary multiple, indicated as the state of the register at the instant under consideration. Of course, the possible different states of the register are 2^N, in so far as each of the two cells can be in one of the two states (0) and (1). At regular intervals of time the register undergoes a very simple process: the content of the register slides to the left one step, such that after the operation has taken place, each cell (or stage) contains the number that was first in the stage to its right. The content of the very first stage at the left exits from play, and the last one is "updated" according to the process of retroaction that follows: in that stage is placed the sum of the data that occupied certain precise stages (we can call them connected stages) before the process began. The sum, to be sure, is the binary sum without carrying. In Figure 13.2 are shown two successive phases of the process, in the case of a register with N = 6 and in which the connected stages have been identified by having arrows pointing away from them toward the summator. Of course these "connected" stages can be changed in a very simple manner: each one is connected to the summator by a switch that can be opened or closed.

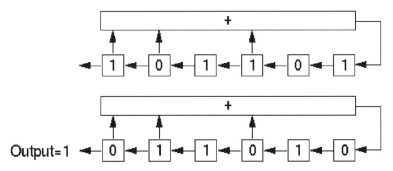

FIGURE 13.2. How a linear feedback shift register operates. Given an initial state of the register (in our case, 101101), the first operation consists in making all the numbers in their places move to the left (so that the first number becomes the first output), and next, filling the last place on the right with the sum (added without carrying) of the numbers which, before the operation, were given an arrow connecting them with the "summator."

We begin by entering N binary numbers (some of which are not zeros), and let the register do its work. We investigate the sequence of zeros and ones which "emerge" from the register to the left and we ask ourselves about its randomness. Clearly, the sequence cannot be random for two fundamental reasons: first of all, the procedure is perfectly deterministic, that is to say, the number that will emerge at a given stage of the process is perfectly and unambiguously determined by the structure of the register (the connected stages) and by the initial input of N numbers. Secondly, since there exist only 2^N distinct stages of the register, the procedure implies that after a certain period, which in the best case scenario will have the maximum length, i.e., it will involve $2^N - 1$ steps (the -1 that appears in this expression takes account of the fact that the register can never be found in the state in which all the stages contain a 0), the register will return to the initial situation and the process will start all over again exactly as before. We considered the problem of the periodicity of the states of the register. Now a question arises: are there registers with a maximum period? The clear answer is yes, and we have very precise mathematical methods for realizing it.

It would be worthwhile to analyze the nature of the randomness that characterizes the output series. It is clear above all that if the series must be a random series of 0's and 1's, then these two numbers should appear with a nearly equal frequency. But this is automatically guaranteed by the fact that the period is a maximum one: indeed, if the states of the register coincide step by step (of course not in the natural order) with all the binary numbers of N digits, that is to say, with all the whole numbers between (000 . . . 001) and (111 . . . 111), it would suffice to point out that if either the first or the last of these numbers is odd, then the total number of odd numbers between 1 and $2^N - 1$ numbers exceeds that of the even numbers in the same interval by 1. Now, in binary notation a number is even or odd according to whether its final digit is 0 or 1, respectively. But the final digits of the register's numbers eventually become the first digits, and thus emerge to form the output series. It follows from this that the number of zeros that come out will be less than the number of 1's only by one, and since, in the case of a register of maximum period with a large N, $2^N - 1$ is a very large number, it can be asserted that the equilibrium between 0 and 1 has been carefully respected. Analogous properties can be shown to hold good for pairs of 0 and 1, for threesomes, and so forth. In fact, the results that we have just now listed provide by themselves an indication that the randomness of the output is in a certain sense only *apparent*. In a genuinely random process, the probability that

321

CHAPTER THIRTEEN

over a very long period the 0's and 1's would appear (to all practical purposes) exactly the same number of times is very close to zero. For these reasons, our series can more appropriately be labeled *pseudorandom*. Nevertheless, for many practical purposes, as long as the period is very long, it can be assumed (and in fact it is assumed) that the output sequence can be considered as random.

At this point, nevertheless, a point of extreme conceptual relevance comes to the fore. Suppose we have a register and we consider the "pseudorandom" output sequence for one period, that is, of the first $2^N - 1$ numbers in binary notation. Now, in our innocence we could be brought to believe that this is the series that we must recognize in order to know how the register functions, or, in cryptological terms, that when Alice and Bob have control over the same register stored with identical numbers, this series represents the key that needs to be discovered in order to violate their secret communication system. But that would be a huge mistake! It would be a simple matter to demonstrate that, given a register with N stages, knowledge of a series of 2N consecutive outputs would allow anybody to determine immediately and without ambiguity all the register's subsequent outputs. To see this, we can reason as follows: We consider the first N outputs. These identify the state of the register at the moment we intercept them. The succeeding N outputs, once we know the initial state, allow us to deduce in a nonambiguous way what elements of the register are connected to the summator.[5] Once again, if we are willing to refer to the case of cryptography based on a random key, in the not unrealistic hypothesis that the spy succeeded in obtaining—let's say, by way of a double agent—both a cryptogram of 2N numbers and the corresponding plaintext (which is equivalent to knowing 2N consecutive outputs of the register), he would then be able to determine the succession of all the outputs with absolute precision, that is, the key of the code. Note, for example, that if N = 100, while 2N = 200, $2^{100} \approx 10^{30}$, that is to say, a succession of a hundred billion billion billion outputs!

Linear feedback shift registers are therefore reasonable systems for the generation of pseudorandom series of 0's and 1's for a variety of purposes (and in fact they are often used in practical scientific applications), but for cryptographical purposes their pseudorandomness (as opposed to the genuine randomness of a perfect key) makes them extremely vulnerable.

But to understand how—apart from cryptographical uses—quantum formalism allows the generation of truly random sequences of numbers, it should be enough to refer to the state (13.3). The reader will have noted how all the terms that appear in the state enter into the superposition with the same coefficient, that is, 1/8. Since the square of this coefficient represents the probability that a certain outcome is obtained in a process of measurement designed to identify in which of the 64 states the system will be found to be, this means that in such a process there is an equal and nonepistemic probability of obtaining any one of the 64 (states = binary numbers) that appear in equation (13.3). Note that the measurement process involved here is extremely simple: it amounts to measuring in which of the two states $|0\rangle$ or $|1\rangle$ each of the 6 quantum bits is found. After the measurement, by returning the system of 6 quantum bits to state (13.3) and subjecting it again to a measurement process, we will carry out a genuinely random choice of one of the 64 numbers between 0 and 63.

13.7. COMPUTATIONAL COMPLEXITY

Another theme of enormous importance in information theory is the computational complexity of certain problems. For our purposes we can define the concept very simply. Let us say we have a problem characterized by a certain number (N) of variables, and we take up the whole family of the same problems with different values for N. A famous example is that of the traveling salesman. The salesman has to visit N clients distributed in a certain region (the United States for example) and the problem consists in determining in what order he should make the rounds to all his customers in order to travel the least possible distance. We will shortly return to the solution of this problem, but first we must clarify the idea of computational complexity. Suppose we have a fixed N: we then ask ourselves, how many steps does a computer have to take in finding a solution?

There are many different degrees of complexity. For our present purpose we will consider only the so-called linear, polynomial, nonpolynomial, and exponential cases. We say that the problem has only linear complexity when the number of steps needed to solve the problem increases proportionally with N. Given P as this number, we will have P = kN, with k a positive constant that we need not specify any further. But we have polynomial complexity when the number P increases as a polynomial of N, for example, $P = aN^3 + bN^2 + c$. Since we are interested in very high values of N, it is clear that the first term of the polynomial now being considered will dominate the other terms. We can say that the

CHAPTER THIRTEEN

polynomial complexity is of the 3rd degree; in general, we say we have a problem of polynomial complexity r, if $P = kN^r$ represents the dominant term for large N's. If we do not have any of the foregoing cases, we can say that the computational complexity of the problem is nonpolynomial. Among this last group those known as having an "exponential" complexity are especially relevant. Without going into details we say that the number of necessary steps increases with the Nth power of a number — and for simplicity's sake we will suppose that this number is 10. This means that $P = 10^N$. In Table 13.1 I have set out how the number of steps necessary to solve the problem grows with N according to the various types of computational complexity.

Without dwelling on this problem I shall limit myself to citing two relevant examples. The problem of the traveling salesman presents a computational complexity of the exponential type, which means that already in the case (relatively common in practice) that he has to visit 50 cities, a computer would have to work out a number of steps equal to 1 followed by 50 zeros before it could find the shortest itinerary. Even for the most sophisticated and powerful computers, this computation is practically insoluble in a reasonable time.

We now pass on to consider a family of problems that present a very peculiar characteristic. We consider two operations that are the opposite of one another. It happens, in certain cases, that while it is relatively simple to carry out a direct operation (typically the relevant complexity is linear or at most polynomial), the inverse operation is extremely difficult, and constitutes a problem with a nonpolynomial computational complexity.

TABLE 13.1.
Showing how the number of steps needed to solve a problem grows with the increase of the number of parameters in the case of the various types of computational complexity of the problem in question.

Type	N =	1	10	100	1,000
Linear	6N =	6	60	600	6,000
Quadratic	N^2 =	1	100	10,000	1,000,000
Exponential	10^N =	10	10 billion	1 followed by 100 zeros	1 followed by 1,000 zeros

The best known and perhaps the most often studied problem of this type is that of factoring the product of two prime numbers. Of course the problem becomes interesting when the number of digits N of the product is very high.

The reader will readily be convinced that it is extremely simple to multiply two prime numbers even in the case when the two factors have a large number of digits, and that any calculator can carry out the operation in a relatively brief time. Nevertheless, the inverse operation—that of moving from the product to the two factors—is not all that simple. Even if it cannot be demonstrated that the inverse operation (the one we are interested in) presents us with what is effectively a nonpolynomial complexity, an algorithm has not as yet been developed that would make it easy to work out. The procedure to be followed is more or less the obvious one, that is, undertaking to divide the product by all the prime numbers, beginning with 2, until we reach a quotient with no remainder.

An example should make the problem clear, and explain the above assertions. We consider the product $127 \times 229 = 29,083$, and we are sure that if we do it on a piece of paper—probably no more than a minute would be needed to find the solution! To go from the product to the two factors, we would only have to divide 29,083 by all the prime numbers between 1 and 127. But there are thirty-four of them. In fact, leaving aside the division by 1 and 2 (which would leave us with 32 divisions), with pencil and paper the problem would really require more than an hour!

This peculiar feature of the problem has led to its adoption as the basis for the most widely diffused and commercialized cryptographic system. Since time does not allow us to enter into the details of the procedure, the idea is in essence the following: the individual who wants to receive coded messages announces what is called the public key of the system, a number of 129 digits (to make explicit reference to the case of the RSA, a commercial system developed by three researchers, Ron Rivest, Adi Shamir, and Leonard Adleman at the Massachusetts Institute of Technology in 1977), obtained by multiplying two prime numbers (both of which have many digits to begin with). The senders of the message follow a procedure of encryption that makes use of the public key. The system is conceived in such a way that the only direct and simple way (from the computational perspective, at least) to move from the coded message to the plaintext requires knowledge of the two prime factors. The addressee can then rapidly decipher the text while any spy would have to factor out the public key. It is important to note that we assume

CHAPTER THIRTEEN

here an almost extreme attitude toward the problem of deciphering, for not only is the public key known to everyone, but anyone also knows how to proceed to decipher the coded message, provided he knows the two factors of the public key: if they are known, no further message will be a problem for him. However—and here is where the original motive for adopting this system comes into play—if the operation (as well known as it may be) needed to decode the message (i.e., finding the factors) takes an extremely long period of time to carry out (let us say a thousand years, even for someone with a very powerful calculator), the users of the system will have no worries about any interceptions or decoding, since these would be impossible in practice.

When the three researchers mentioned above introduced their system, they challenged the scientific community to find the two factors of their number, known as RSA 129. Nobody had met the challenge until 1995, when 1,600 computers around the whole world, connected by the Internet, finally succeeded. The task was not easy, even with a calculating power of this order, since it took eight months to reach the solution. It is easy to show that if we use the algorithms presently in use for factorization, the problem presents exponential difficulties of computation. This means it would be enough to bring the number of 129 digits to 500 digits, for example, to be able practically to exclude the possibility that the code would be violated in a passage of time much longer than one would be using that particular cryptographic method or that specific public key.

But why have we taken so much time for the description of computational problems? What connection does this have with the possibility of using quantum computers? Here is a real "scoop" from the world of quantum information theory. In 1995, Peter Shor, a programmer at AT&T Bell Laboratories in New Jersey presented a list of "operations" that could be accomplished by using a quantum computer but not a classical one, and showed how these operations can be combined in an ingenious way to solve in a very short time precisely the problem of the factorization we have just been discussing. In Shor's example, the computer would require about 100,000 quantum dots working coherently (in the quantum sense) for a certain stretch of time, a system very far from any practical feasibility. But after Shor's work, Arthur Eckert and Adrian Barenco discovered a remarkable way to simplify the system. A computer that could preserve the needed coherence among 2,000 quantum dots of the type considered above could solve the problem of the RSA 129 in about 8 seconds! An algorithm that makes use of the possibilities of a quantum computer, and would likewise easily solve the problem of the traveling salesman, has not yet been worked out, although Shor hopes he will succeed in finding the solution.

13.8. COMPRESSION OF DATA

A preeminent object of research in the field of information theory is the problem of determining the maximum possible compression of data. With his famous theorem, Shannon himself provided an answer for the classical case in the ideal hypothesis of a non-noisy channel of information. To understand what we are talking about we can consider the following specific problem: let us say someone wants to rewrite *Moby Dick* using the binary alphabet (i.e., only 1 and 0) instead of the English alphabet. As we know, a reasonable method would be to codify every letter that appears in the novel by associating it with a binary number, and we end up with a very long string of 0's and 1's. Suppose we now want to "economize" as much as possible on the length of the coded text. Of course we would have to make some selections, such as, for example, codifying the most common letters like E—and also the spaces and the punctuation signs—with short numbers (of one or two digits), while giving longer numbers to letters Z and Q. In fact, a little analysis shows that further steps in compression can be made by choosing whole groups of letters that frequently appear to be represented with suitable binary numbers.

And here is where Shannon's theorem comes into the picture: he asserted that there is an impassable limit to compression. The theory is based on the idea of information entropy. We cannot spend a long time on this fascinating problem but will limit ourselves to reporting Shannon's conclusion, derived from very general hypotheses (that is, excluding the most peculiar cases): the limit to the compression of the novel is about 2.8 bits per letter.[6]

Shannon's theorem is based on classical information theory. In 1994, Benjamin Schumacher succeeded in showing that even this fundamental theorem, known as one of the milestones of information theory, has to be reconsidered when the methods of codification based on quantum mechanics become available. In particular, by resorting to such methods the limit to the number of bits (quantum bits, of course) necessary for the coding of the letters of a message is lower than the one just mentioned.

CHAPTER THIRTEEN

13.9. The Experimental Possibilities

Of course, any system of storage, transmission, and elaboration of data is subject to error, whether classical or quantum. But the situation that concerns us is particularly delicate thanks to one of the remarkable characteristics of quantum systems, namely, their entanglement. If a system of this kind (such as our collection of 2,000 quantum bits) interacts with some other system (typically, the environment) the one will inevitably be entangled with the other at the end of the process. Now, since in the quantum case the functioning of the entire system is strictly based on the quantum coherence among the various constituents, this process can be devastating. In fact, if there is a system in the superposition of two states that interacts with another system, giving rise to an entangled state, and if afterwards this second system is ignored (this is unavoidable when the system is so complex as the environment), then the coherence between the two original quantum states is irreversibly lost. It is like moving from a quantum superposition, the famous "+," which we have been studying, to a mixture, that is, to an "either . . . or," typical of classical situations. And correspondingly, the quantum computer would not be in any sense more efficient than its classical analogue.

The experimental challenge is located fully in the conflict between the time that the computer requires to carry out the calculation and the lack of coherence induced by the inevitable interactions with the environment. Profound studies in this field are now under way. Certainly, there exist elementary quantum circuits that operate at speeds of the order of milliseconds, such that it is possible for them to maintain the coherence between the involved states for one minute. But the field is still in too rapid an evolution to make any reasonable predictions. On the other hand, recent technological progress in cold lasers, thermic isolations, and quantum optics in general provide real hope for development in these fascinating new fields.

13.10. Quantum Teleportation

Before this chapter is concluded, it will be convenient to take advantage of the formulas we introduced in section 13.5 to illustrate another science fiction possibility offered by the quantum behavior of microsystems, the possibility known as the teleportation of a quantum state. This theme is of lively interest for the problems we have been considering, insofar as it

QUANTUM COMPUTERS

is one of the effects made use of by quantum computers, but, as the reader will understand, it has a very general conceptual interest. The idea can be formulated simply as follows: in the usual situation, in which Alice and Bob have at their disposal a common source of particle pairs in their entangled state (8.10), which we will write in the form (13.4), Alice, by performing a series of operations in her laboratory and transmitting a classical message to Bob (such as by a radio signal) can bring it about that Bob can have a situation identical to hers in his laboratory. We underline immediately three specific characteristics of the process:

1. Alice can "teleport" her own state to Bob even in the case where she does not know what state is in question (a gift in a closed box).
2. The teleportation process destroys Alice's system, so that, in the literal sense, a precise system disappears from a certain region and is "reconstructed" in a distant one (the message serves only to explain to Bob what he has to do to carry out the reconstruction).
3. The teleportation requires communication between Alice and Bob and thus cannot take place at speeds greater than light.

The idea is extremely simple. We suppose that Alice has a photon, in her laboratory, in a generic state that she herself does not know exactly but which we write in the form (13.1), which we can repeat here, associating the index T (for teleport) with this constituent:

$$|\Psi\rangle = \alpha |T,0\rangle + \beta |T,1\rangle. \tag{13.5}$$

We further suppose that Alice and Bob have the usual state (13.4) in which we highlight the two indices that characterize the constituent near Alice and the constituent near Bob:

$$|\Phi\rangle = \frac{1}{\sqrt{2}} \big[|A,0; B,0\rangle + |A,1; B,1\rangle\big]. \tag{13.6}$$

We thus have a system of three photons of which two (the ones denoted by T and A) are in Alice's laboratory and the third (B) is in Bob's. We write the state of the three photons by carrying out explicitly the product of the two preceding states:

$$|\Psi\rangle|\Phi\rangle = \frac{\alpha}{\sqrt{2}} \big[|T,0\rangle|A,0; B,0\rangle + |T,0\rangle|A,1; B1\rangle\big] \\ + \frac{\beta}{\sqrt{2}} \big[|T,1\rangle|A,0; B,0\rangle + |T,1\rangle|A,1; B1\rangle\big]. \tag{13.7}$$

We now rewrite the state, highlighting Alice's photon pair:

$$|\Psi\rangle|\Phi\rangle = \left[|T,0\rangle|A,0\rangle\right]\frac{\alpha}{\sqrt{2}}|B,0\rangle + \left[|T,0\rangle|A,1\rangle\right]\frac{\alpha}{\sqrt{2}}|B,1\rangle$$
$$+ \left[|T,1\rangle|A,0\rangle\right]\frac{\beta}{\sqrt{2}}|B,0\rangle + \left[|T,1\rangle|A,1\rangle\right]\frac{\beta}{\sqrt{2}}|B,1\rangle. \quad (13.8)$$

At this point we note that the states of Alice's pair are really the four states $|0,0\rangle$, $|0,1\rangle$, $|1,0\rangle$, $|1,1\rangle$ that appear on the right side of equations 1–4 appearing in section 13.5 above. The four relationships written there can be solved and allow us to express these four states in terms of the four states $|\Psi^+\rangle$, $|\Psi^-\rangle$, $|\Phi^+\rangle$, $|\Phi^-\rangle$, which, because they are mutually exclusive, are distinguishable by a process of measurement. If we follow out this operation we have

$$|\Psi\rangle|\Phi\rangle = \frac{1}{2}|\Psi^+\rangle\left[\alpha|B,0\rangle + \beta|B,1\rangle\right] + \frac{1}{2}|\Psi^-\rangle\left[\alpha|B,0\rangle - \beta|B,1\rangle\right]$$
$$+ \frac{1}{2}|\Phi^+\rangle\left[\alpha|B,1\rangle + \beta|B,0\rangle\right] + \frac{1}{2}|\Phi^-\rangle\left[\alpha|B,1\rangle + \beta|B,0\rangle\right] \quad (13.9)$$

We now proceed in this way: Alice performs a measurement designed to identify which of the four states she finds:

1. If she gets the result $|\Psi^+\rangle$, she sends Bob a message saying, "Don't do anything to your photon." Recalling that the postulate of the reduction of the wave packet implies that the state of the system as a whole is $|\Psi^+\rangle[\alpha|B,0\rangle + \beta|B,1\rangle]$, it follows that photon B is in the original state of Alice's photon.
2. If her outcome is $|\Phi^+\rangle$, she asks Bob to carry out what we designated as operation 2 in section 13.5, which transforms the state $[\alpha|B,1\rangle + \beta|B,0\rangle]$ into the state $[\alpha|B,0\rangle + \beta|B,1\rangle]$, which would once again be what Alice has.
3. If she gets $|\Psi^-\rangle$, she asks Bob to carry out operation 3.
4. If she gets $|\Phi^-\rangle$, she asks Bob to carry out operation 4.

As will be quickly verified, in every instance Bob has "transformed" the state of his photon into the state Alice's photon had at the beginning of the process: her photon has literally become teleported into Bob's space. We also note that Alice now has a definite state of two photons and has "lost" the photon she "gave" to Bob.

CHAPTER FOURTEEN

Systems of Identical Particles

> As far as we know today, one electron is identical to another electron, one proton to another proton, and so on. Think of a country where all men were monoovular twins and looked all alike. How can you talk of Peter and Paul if you cannot distinguish the one from the other and from all the other men? Of course, you can ask, "Are you Peter or Paul?" But two electrons would give you no answer! Nor can two electrons bear a red or a blue ribbon respectively on their arms.
> —*Giuliano Toraldo di Francia and Maria Luisa Dalla Chiara*

IN THIS CHAPTER I would like to clarify the special role played by the identity of microsystems within quantum formalism. The study of systems that contain identical constituents has enormous practical and conceptual significance: for example, it is really the identity of electrons that makes it possible to explain covalent chemical bonds (such as what makes two atoms of hydrogen capable of forming an H_2 molecule). In an analogous fashion, the regularity of atomic properties which gives rise to their ordered systematization in the periodic table of elements, and the fact that the statistical behavior of a complex of identical systems depends in a crucial way on their spin values, are both immediate consequences of this identity.

But aside from the extremely significant practical implications, the problem of the identity of microsystems presents a huge conceptual and epistemological challenge, quite naturally connected to the debate about Leibniz's doctrine of the individuality of what is indiscernible. We can sum up the position of this great thinker in his own words as follows: "Nowhere in nature can be found two entities so exactly alike that some inner difference cannot be found."

I will now illustrate, in the context of quantum formalism, the specific implications of the fact that, as far as we know (and, I would add, as far

CHAPTER FOURTEEN

as is shown by all the significant consequences of the hypothesis), every electron is identical to any other electron, every proton to every other proton, and so on.

14.1. IDENTIFICATION OF INDISCERNIBLES IN CLASSICAL AND QUANTUM PHYSICS

For simplicity's sake, we can consider the case of a physical system that comprises two elementary identical constituents, such as, for example, two particles of the same type, that is, two electrons or two protons, mesons, etc. First of all, we need to be precise about what we mean when we say that two objects of this kind must be considered identical. The phenomenology of elementary particles, as well as the experiments we can imagine carrying out on them, leads us to conclude that all the intrinsic physical properties that characterize them are exactly identical: all the electrons, protons, mesons, etc., of the entire universe have exactly the same charge, mass, spin, and every other quality. If I were to present you with a specific electron to investigate (admitting, as we will have to ask ourselves soon, that this assertion has some meaning) you would not be able to identify any physical process or carry out any measurement that would permit you to distinguish this electron from any other one, independently from its past "history," or from the processes in which it has been involved, or from the production mechanism that has generated it (this electron could have its origin in a process whereby a photon created an electron-positron coupling, or in the decay of a neutron into a proton, an electron, and a neutrino). Any two electrons whatsoever are absolutely and fundamentally indistinguishable at the individual level. Of course, other characteristics that even Leibniz took into account can be present, and can be used to differentiate these indistinguishable entities: for example, two identical particles can be differentiated by their position in space.

This problem also presents itself in a genuinely classical schema; and within such a conceptual framework, nothing in principle prohibits us from considering two systems to be absolutely identical. However (and here is something that will offer us the opportunity to discuss the very heart of the problem), since two identical material entities can be distinguished in such a scheme because they are in different spatial regions, we would naturally ask, is it possible to anchor the individuality of indiscernibles in spatiotemporal characteristics?

SYSTEMS OF IDENTICAL PARTICLES

Let us consider two identical classical particles with different positions at a certain instant of time, moving under the influence of a field of forces. At the beginning of the process there is nothing to distinguish the two entities except their position. Despite this fundamental fact, it is nevertheless possible to identify the particles; using the language of the quotation at the beginning of this chapter, we can decide to call one of them Peter and the other one Paul. We ask ourselves, is this attempt to distinguish the indistinguishable logically correct within the theoretical scheme under consideration? The answer is "yes." In fact, even if at any subsequent instant the mere observation of the physical characteristics of one or the other identical system does not in itself permit us to assert that this particle here is the one that we *arbitrarily decided to call Peter or Paul*, knowledge of the law of dynamics (Newton's equation) that governs the propagation of material bodies guarantees us that each of the two particles has followed a very precise trajectory that can (in principle) be determined without ambiguity. This fact makes legitimate and appropriate the assertion that the particle that is here is precisely the one we had decided to christen Peter (or Paul).

The reader will not have difficulty in understanding the radical difference that characterizes the corresponding quantum situation. The state of the system that interests us now will be described by a wave function $\Psi(r_1, r_2)$, that depends on the coordinates r_1 and r_2 of the two particles.[1] We suppose that the initial situation is ideal for our purposes (that is, such as to allow the individuation of the constituents, despite their identity), and this means that we suppose that the wave function in question is the product of the two functions $\phi_A(r_1)$ and $\chi_B(r_2)$ that are different from zero only in the two spatially separated regions A and B (see Figure 14.1) where the regions are distinguished as gray (A) and striped (B):

$$\Psi(r_1, r_2) = \phi_A(r_1) \chi_B(r_2), \tag{14.1}$$

and it will thus be legitimate to decide to designate as Peter (or Paul) the particle in region A (or B).

We now have to take into account the fact that, as we have already said, wave functions spread out, even in the case when the particles are not subjected to any field of forces and do not interact with each other, and that means that they are different from zero in regions that become increasingly more extended with the passage of time.[2]

Let us consider therefore the region C, representing the intersection of the regions "invaded" by the evolution of the original functions. In such

CHAPTER FOURTEEN

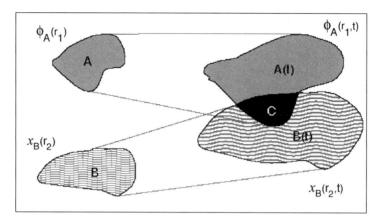

FIGURE 14.1. The wave functions corresponding to the two identical particles are initially different from zero in the separated regions A and B, but, by evolution in time, become superposed (in the figure $A_{(t)}$ and $B_{(t)}$ indicate the spatial regions where at time t the stated wave functions are different from zero). When in a process of measurement a counter reveals a particle in the shared region C, it is in practice impossible to determine, and in fact conceptually illegitimate to assert, that the located particle was the one originally in region A or originally in B.

a region there is a nonzero probability of finding a particle in a position measurement. It should now be clear to the reader that there would be no sense in trying to "reconstruct" the "trajectories" of the two particles, as we did in the classical situation of the two particles we dubbed Peter and Paul. In fact, because of the uncertainty principle which prohibits us from attributing to the particles position and velocity at the same time, the very concept of a trajectory loses all meaning in a quantum context. The sole legitimate assertion to be made in the present case is to say that there exists a certain probability of finding an electron in one or other spatial region, and, correspondingly, the sole legitimate operation is that of placing a detector in that region and ascertaining whether or not it registers the presence of one of the two particles: asking if this particle was originally in A or B, that is, whether it was Peter or Paul, is incompatible with the conceptual structure of the theory. As was fittingly stated in the opening quotation of this chapter, we cannot put any labels on elementary particles or "put a red or blue ribbon on them" to bring about their individuation.

14.2. IMMEDIATE IMPLICATIONS OF IDENTITY

The most logical attitude to assume in regard to this problem is to overhaul its formulation in some way. Once we have understood that it is impossible to individualize the indistinguishables, we are led to assuming the impossibility of such an operation itself as a criterion of identity: two elementary particles will be proclaimed identical when we must recognize that no physical procedure exists that permits us to distinguish them from each other, or in other words, when all the physical implications about the system in question are unchanged when we imagine "switching" the constituents. I do not want to take the time to explain how this problem is treated in a mathematically suitable way, but I will limit myself to stating some of its formal implications. Above all, the adoption of this point of view permits us to conclude that the observables relative to a system of two identical constituent systems must depend symmetrically on the constituents themselves: for example, for a system of this type it would not be legitimate to ask what the probability would be of finding the "initial" particle Peter (or Paul) in a certain infinitesimal volume, but only what the probability would be of finding one of the two particles in that volume. In turn, this formal fact has precise consequences for the wave functions of the composite system: they will have to have precise symmetry properties for the exchange of the two constituents. By this expression it is meant that the wave function $\Psi(r_1, r_2)$ will either have to stay unchanged (and in such a case will be called symmetrical) or will have to change its sign (and then will be called antisymmetrical) when all the "coordinates" of the two constituents are exchanged:

$$\Psi(r_1, r_2) = \pm \Psi(r_2, r_1). \tag{14.2}$$

We now can ask ourselves: what physical fact determines the symmetry of the wave function? Here, nature gives us a clear answer (usually indicated as the spin statistics connection): with respect to the exchange of all the variables (in general, therefore, we will have to exchange not only the spatial variables explicitly indicated in equation [14.2], but also the spin variables) the symmetry character is governed just by the spin value of the identical constituents. Systems of particles with integer spin such as photons, π mesons, etc. (called bosons because they obey Bose-Einstein statistics) have to be described by wave functions that are symmetrical under the exchange

CHAPTER FOURTEEN

$$\Psi_{boson}(1,2) = +\Psi_{boson}(2,1) \qquad (14.3)$$

where we have comprehensively indicated by the symbols 1,2 the totality of variables (spatial and spin) of the two constituents. On the other hand, particles with half-integer spin such as electrons, protons, etc. (called fermions because they obey Fermi-Dirac statistics) have to be described by wave functions that are antisymmetrical under the exchange:

$$\Psi_{fermion}(1,2) = -\Psi_{fermion}(2,1). \qquad (14.4)$$

14.3. Physical Consequences of Identity

The most direct and relevant consequence of identity in the case of particles with half-integer spin is known as Pauli's exclusion principle. This principle says simply that two such particles cannot have exactly the same spatial wave function and the same spin state. In fact, if we designate such functions as $\phi(r)$ and $\chi(s)$, the function of the system will be

$$\phi(r_1)\chi(s_1)\phi(r_2)\chi(s_2) \qquad (14.5)$$

which of course does not change (i.e., it is symmetrical) if we exchange r_1 and s_1 for r_2 and s_2.

I shall limit myself to pointing out a few important consequences of this fact. We can think of a system of many electrons in a certain field of force not dependent on spin and, in first approximation, we disregard the repulsion between the electrons caused by the fact that they have the same electric charge. We then ask ourselves in what states the various electrons would "locate" themselves if it weren't for Pauli's principle? Clearly, the system in question (like all systems in nature) tends to settle at the lowest state of energy possible. This state would have to be one where each electron is found in the lowest state of energy relative to the force field in which they are immersed (such as, for example, the force field of the atomic nucleus to which they are bonded). However, in the present case, this state of a single electron corresponds to a precise spatial wave function, which we can call $\Phi_{ground}(r)$. What will happen? The first electron will end up in the state we are considering now and will have a certain spin function—for example, one that corresponds to spin up along the z axis, which we will denote as $\alpha(s_1)$. The second electron can then be found in the same state $\Phi_{ground}(r)$, but will have to have a different spin state, in accordance with Pauli's principle, which we identify as

$\beta(s_2)$, or spin down along the z axis. At this point we consider the third electron and we ask, can this electron also be found in the state of lowest energy, that is, $\Phi_{ground}(r)$? Well, of course not, because then it would have to have a spin function different from those of the first two electrons; but there are only two possible spin functions and they are both "occupied" by the first two electrons.[3] The third electron will therefore have to have a different spatial wave function. The best one with respect to energy corresponds to the state closest to the fundamental one, that is, to the next excited state: to use Bohr's language, the third electron will locate itself on the second of the permitted orbits. This electron would not be able to fit on the first orbit as the other two can. Accordingly, we understand that the state of lowest energy of a system of fermions cannot correspond to the fact that all of them are individually in the state of the lowest possible energy, because a situation like that is incompatible with Pauli's principle. We also understand how this third electron is, in a precise sense, less closely bound to the nucleus than the other two. The argument as developed this far can be repeated with reference to the various elements that have different nuclear charges and therefore different numbers of electrons. The only relevant modification derives from the fact that the distinct spatial functions associated with a determinate energy level increase with increase of the number n that characterizes the different orbits. In particular, with the first level (n = 1) only one spatial wave function is associated, while the second has four. Since each of these has two possible states of spin, the second orbit can contain as many as eight electrons.

Allow me to take a few moments to show how this fact determines the chemical properties of the various elements, typically their electronic affinity (the propensity of an element to give away or capture electrons), and to show how this implies the regularity of the periodic chart of the elements.

Let us take a hydrogen atom. Its nucleus has a charge of 1 and one electron is orbiting around it. In the ground state this electron will be located on the orbit closest to the nucleus. Let us move to the second element, helium. The nucleus of a helium atom has a charge of 2, and two electrons are in the orbit closest to the nucleus (different, of course, from the one before because the charge is different). The two electrons of helium are equally bound and therefore this element has no special propensity to lose or gain a third electron. Atoms like this are chemically inert and are known as the noble gases. But the situation changes when we come to the next element, lithium, with a charge of 3, and three electrons in orbit

337

CHAPTER FOURTEEN

around its nucleus. As we described before, two of these will arrange themselves upon the orbit closest to the nucleus, while the third will necessarily be situated in the next orbit, further away from the nucleus and less closely bound to it. And this is the reason why lithium has a certain propensity to "lose" the unpaired electron, that is, the one less tightly bound. If we bring to mind what was mentioned above, that in the second orbit eight electrons can be found, we can pass from lithium to beryllium (which has *two* electrons less closely bound than its first two), and so on, until we reach fluorine. This element has a charge of 9, and in its fundamental state it will have two electrons in the first orbit and seven on the second. Since the second orbit can contain eight electrons, we understand why fluorine has a certain propensity for "capturing" an electron. We now make two further steps: we increase the nuclear charge by 1 and arrive at neon (another noble gas) with charge 10, and enough electrons to completely fill the first two orbits. The next element is sodium, with charge 11. How are its electrons arranged? Of course, ten will occupy the first two levels but there is no further possibility for the last electron to have a spatial wave function corresponding to the same energy and be different from those already occupied and corresponding to the first two levels. In simpler terms, the eleventh electron will become part of the system by occupying an orbit further from the nucleus, and consequently will be less closely bound to its companions, and will be more readily "lost" by the nucleus. Here then, is where the periodicity of the various elements becomes apparent: sodium has properties very similar to those of lithium, and the same goes for the rest.

We now have grasped the essential role of the exclusion principle in determining the regularity of the periodic chart and the electron affinity of the various elements. This affinity for losing or acquiring electrons in turn plays an important role in the formation of ionic bonds, those bonds which lead (for example) an atom of fluorine and an atom of sodium to combine to form a molecule of sodium fluoride: Na^+–Fl^-. But this bond is fundamentally "classical," since it is caused by the Coulomb attraction between opposite electrical charges.

But in nature there is also another, and very important, type of chemical bond: the covalent bond, and the most characteristic example of this is provided by the formation of a molecule from two equal atoms, such as the hydrogen molecule H_2. The role of the identity of constituents is absolutely essential for this type of bond, which finds its only explanation within quantum formalism as a direct consequence of the exclusion principle. Because of technicalities it would be neither convenient nor even possible to enter into the details of this very relevant problem, so I will limit myself to the essentials.

In order to determine the state of minimum energy of a hydrogen molecule, we can attempt to approach the problem in two ways. First, the calculations can be carried out without taking into account the exclusion principle, or better, without stipulating that the wave function of the system be antisymmetrical by the exchange of electrons. It turns out that the two atoms can be bound, but the energy of the bond is openly in contradiction with experimental data (the molecule would have a bond energy 1/10 of what is actually measured in the laboratory). If we then consider the same problem mathematically, but impose the requirement of antisymmetry to the wave function, we obtain a bonding energy in excellent agreement with experiment. The process can be described by saying that the requirement of antisymmetry generates the terms that contribute to the bond, and these terms derive from the possibility of "exchanging" the two electrons. These terms are comparable to the effect of particular forces without classical analogues, called exchange forces.

Before I conclude this section it is essential to point out some important implications of the identity of constituents for statistical mechanics, especially for the behavior of ensembles of many identical systems. To understand these significant effects it will suffice to think of a "gas" of bosons or fermions at temperatures very close to absolute zero. At low temperatures, of course, any system is at its lowest energy. While in the case of bosons this state is characterized by the fact that all the constituents end up in the lowest-energy state, for a system of fermions, as we have already discussed, this is not possible since it would imply a violation of Pauli's principle. It follows that these two types of gas have radically different behaviors at very low temperatures. Many readers have heard of the peculiar properties of helium at such temperatures, which is nothing else than the consequence of the fact that the helium atom, when considered (as it should be in this context) as an elementary system, has spin zero and is therefore a boson.

14.4. Identity and Entanglement

Let us consider a system of two electrons in a factorized state:

$$\Psi(1,2) = \Phi(1)\Xi(2). \qquad (14.6)$$

CHAPTER FOURTEEN

A state like this does not satisfy the requirement of antisymmetry for the exchange of the two electrons, which must be satisfied since they are identical particles of half-integer spin. This means that, instead of the state (14.6), we should consider the state

$$\Psi_A(1,2) = N[\Phi(1)\Xi(2) - \Phi(2)\Xi(1)], \qquad (14.7)$$

where N is a suitable constant of normalization.[4] As we already discussed in detail in chapter 8, the state (14.7) is an entangled state of two constituents, and usually the entanglement has some relevant conceptual consequences for the possibility of attributing definite physical properties to the constituents. Here, then, a very important conceptual question arises: since the identity of the elementary constituents of every physical system requires that the relative wave functions cannot be factorized,[5] can it be said that the identity by itself brings about all the peculiar consequences of entanglement, and thereby the impossibility of attributing just any property to the constituents of every physical system and in particular to the elementary building blocks of the universe?

The question is a legitimate one and has been powerfully expressed by the French physicist Jean Marc Levy-Leblond: "I would like to remind you that there is a universal correlation of the EPR type which we do not have to cleverly set up, and which we cannot avoid but which (and that is the question) we may very often forget; it is simply the total antisymmetrization of a many-fermion state, which correlates the electrons of my body with those of any inhabitant of the Andromeda galaxy." The observation is certainly very incisive and appropriate: the identity of the elementary constituents requires in practice that one cannot consider states which are factorized with respect to the elementary constituents of the universe such as electrons, protons, etc. However, at the same time, the affirmation is seriously misleading and thus requires some explanation. What has to be emphasized is that, in itself, the requirement of symmetry or antisymmetry of a composite system's state does not automatically require the peculiar consequences associated with entanglement. In order to understand this important fact, let us go back for a moment to a case of *distinguishable* constituents and remind ourselves of the difficulties we encountered when dealing with entangled states.

Consider, for example, the system of an electron and a proton. We indicate with r_e, r_p, s_e, and s_p, respectively, the spatial and spin variables of the two constituents. We suppose that the system is in a particular factorized state:

SYSTEMS OF IDENTICAL PARTICLES

$$\Psi(r_e, s_e, r_p, s_p) = \phi_A(r_e)\alpha(s_e)\chi_B(r_p)\beta(s_p), \tag{14.8}$$

where the states $\phi_A(r_e)$ land $\chi_B(r_p)$ are the ones considered at the beginning of the chapter (see equation [14.1]) which are different from zero only in the separated spatial regions A and B of the figure, and, as set forth above, the states $\alpha(s_e)$ and (sp) are the spin states corresponding to spin up and spin down along the z axis for the electron and proton, respectively. State (14.8) allows us to attribute definite properties to the two constituents. For this state we are allowed to say that there is an electron in region A and that this electron has spin up along the z axis, and that there is a proton in region B with spin down along the same direction. Now, let us say that, instead of state (14.8), we were to consider an entangled state such as the following (for reasons of simplicity we will suppose that only the spin variables are entangled in the manner of an EPR situation):

$$\Psi_{entangled}(r_e, s_e, r_p, s_p) = \frac{1}{\sqrt{2}} \phi_A(r_e)\chi_B(r_p)[\alpha(s_e)\beta(s_p) - \beta(s_e)\alpha(s_p)]. \tag{14.9}$$

Now, while it is still legitimate to assert that there is an electron in A and a proton in B, it is not possible to make any assertion about the spin properties of the two constituents. In fact, as in the EPR case, the probabilities, in a process of measurement along any arbitrary direction, of getting the outcome that the electron (or proton) has spin up (or spin down) are all equal to 1/2. This is the essence of entanglement: the loss (in general) of any property of the entangled constituents.

Suppose we now consider the analogous situation in the case of identical particles, say two electrons, and we abandon the specification of the variables e and p (for electron and proton) and substitute for them the numbers 1 and 2 (indicating the two electrons). We end up with the wave function:

$$\Psi(r_1, s_1, r_2, s_2) = \phi_A(r_1)\alpha(s_1)\chi_B(r_2)\beta(s_2). \tag{14.10}$$

This function is not acceptable for our system because it is not antisymmetrical for the exchange of the variables of the two particles. It can easily be made antisymmetrical by subtracting from it the same state—in which, however, the indices 1 and 2 have been exchanged—and normalizing it as follows:

$$\Psi_{antisymm}(r_1, s_1, r_2, s_2) = \frac{1}{\sqrt{2}}[\phi_A(r_1)\alpha(s_1)\chi_B(r_2)\beta(s_2) - \chi_B(r_1)\beta(s_1)\phi_A(r_2)\alpha(s_2)].$$

341

$$\tag{14.11}$$

Of course, the state is not factorized. However, we can ask ourselves: do properties exist that can be considered as objectively possessed by the system (that is, independently of our carrying out a measurement process on it)? It is not difficult to be convinced (so long as we use the expression [14.11] and keep in mind the rules of quantum formalism) that the state in question guarantees us that in a measurement intended to check whether there is a particle in A, there is a 1/2 probability of finding there the electron that we have named 1 and with spin up, and alternatively, a 1/2 probability of finding in A the electron we have named 2, again with spin up. Analogously, the state in question implies that in a process of measurement in B, one will *certainly* find an electron there and it will *certainly* have spin down. The conclusion is clear enough: the entangled state (14.11) is just as meaningful and sensible from the physical point of view as the state (14.10). The sole difference between the two comes from the fact that if the state were the one given by (14.10), we would be justified in saying that the electron in A is the one we have dubbed as 1. But the interesting physical fact is still true, namely, that the state guarantees that in A there is an electron with spin up and in B an electron with spin down.

This example should make clear to the reader the fact that entanglement induced by the pure and simple demand that the states of systems containing identical constituents have definite symmetry, is absolutely benign. If a state of identical particles allows precise assertions about properties possessed by its constituents, making it symmetrical or asymmetrical does not render the assertions in question inappropriate or illegitimate. The only difference comes from the fact that it is not legitimate to assert which one (Peter or Paul) of the identical constituents is found in a certain position and has certain properties. I do not want to be pedantic, but it seems a good idea for me to underline strongly the following conclusion: the only information that we lose in passing from state (14.10) to state (14.11) is the information about which electron is found in A or B (an unimportant loss, although it is necessary, given that we have already understood that it is not possible to distinguish identical constituents individually), but the truth remains that in both cases there is an electron in A with a precise spin (up) and another in B with spin (down).

With reference to the incisive phrasing of Levy-Leblond I must point out that the entanglement of electrons of my own body with those of the inhabitants of the Andromeda Galaxy has as a unique consequence that it is no longer legitimate to say that these are "my" electrons and those are "theirs." But, thanks to the effect of pure and simple antisymmetrization, nothing changes with regard to the objective fact that here on earth there exists a being that coincides with me, and on Andromeda there is another one that would be very different.

The situation is quite otherwise when the point of departure is a state that, besides not satisfying the demands of symmetry, presents the peculiarities of a state that is genuinely entangled from the beginning. In other words, as the reader will easily understand, when the point of departure is a state analogous to that of equation (14.9) and it is made antisymmetrical, the embarrassing fact remains that no longer can any certain assertion be made about the result of a spin measurement of a particle in A along any arbitrary direction. Once again the probabilities of finding spin up or down along any arbitrary direction of the particle in one of the two regions are equal, and equal to 1/2, as in the case of EPR. The constituents do not have definite spin properties, but this is not because of their identity, but rather because the initial state was "genuinely" entangled.

In conclusion: from the point of view of the possibility of attributing properties to constituents of composite systems, their identity does not bring any complication. Apart from the impossibility of individualizing the constituents, when it comes to individual particles, "unproblematic" (i.e., nonentangled) states of systems with distinguishable constituents remain "unproblematic."

CHAPTER FIFTEEN

From Microscopic to Macroscopic

> There is a fundamental ambiguity in quantum mechanics, in that nobody knows exactly what it says about any particular situation, for nobody knows exactly where the boundary between the wavy quantum world and the world of particular events is located. That for me is the problem of quantum mechanics. It is not a problem in practice—in practice, we always can take this boundary judiciously so that moving it a bit one way or the other doesn't much matter. But every time we put that boundary in— we must put it somewhere—we are dividing the world arbitrarily into two pieces, using two quite different descriptions, one on one side and one on the other.
> —*John Stewart Bell*

AFTER ANALYZING the structure of quantum mechanics, describing its unparalleled successes, and illustrating the conceptual revolution it implies, I have given some examples of the surprising behavior of microsystems. It is now time to confront a serious question: does quantum mechanics, a theory that presents itself as "fundamental" (that is, as a theory that grasps at least some portion of the truth) really enable us to "understand" or "explain" reality, or does it rather amount merely to a series of "recipes" for describing the outcomes of our experiments? In this chapter we will encounter the important epistemological problem that Abner Shimony has appropriately summed up as *closing the circle*, a happy expression which he uses to characterize, among other things, the possibility of elaborating a coherent vision of the world that is compatible both with the behavior of microsystems and with the data of our everyday experience of macroscopic systems.

The problems we are going to confront in this chapter represent a field of investigation that is still wide open to development, after more than three-quarters of a century of very lively debate. Consequently, this final part of the book will differ from the foregoing parts in one essential respect: while there exists a certain agreement that could be called universal with regard to certain characteristics we have identified as typical of *microsystems* (including, in particular, nonlocality and nonseparability), nevertheless when it comes to the question of how to overcome the difficulties encountered in describing *macrosystems*, the positions are still rather diverse. In keeping with my ambition to present a general outline (within the limits suitable for a book of this kind), and to present as unprejudiced a view as possible of the debate that still rages over the conceptual basis of the theory, in what follows I shall first make clear the exact points where the difficulties lie, and then review in turn all the positions that are of abiding interest today.

In all honesty I cannot omit to mention that there does exist a tiny minority of scientists who think there are no conceptual problems with the theory at all. Fortunately, this (in my view) superficial attitude is gradually subsiding, while an almost unanimous consensus is taking shape among the experts in the field that the theory in its orthodox interpretation does not permit the elaboration of a coherent vision of reality, and that this theory has still to be supplemented, given alternative interpretations, or simply changed, before it can be "taken seriously" as a fundamental theory of the physical world.

15.1. THE SUPERPOSITION PRINCIPLE AND MACROSCOPIC SYSTEMS

When we begin to discuss the conceptual difficulties that quantum formalism encounters in attempting to explain macrosystems, it will help to make a certain qualification. Of course, nobody pretends that quantum mechanics constitutes the *definitive* framework for understanding reality: like any other scientific theory, it will surely be superseded someday, and eventually its field of application will find certain limits. Actually, the formalism we have been presenting is fundamentally nonrelativistic (as is well known) and therefore had to be modified in order to make it compatible with the inescapable requirements of the theory of relativity. The resulting marriage between these two pillars of twentieth-century science has already been fruitful: Dirac formulated a first relativistic generalization of Schrödinger's equation for the electron that led, among other things, to the discovery of antimatter, one of the most important discoveries of our time. But that was not enough: even Dirac's formalism for the electron and the analogous relativistic equations for the other types of elementary particles present serious problems of consistency and have called for the elaboration of modern quantum field theories, a fascinating topic we cannot enter into here.

And yet this does not affect the problems we are going to discuss now, or the conceptual difficulties I am now trying to bring into the open. Indeed, as we shall see, all these difficulties derive exclusively from the superposition principle, or the linear character of the theory, a formal aspect that characterizes even the generalizations just mentioned. As I have repeatedly pointed out, the enormous mass of experimental data accumulated over an entire century of research about microsystems has confirmed beyond any reasonable doubt that the principle of superposition represents a fundamental characteristic of natural processes. Because of this, in none of the generalizations undergone by the formalism from the 1930s until the present has the linearity of the formalism ever been put in doubt. Consequently, the validity of this principle has assumed the conceptual and logical status of a definitively acquired fact, which should constitute a fundamental ingredient of every conceivable future development.

Among the many implications of this, by far the most important one for what we are considering in this final part of the book is that it allows the occurrence of superpositions of states corresponding to definite and diverse properties for a physical observable. We know very well that in such a case the theory, if considered complete, does not allow us to think that the observable in question objectively possesses a precise value, but implies that, in a genuinely nonepistemic way, we can speak only of the probabilities of outcomes that will follow the effective execution of a measuring process. The most characteristic example, which has constituted a kind of leitmotif for most of the preceding chapters, is that of a photon in the state

$$|45\rangle = \frac{1}{\sqrt{2}}\big[|V\rangle + |H\rangle\big]. \qquad (15.1)$$

As we know, this is the superposition of states that correspond to definite, mutually exclusive and exhaustive properties relative to a test for vertical polarization (the state $|V\rangle$ corresponding to the certainty that the photon will pass a test of this kind, while the state $|H\rangle$ corresponds to the certainty that it will fail the test). Such a state does not permit us to think in the least that, even in a way unknown to us, the photon will in fact have certain characteristics that determine its fate a priori once it is subjected to the considered test. I have spent some time recalling a few well-known facts, in order to emphasize once again how the superposition principle plays an absolutely fundamental role: it implies indeterminism and the incompatibility of some properties[1]: in brief, it incorporates all the most significant aspects of the conceptual revolution imposed by the theory.

But even though every step of the investigation into the microcosm, every new field of research that has been opened, provides us with ever more certain confirmation of this principle, nevertheless, ever since the very first moments of the debate about formalism, the principle has represented an insurmountable obstacle for any attempt to reconcile the theory with the data of our daily experience. The reasons for this state of affairs will be obvious when we consider a situation analogous to that of equation (15.1), but with reference, not to a microsystem and a microscopic property, but to a macrosystem and a macroscopic observable.

Let us consider a system of this kind, S. Of course, our immediate experience tells us that this system can possess various macroscopic properties. This chair, for example, can be found in a certain location A or in another location B: let us say, in this or that corner of my office. The theory assures us that any individual physical system is associated with a precise state vector. We can indicate by $|here\rangle$ and $|there\rangle$ the two states in question for our chair.

Let us argue, for now, on purely formal grounds. The theory implies that for a physical system if a certain state is possible and another state is possible, then even their linear combination is possible. But this means that according to the formalism the chair can be found (in principle, for now: we will see later whether this can occur in practice) in a state analogous to that of the photon considered above:

$$|?\rangle = \frac{1}{\sqrt{2}}\left[|here\rangle + |there\rangle\right]. \quad (15.2)$$

The reader who has been assimilating the principles of the theory cannot but be embarrassed by an attempt to give a meaning to the state in question (hence the question mark I put there). The reader knows very well that the + that appears between the two terms of the sum plays an absolutely peculiar role, and that, in particular, it cannot be substituted by an assertion that contemplates the alternative "either . . . or." In fact, unless limits are placed (limits which the theory has never identified) on the possibility of carrying out any type of measurement on a system, it can be said that the state $|?\rangle$ has some physical properties that are incompatible with the assumption that the state is either $|here\rangle$ or that it is $|there\rangle$. What meaning can there be in a state that makes it illegitimate to think that our chair is *either* here *or* in some other place? To what macroscopic situation does this correspond? What sense can the assertion have (which is inevitable if we do not want to deny that chairs are also physical systems subject to quantum laws) that in the state under consideration only potentialities exist about the location of the chair, potentialities that cannot be realized, unless we carry out a measurement of position? How can it be understood that, attached to these potentialities, is a *nonepistemic* probability that in a subsequent measurement of position the chair will be found here or there (which is equivalent to asserting that, before the measurement was carried out, the chair could be neither here nor there, nor in both places, nor in neither place)?

The conclusion is obvious but unsettling: if we admit that the theory (or rather, simply, the superposition principle) has universal validity, and thus governs as well the behavior of macrosystems, it is inescapable to accept that, in principle, macrosystems too can be found in superpositions of states corresponding to precise and different macroscopic properties, with the consequence that these systems cannot legitimately be thought to possess any of these properties. In our specific case, a macroscopic object "is capable of not having the property of *being in some place.*"

For the moment I have pointed out how the theory, by its own formal structure, forces us to confront a serious conceptual problem as long as it is also taken to govern the behavior of macrosystems. In the following section, I intend to show how the disturbing states just considered are not only possible in principle, but in fact must crop up (to say the least) in the course of a measurement process. In other words, they must inevitably be taken into account if we assume that there exist reliable apparatuses to obtain information about microsystems. But before discussing this important theme it seems a good idea to present in summary form a typical objection of those who think that the theory presents no problems, and to show in reply why the objection does not hold.

. . . . .

In developing the preceding argument, I deliberately chose an oversimplified form of notation. In fact, the specification of such a complicated state as that of a macrosystem by focusing only on its spatial location is obviously reductive. But that is true not only in quantum mechanics: it is clear that even in a classical context there is an enormous variety of states for the chair that correspond to what we mean by saying that "the chair is here"—states that can differ by an infinity of particulars (for example, by the precise positions and velocities of the 10^{24} molecules that constitute a system of the kind). It is therefore fitting to take this into account and to adopt a new notation, substituting for the $|here\rangle$, $|there\rangle$ of the preceding formulas new expressions like $|here, \ldots\rangle$, $|there, \ldots\rangle$. The added dots are meant to emphasize how the specifications included in Dirac's symbols are not sufficient to characterize the chair's situation completely. Instead, I will go further: just as it would be excessive to try (in practice) to determine perfectly the state of a chair even in the classical case, I have no difficulty admitting that it is impossible (still from the practical point of view) to determine perfectly the quantum state of a chair. But this has no effect on the fact that the states in question correspond to diverse macroscopic properties.

Of course, the analogue of the embarrassing equation (15.2) is now the expression

$$|?\rangle = \frac{1}{\sqrt{2}}\left[|here, \ldots\rangle + |there, \ldots\rangle\right]. \tag{15.3}$$

Some writers, who maintain that the theory does not encounter any difficulties in dealing with macrosystems, argue that a state like the one of equation (15.3) need not be considered because it can never occur. The reason for this would derive from the fact that the dots are there to show that we do not know the state of the system exactly and that we cannot completely control it. It would follow from this that in fact the two states that appear in (15.3) are fundamentally ambiguous and that in reality we do not know exactly what we are superposing; the ensuing implications would then be irrelevant or could not be legitimately derived. Now I must explain that this argument is erroneous. In the case of the superposition of two quantum states that correspond to different properties for a specific observable, ignorance of the exact form of the state with regard to the other variables that are independent of the ones related to the observable in question, has no effect upon, and does not in any way change the fact that it can be asserted that the state describes a system that cannot be thought to possess any precise property relative to that observable.

To understand this point, we can consider the case of the state of a particle with spin concerning which nothing is known about its position, but it is known that the spin is a superposition of spin up and spin down along the z axis:

$$|\Psi\rangle = \frac{1}{\sqrt{2}}\left[|z\ up\rangle |\phi(r)\rangle + |z\ down\rangle |\chi(r)\rangle\right]. \tag{15.4}$$

The ignorance about the position means that the precise form of the spatial functions $\phi(r)$ and $\chi(r)$ is unknown, apart from the fact that they are different from each other. Nevertheless, this does not change in the least the conclusion that it is not permissible to assert that the particle has definite properties concerning the projection of its spin along the z axis: for it there exists a nonepistemic probability (which means that it is illegitimate to think that in fact the particle possesses one of the two alternative properties) of obtaining the result "the spin is up" or "the spin is down" when a measurement is made for that observable.

. . . . .

15.2. The Quantum Theory of Measurement

To illustrate how inevitably we are confronted—once the limitless validity of the principle of superposition is assumed—with the practical possibility of occurrence of states like (15.2), let us now take a look at the thorniest and most controversial problem of the theory, the process of measurement. The reader needs to be warned that the problem cannot be avoided, and that it inevitably involves macroscopic systems. In fact, owing to the huge difference of scale between human beings and the microscopic systems we want to study, any attempt to obtain information about them requires a process of amplification that strictly correlates the microscopic properties to situations that are macroscopically perceivable and hence distinguishable to our perception. The problem is traditionally approached with reference to what is technically known as von Neumann's ideal measurement process, named from the scientist who first formulated it in precise terms.

For this purpose we can refer to experiments of the kind discussed in chapters 3 and 4, in which photons with definite states of polarization were sent into a birefringent crystal, and we enrich the analysis by including the dynamics of the detecting apparatus. In the discussion of those chapters we simply said that "the detector detects (does not detect) the arrival of a photon," insofar as this was the only relevant information for the analysis we were then making. Now, however, we must inquire into the precise sense of this assertion and investigate the physical processes to which reference is made in the cases of concrete experiments carried out in laboratories. A reasonable and sufficiently realistic description of what happens (as pointed out in chapter 6) would be the following: the purpose of a measurement is to infer something about the system being measured from the outcome of an appropriate experiment. It follows that the interaction between microsystem and measuring apparatus should bring about a macroscopic change of the apparatus in such a way that, by observing the state of the apparatus after the process, we would be able to obtain the desired information. Instead of what we did in the foregoing chapters, i.e., consider two different detectors placed, respectively, on the ordinary and extraordinary beams, this time we will imagine the apparatus as a box (Figure 15.1) with two regions on it, equally sensitive to incoming photons, and we will suppose that the box includes an instrument with a macroscopic pointer that can be in three positions, designated as −1, 0, and +1. The zero position of the pointer is the state of the apparatus before it has interacted with a photon and can be designated as "the state of the apparatus ready to carry out a measurement." We will also have to suppose that the interaction of the apparatus with the photon is such as to permit us to recognize from the final position of the pointer, if the photon struck the apparatus in the spot L (for lower), corresponding to the ordinary ray (+1), or the spot U (for upper), corresponding to the extraordinary (−1).

To take up the problem it will be convenient to analyze first of all the case in which a vertically polarized photon is sent into a birefringent crystal placed in front of the apparatus. As we know, the photon will be propagated along the ordinary ray and will end up striking the region L of the apparatus. Alternatively, we will consider a horizontally polarized pho-

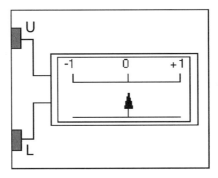

FIGURE 15.1. Schematic representation of a measuring apparatus with the pointer in position 0, which indicates that it is ready to register the arrival of a photon either in the upper shaded region U or in the lower shaded region L. In response, the pointer of the apparatus will move, once the interaction takes place, to position −1 or +1, respectively.

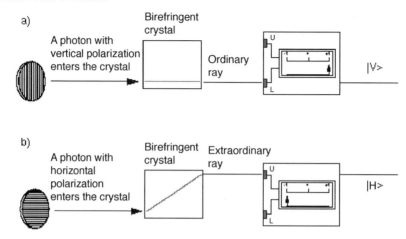

FIGURE 15.2. Schematic representation of the effect obtained by activating an ideal measuring apparatus by states having the precise polarizations that the apparatus is set up to reveal.

ton that we know will with certainty follow the extraordinary ray and will end up striking region U of the apparatus (Fig. 15.2a,b). We designate as $|A_0\rangle$, $|A_{+1}\rangle$, and $|A_{-1}\rangle$ the states of the apparatus in which the pointer points, respectively, to the three indicated positions. For simplicity we will suppose, with von Neumann, that the photon passes through the apparatus without changing its state.

The quantum evolution of the two states can be represented as follows:

$$|V, A_0\rangle \Rightarrow |V, A_{+1}\rangle, \tag{15.5a}$$

$$|H, A_0\rangle \Rightarrow |H, A_{-1}\rangle, \tag{15.5b}$$

in which the final states describe, respectively, a vertically polarized photon correlated with the state in which the pointer of the apparatus indicates +1, and a horizontally polarized photon correlated with the state in which the apparatus indicates −1.

We are now prepared to show how the problem of the quantum theory of measurement arises. First, since the two processes described above involve physical systems that are completely normal for the theory in question (the apparatus is composed of electrons, protons, neutrons, etc.), we must inevitably assume that the whole evolution symbolized by the arrows in the formulas above is nothing other than the Schrödinger evolution for the system "photon + apparatus." But the fundamental charac-

FROM MICROSCOPIC TO MACROSCOPIC

teristic of quantum evolution, once again, consists in its linear character: the evolution of a state which is the linear superposition of two initial states is the same as the superposition of what has evolved out of the same initial states. We now prepare an initial state that is a linear combination of the two initial states considered above: for this purpose it is enough to send a photon with a polarization plane of 45° into a birefringent crystal. Then, for the initial state we have the following:

$$|\text{Initial}\rangle = \frac{1}{\sqrt{2}}\big[|V\rangle + |H\rangle\big] \bullet |A_0\rangle = \frac{1}{\sqrt{2}}\big[|V, A_0\rangle + |H, A_0\rangle\big]. \quad (15.6)$$

Now this, thanks to the linear character of the dynamics, will evolve into the same linear combination of the two final states of the equations (15.5a,b):

$$|\text{Initial}\rangle \Rightarrow \frac{1}{\sqrt{2}}\big[|V, A_{+1}\rangle + |H, A_{-A}\rangle\big]. \quad (15.7)$$

The final state (Figure 15.3) is a superposition of two states that are macroscopically different, since in the first of the two, the macroscopic pointer of the apparatus points to +1, and in the second it points to −1. The argument shows in a simple way how to realize a state of the type considered in the preceding section.

A few observations:

1. As discussed in Section 15.1, it is extremely problematic to find some physically intelligible meaning for state (15.7).

2. This state is an entangled state of the photon and the apparatus and as such it is not legitimate to attribute definite polarization properties to the photon or definite positions to the pointer of the apparatus.

3. The treatment is oversimplified. The states of the apparatus that appear in the preceding formulas concern a macroscopic system, and, as we have already pointed out, the specification for the position of the pointer is not enough to characterize it completely. To be rigorous we would have to do what we did at first, namely, add points that would refer to all the other degrees of freedom of a system as complex as this. Nevertheless, as observed in the preceding section, this does not have any consequences for what concerns the fact that any assertion about the position of the pointer is illegitimate, and it is even not permitted us to think that it has a precise position. The relevant physical implications of the enormous complexity of the apparatus will emerge when we pose the problem of how to "verify in the laboratory" that in fact, at the end, there is a su-

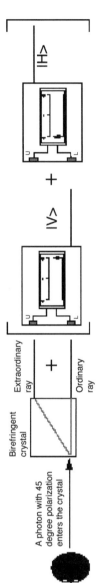

FIGURE 15.3. In the case when the apparatus is set off by a photon with 45 degree polarization that is the superposition of the states that it is programmed to register, and under the assumption that the measuring process is governed by the linear laws of the theory, one must conclude that the final state does not correspond to an apparatus with the pointer in a definite position. The quantum mechanical ambiguity associated with the superposition becomes, as it were, transmitted to the macroscopic system.

perposition of macroscopically distinguishable states. But this does not change in the least the objective "given" that such a superposition must be considered present whenever it is assumed that we have at our disposal an apparatus that reliably allows the identification of vertical or horizontal polarization of a photon, and that the interaction between the measured system and the measuring apparatus is a process that obeys the general laws of the theory.

4. As already mentioned, the difficulties that appear do not derive from our idealizations or simplifications introduced into the interpretation. This has been emphasized repeatedly in the literature by John Stewart Bell and Bernard d'Espagnat among others. To my knowledge, the most mathematically rigorous demonstration of the fact that the very possibility of using measuring apparatuses with a high degree of reliability implies the indefiniteness of some of their macroscopic properties, appears in a paper which I recently wrote in collaboration with my doctoral student A. Bassi.

15.3. Quantum Evolution and the Reduction of the Wave Packet

I would now like to direct the reader's attention to the way the orthodox interpretation gets around the difficulties we have outlined in the foregoing sections. Why would that interpretation not have to face the embarrassing occurrence of superpositions of states that are distinguishable at the macroscopic level? In order to understand this point, we need to recall that the theory in its general formulation includes a postulate that becomes operative every time a process of measurement is carried out, and that is the postulate of the reduction (or collapse) of the wave packet. Before we carried out our analysis, every reader, when faced with the situation summarized by the equation (15.7), would have replied correctly about the ultimate situation of the experiment, asserting that in an entirely casual and unpredictable way there would occur *either* the situation with the photon polarized vertically and the apparatus registering +1, *or* the situation in which the photon is polarized horizontally, with the apparatus showing −1. And this is in fact what we would experience if we actually looked at the apparatus. No physicist would be embarrassed about this: going into the laboratory, and looking at the pointer on the gauge, he (or she) would find it definitely in one position or the other.

CHAPTER FIFTEEN

It is consequently very tempting to say that the problem we are discussing is a pseudoproblem: the theory has already given us a solution that does not cause any embarrassing situation, and our own experience confirms the correctness of such an assessment. And so what can be wrong with this simple, clear argument? The problem emerges from the analysis itself: the application of a principle (linearity), whose unlimited validity is assumed by the theory, to the description of the measuring process, leads inevitably to the conclusion that the final state of system + apparatus is the embarrassing state that appears on the right side of equation (15.7), while the postulate of the reduction of the wave packet says that the final state is something else—namely, one of the nonproblematic terms of the sum (for each of which it can be said that the pointer of the apparatus is in a definite position). The conclusion is clear: the postulate of the reduction of the wave packet logically contradicts the hypothesis that the evolution of the system we are investigating is in fact governed by quantum mechanics. In other words, the theory is not in a position to explain how it could ever come about, in an interaction between a system and a measuring apparatus, that the peculiar process leading to a definite outcome could emerge—that is, how an apparatus could ever behave in the manner it is supposed to do!

It is important to emphasize the radical differences in the description of the measuring process when we use the evolution equation of the theory or when we resort to the postulate of the reduction of the wave packet. As we have repeatedly observed and as is clearly illustrated by the equations (15.6) and (15.7), the quantum evolution is perfectly deterministic and linear: the initial state determines the final state without any ambiguity, and the sums of the initial states evolve into the corresponding sums of the final states. By contrast, the reduction of the wave packet is a fundamentally stochastic and nonlinear physical process: in general, the outcomes of measurement are unforeseeable, and since the relative probabilities depend upon the square of the wave function, phenomena of interference present themselves, which in turn imply, as we know well, that the probabilities for the various possible outcomes in the case of a superposition are not the sum of the probabilities associated with the terms of the superposition itself. Furthermore, while the quantum dynamics is perfectly reversible, as in the classical case, the process of the reduction of the wave packet is fundamentally irreversible.

We can ask ourselves then: does this singular fact, this internal inconsistency of the theory represent in itself an insurmountable difficulty? Of course not: it simply leads us to conclude that we have to allow for the fact that there are systems in nature that do not obey the laws of the theory. Like any physical theory, even quantum mechanics will have a limited field of application. And this represents in a certain sense (although many ambiguities remain) the orthodox position: two principles of evolution must be adopted: one governing all the processes that involve interactions between microsystems which is described by Schrödinger's equation; the other which is to be used to describe measurement processes and is accounted for by the postulate of the reduction of the wave packet. The problem that remains, and which is of no small account, is that of succeeding in identifying in a nonambiguous way the boundary line between these two levels of the real, which require two essentially different and irreconcilable physical descriptions.

We have now reached the point where we can face the so-called problem of the macro-objectification of properties: how, when, and under what conditions do definite macroscopic properties emerge (in accordance with our daily experience) for systems that, when all is said and done, we have no good reasons for thinking are fundamentally different from the microsystems of which they are composed? To appreciate fully the relevance of this question, and to prepare ourselves to analyze the multiple and interesting proposals advanced as a way of getting around such difficulties, it will be useful to begin by deepening our analysis.

15.4. THE VON NEUMANN CHAIN

As we once pointed out, John von Neumann who was the first to give a lucid and rigorous formulation to the problem we are now considering, also clarified one of its benign aspects. Let us consider again state (15.7) and suppose that this is truly the state that occurs in nature at the end of the process, and let us ask ourselves: does the expression on the right side of the equation represent the conclusion of the measuring process? The state describes a physical system, the apparatus A, which does not have a definite physical property as far as concerns the position of its (macroscopic) pointer. Now suppose that at this level we have still not reached the limit of the field of applicability of the theory, and that the final state must be taken seriously as that which correctly describes the physical situation. What is left to do? Clearly, we need to ask ourselves, in what actual position is the pointer? To answer this question, instead of analyzing the mathematical properties of the system, we must proceed to carry out a new measurement to determine the position of the pointer. Let us sup-

CHAPTER FIFTEEN

pose we have at our disposal an instrument capable of performing this task. We call it B, and once again we consider three possible states $|B_0\rangle$, $|B_{+1}\rangle$, and $|B_{-1}\rangle$ which correspond, respectively, to the situation in which the instrument is ready (at the beginning) to measure the position of pointer A; the situation in which, in the measurement, pointer A has been found to be in position +1; and (finally) the situation in which, in the measurement, pointer A has been found in position −1. In place of equation (15.5), then, we now have the following equations in two steps:

$$|V, A_0\rangle|B_0\rangle \Rightarrow |V, A_{+1}\rangle|B_0\rangle \Rightarrow |V, A_{+1}\rangle|B_{+1}\rangle, \quad (15.8a)$$

$$|H, A_0\rangle|B_0\rangle \Rightarrow |H, A_{-1}\rangle|B_0\rangle \Rightarrow |H, A_{-1}\rangle|B_{-1}\rangle, \quad (15.8b)$$

with an obvious meaning. In the first case, the photon has entered the apparatus, bringing the indicator to the position +1, and the next measurement of the pointer's position tells us, in essence, that the apparatus B has found the pointer of A in the position shown (and the same holds for the second case.)

Of course, the equations (15.8) imply that the final state in the case of an initial photon polarized at 45 is

$$|\text{initial}\rangle \Rightarrow \frac{1}{\sqrt{2}}\left[|V, A_{+1}, B_{+1}\rangle + |H, A_{-1}, B_{-1}\rangle\right], \quad (15.9)$$

which is, again, a superposition of states with different macroscopic properties (that now also involve the apparatus B) and a more complicated entanglement among the three physical systems under inspection.

We can ask ourselves again: is this the end of the story? And once again we will say that it is not, that it is necessary to carry out a new measurement to ascertain the position of B, and so on. This is how von Neumann's chain arises, and how it gradually involves an ever larger number of physical systems, the ultimate system being the observer himself. Each system of the chain is prepared to ascertain the state of the one that precedes it in the hierarchy:

$$|\text{initial}\rangle \Rightarrow \frac{1}{\sqrt{2}}\left[|V, A_{+1}, B_{+1}, \ldots, Z_{+1}\rangle + |H, A_{-1}, B_{-1}, \ldots, Z_{-1}\rangle\right]. \quad (15.10)$$

Equation (15.10) brings out an extremely important characteristic of the process. If, on the one hand, it seems that the situation is getting more and more complicated and that the entanglement is spreading like a contagious disease, involving an ever larger number of physical systems, on the other hand, all the systems that participate in the process end up having perfectly correlated indices: in each one of the two terms of the superposition the apparatuses A, B, . . . , Z are in states in which their pointers are either all at +1 or all at −1. This means that if a stage is reached along the chain (it does not matter which one, so long as it precedes or coincides with the act of conscious perception on the part of an observer) in which quantum mechanics effectively ceases to be valid and recourse needs to be made to the reduction postulate, then, regardless of the level where this happens, the state on the right side of the equation (15.10) will be replaced by one of the two terms, and for each one of these there is a perfect coherence of all the outcomes. Let us suppose that the reduction brings us to the first state of the superposition: then it will be legitimate to say that *the photon is vertically polarized, apparatus A has its pointer in the position corresponding to this outcome, and apparatus B confirms the preceding assertion about the position of A's pointer, and so on.* All of a sudden, once the reduction has taken place, everything makes sense again, and all the physical systems that were involved in the process end up in perfectly definite macroscopic states, exhibiting perfectly consistent properties.

There is no doubt that the process just described has great conceptual relevance and that it is very helpful for at least two reasons. First, if there are outcomes, they do not lead to any contradiction between the situations of all the systems involved. In particular, it is possible to include in the chain an apparatus R, for example, that would record the outcome on a slip of paper, as well as a conscious observer Z. When Z perceives, by watching apparatus Y, that the polarization is vertical, he then goes over to see what the apparatus R had recorded on the paper, and finds written, "the photon has vertical polarization." Secondly, the consideration of the chain suggests a naturally *pragmatic* solution to the problem of macro-objectification. Since, as we shall soon see, distinguishing the linear superposition from the statistical mixture of its terms (that is, the fact that one or the other of the events associated with these states actually occurs) is in practice more difficult, the more complex the system under investigation, it will be possible to resort to the rapid coupling or interaction of apparatuses with the environment (from which they can never be perfectly isolated) to conclude that everything happens as if the reduction process actually took place, even if in fact it never happened, that is, even if the theory had unlimited validity.

However, I have to conclude that this escape route (as attractive as it may be from the practical point of view) does not represent an acceptable solution on the logical and epistemological level. It is equivalent to admitting that when we take the theory seriously as regulating all the processes of the universe, situations occur to which it is not possible to attribute any acceptable meaning and which contradict our sensible experience with macro-objects. On the other hand, for purely contingent reasons, everything happens as if we could replace the exact equations of the theory with approximate equations that bring about states that have meaning and lead to outcomes in accord with our experience. But in the history of science it has never been the case that a theory that was correct, but without meaning, acquired meaning as the result of an approximation. It is not by chance that a thinker with von Neumann's rigor immediately understood that this escape route is not logically sustainable, and consequently he preferred to support the idea which we will analyze in the following chapters, an idea later endorsed by Eugene Paul Wigner (and to a certain extent more or less consciously by the various proponents of orthodoxy), that the breaking of the chain occurs precisely in correspondence with the act of conscious perception on the part of the observer. The solution is legitimate when it is accepted that because such an act is not a physical process of the same type as those that take place in our laboratories or in the phenomenal world, there would consequently be no reason why it should be governed by the theory. We will return to this point in due time. For the present, I would like to come to a close by reporting that this position has driven many an author to assert that quantum mechanics has "undone the Copernican revolution" by placing man at the center of the whole of reality again, since he and only he can determine that actualization of potentialities that characterizes reality as we perceive it.

15.5. Difficulties Involved in Testing Macroscopic Superpositions

In the preceding section we made it clear that the proliferation of the number of systems that become entangled (as a consequence of the inevitable interaction between the system being measured and the measuring device and of this with its environment) seems to open some pragmatic escape routes from the impasse of the occurrence of linear superpositions of macroscopically distinguishable states. The reason for this lies in the fact that, the more complex the system, the more difficult it is to carry out an experiment that would bring into the open that notorious plus sign (+) that is so characteristic of superpositions. In this section I will list a number of facts that illustrate this specific aspect of the problem.

First, we have to remember that the theory we are considering is fundamentally probabilistic. That implies that to test it we need to consider an "ensemble" of systems prepared in an identical fashion, subject them all to the same measurements, and record the relative frequencies of the various outcomes.[2] If the theory is correct, as the number of systems in the ensemble grows, these frequencies (according to the law of large numbers) will steadily approach the theoretical probabilities.

With reference to the process analyzed in Section 15.3, let us consider again the first link in the von Neumann chain. As we have seen, if we assume that the theory also governs the unfolding of the measurement process, once the system-apparatus interaction is complete, all the members of the ensemble will be found in the embarrassing state on the right-hand side of equation (15.7):

$$|\text{final}\rangle = \frac{1}{\sqrt{2}}\left[|V, A_{+1}\rangle + |H, A_{-1}\rangle\right]. \qquad (15.11)$$

On the other hand, if one assumes that the measurement process induces the wave-packet reduction, we will have to conclude that around one-half of the systems will actually be in the state $|V, A_{+1}\rangle$, while the rest will be in state $|H, A_{-1}\rangle$. We can call the ensemble just now described E_{red}, that is, the one for which half the apparatuses show a + (and the related photon is vertically polarized), and for which half show a − (and the related photon is horizontally polarized).

Now, supposing that the final situation is one where all the members of the ensemble are in state (15.11), let us take a second step, that is, carry out a measurement to see how many apparatuses register a + and how many a −. Let us also suppose that at this stage the reduction actually takes place, and we ignore the second apparatus B, concentrating all our attention on the system (photon) + (apparatus A). Since the two states in (15.11) have equal coefficients (that is, $1/\sqrt{2}$), the theory tells us there will be equal probabilities of obtaining the two outcomes and that therefore, at the end of it, we will find ourselves in exactly the same situation as we described before, when we had assumed that the reduction took place at the first stage.

What meaning can be given to the analysis just described? Simply this: that if the only types of experiments we consider are experiments intended to detect how many apparatuses show a + and how many a −, then the state (15.11) and the mixture E_{red} are indistinguishable. These

361

CHAPTER FIFTEEN

measurements (analogously to those intended to indicate if the photons are polarized horizontally or vertically) do not permit us to distinguish an ensemble in which all the systems are in state (15.11) from one in which one-half of them are in the first of the superposed states, and one-half in the second.

Accordingly, in order to be able to distinguish these two cases, that is, to be able to provide an experimental answer to the question, "Is the process we are studying governed by the linear evolution equation that is characteristic of the theory, or by the reduction mechanism?," we will have to have recourse to more refined measurements. I will now mention a whole series of measurements that likewise do not permit us to distinguish state (15.11) from the mixture E_{red}. In Appendix 15A those readers who are interested will find some simple illustrations of the facts I am about to list:

It is impossible to distinguish the two situations that interest us in the following ways:

- By any type whatsoever of measurement involving only one of the constituents that enter into (15.11). In other words, if measurements are carried out that involve properties of only the photon (for example, measurements of polarization along arbitrary planes) or of only the apparatus, the probabilities of their outcomes will be equal in the two cases considered.
- By measuring correlations between pairs of observables for the photon and for the apparatus, if the observations referring to at least one of the constituents are compatible with the values that characterize the superposed states. Or stated more simply: if the observables that are measured on the microsystem or on the apparatus turn out to be compatible with the polarization involved or with the observable "position of the apparatus's pointer," respectively, then the probabilities of the outcomes are the same for the state (15.11) and the mixture E_{red}.
- What has been said for the first stage of the chain holds for any later stage whatsoever; in other words, in the general case, it cannot be shown if there is effectively a + on the right-hand side of the equation (15.10), if measurements of correlation are not carried out that involve all the systems (photons A,B,C, . . . , Z) that appear, and if, even for one only of these systems, we limit ourselves to observables that are compatible with those indicated (we should recall that, for example, C can be an observable that answers the question: Did the computer write "vertical" or "horizontal"?)

The conclusion should be clear enough. In order to be able to distinguish an ensemble in which every system is in state (15.10) from a corresponding ensemble in which every system has the precise properties associated with the individual states appearing in (15.10), measurements of correlation need to be carried out among all the systems involved, and moreover of correlations between observables that are incompatible with those associated with the terms of the superposition. It will not be surprising therefore that our study of the von Neumann chain and its extension to a state which calls into play an entanglement with the entire surrounding environment would quickly cause an explosive increase in the difficulty of experimentally distinguishing the state from the mixture. In particular, to obtain this result would require carrying out measurements of correlation between the apparatus and, for example, every molecule of the surrounding air with which the apparatus has interacted, with the systems that have interacted with these systems, and so on. This is impossible in practice.

But I do not want to be misunderstood: I have tried to point out the practical difficulty of distinguishing these two situations, but this does prevent the conceptual problem from remaining unchanged. Even if the distinction is experimentally very difficult, it is nevertheless possible in principle. The theory itself guarantees that there exists some physical quantity for which the predictions of the state and of the mixture are arbitrarily diverse. And here emerges the ultimate problem: if the theory has universal validity, it brings about situations to which we do not know how to attribute any meaning; if it has limited validity, so that at a certain stage the linear nature of the evolution disappears and the reduction of the packet emerges, then it is necessary to identify the line of demarcation between systems that have to be described by different sets of laws.

15.6. The Ambiguous Boundary

What, in essence, is the crucial problem we are facing? As we repeatedly emphasized in the early chapters of the book, the theory has been formulated in such a way as not to speak in general of properties possessed by systems but only of the probabilities of obtaining certain outcomes if measurements are carried out that intend to identify the values of the properties we are interested in. But if the theory has a universal validity an endless process begins: in order to ascertain the properties of a system we will have to make it interact with an apparatus, and this will react differently according to the potentialities of the system (except in the case

CHAPTER FIFTEEN

where the system is in a state yielding a certain measurement result), and then in their turn the potential outcomes of the measurements will not be realized, and will become real only if a measurement is carried out to ascertain them, and so on, until what end is reached? This reasoning can in principle be extended until it involves the entire universe, and it implies that even the universe would have only potentialities. But then the question is, "Who measures the universe, and brings it out of the limbo of the potentialities?"

It is time to mention another fact, until now only treated marginally, which makes the problem even more serious. It is absolutely true (unless we completely change our perspective and adopt some rather fascinating but science-fiction positions to be discussed later) that measurements have outcomes and therefore that at a certain stage a reduction actually takes place, i.e., that a passage from the potentialities to the actualities occurs. Indeed, for every process of measurement that interests us, at a certain level there must certainly intervene some observer whose perceptions are definite. I would like to call the attention of the reader to the fact that if we think, for example, that the last link in the von Neumann chain is a conscious observer, then the two states $|Z_{+1}\rangle$ and $|Z_{-1}\rangle$ that appear in our equations are simply abbreviations for states that could be expressed more appropriately as follows:

|the conscious observer sees a macroscopic gauge that points to $+1\rangle$ and
|the conscious observer sees a macroscopic gauge that points to $-1\rangle$,

or alternatively

|the conscious observer reads the word VERTICAL on the computer screen$\rangle$ and |the conscious observer reads the word HORIZONTAL on the computer screen$\rangle$.

We do not have any sensate experience of a situation where there are the potentialities of having read VERTICAL and of having read HORIZONTAL instead of a very precise actualization of one of these alternatives. We know very well that every time we watch one of the computers of the ensemble we read something definite and we do not end up in a state of mental confusion in which we have "potentially" read both expressions, leaving it ambiguous what we have actually seen.

This simple observation tells us that at a certain point between the level of microscopic events that are doubtlessly governed by the principle of superposition (microsystems show the effects of interference) and the level of the perception on the part of a conscious observer, the linearity has to be violated. Where do we locate this boundary? This is a crucial question for the theory, the problem that, as Bell says in the quotation given at the beginning of this chapter, makes it impossible for anyone to know exactly what the theory is saying about any specific situation.

Some qualifications are in order. When we speak of a precise boundary, nobody is pretending that there exists some perfectly defined criterion of demarcation that allows us to say that up to here quantum mechanics is valid; from this point on, the reduction of the wave packet takes over. By a "precise boundary" is meant that the theory ought to contain at least some parameter that defines a scale that in turn would permit us to evaluate when it is legitimate to use the linear equations, and when it is only an approximation to use them, and when it is just plain erroneous to use them. An example should clarify this idea. Let us consider Newtonian mechanics and the theory of relativity. In this case it is easy to identify a precise parameter: the universal constant that is so characteristic of the theory of relativity, namely, the speed of light. This serves extraordinarily well for defining the (limited) area of validity of classical mechanics. If a body has a velocity much less than the velocity of light, Newton's theory is appropriate, but as the velocity increases it becomes increasingly less precise and fails completely when it comes to describing bodies that travel at velocities nearing that of light. We then can ask ourselves: what parameter plays a role in quantum mechanics analogous to the speed of light in classical mechanics? The simple and clear answer is that nobody has as yet been able to find it. The foregoing arguments might seem to point toward the number of particles as a possible candidate, but that does not work: there are macroscopic systems that require a quantum treatment to be correctly described. Superconductors could be mentioned, which show typical "tunnel effects" involving a macroscopic number of constituents. But it must be emphasized that to account for the internal structural properties of even a simple macroscopic crystal, or the behavior of electronic chips, or the functioning of transistors, etc., a quantum treatment and the principle of superposition are indispensable.

I would like to conclude this section by underlining the essential role that the ambiguity about the "boundary" has played in the debate over the conceptual basis of the theory. I will then sum up the matter through reference to a stimulating picture drawn by John Bell for one of his last lectures: an image that has the advantage of going right to the core of the problem.

As regards the first point, it should be enough to recall the debate between Bohr and Einstein that we analyzed in detail in chapter 7. The

CHAPTER FIFTEEN

reader may recall the escape strategy Bohr used to defend his position against Einstein's observations that a precise measurement of the state of a macroscopic object (see Figures 7.4 and 7.6) would lead to a violation of the uncertainty principle. Bohr's point consisted in asserting that it is only decisive that, in contrast to the proper measuring instruments [?], these bodies [i.e., moving diaphragms or pointers of an apparatus balance], together with the particles, would in such a case constitute the system to which the quantum mechanical formalism has to be applied. But what, in Bohr's view, would make the diaphragm or the pointer different from other systems used to determine the states of these objects—that is the real mystery. We have indicated that if Einstein had insisted on his requirement of attributing definite properties up to an appropriate point (of what we can call the von Neumann chain) he would have constrained Bohr to accept that the entire universe requires a quantum treatment. While discussing this debate previously, I referred in passing to the fact that the same von Neumann and Wigner were led to relocate the boundary between quantum and classical, between reversible and irreversible, at the act of perception on the part of a conscious observer. But it must be noted that even this last solution is not without ambiguity. In fact, the problem is simply transferred to the problem of defining precisely what is meant by a conscious observer, a concept the present state of our knowledge does not permit us to define in an unambiguous way.

I would now like to comment on the picture drawn by J. S. Bell and presented by him at two conferences: the first at the University of California at Los Angeles in March 1988 on the occasion of the seventieth birthday of the Nobel Prize winner Julian Schwinger, the second at Trieste in November 1989, on the occasion of the twenty-fifth anniversary of the founding of the International Center for Theoretical Physics by the International Atomic Energy Agency at Vienna. Bell analyzed the process of diffraction, through a slit, of a beam containing a certain number N of electrons, and the formation of the image on a photographic film placed behind the slit. He observed that it makes no sense to treat the electrons as if they were punctifom bodies that follow precise trajectories; they must be described by means of a diffracting wave function (Figure 15.4).

Since the number of the electrons is very large but still finite, what we see on the film will be a series of N black dots that correspond to the positions in which they are, so to speak, "revealed" by that object, which can be thought of as a measuring device. In making this assertion, we have made a logical jump from the language of wave functions, and of the potentialities of a microsystem, to the language of the reality of the

FROM MICROSCOPIC TO MACROSCOPIC

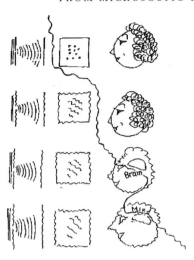

FIGURE 15.4. Where are we to place the boundary between the vague microscopic world and the precise world of our sensory experience? Here is an image John Bell proposed to illustrate this deep question.

dots. But we know very well (recall Figure 3.10) that because the wave function of each electron is appreciably different from zero at all the points of the figure of diffraction, in reality, if we were to treat the photographic film as a quantum system (and what would be wrong in doing so?), we would have a linear superposition of states. In one of them, for example, each of the N electrons is in a precise position among the infinite that are possible, and the film is in the state in which the activated grains of silver bromide are really those that correspond to the positions indicated. But there exists an infinity of other possible outcomes of this process, each one of which would correspond to a different distribution of the N electrons in the central region of the spot and to a different collection of activated grains of the emulsion. If we were truly interested in describing the process, we would have to elaborate the photographic process in detail, and look at the entire electron + film system as a genuinely quantum system. This would mean that we could no longer speak of spots in precise positions, but only of the potentiality that the film would be exposed at certain N points rather than at others. We have dislodged the boundary from the microcosm of the electron beam to the macrocosm of the film, as shown in the second stage of the figure. But we cannot stop here. When we watch the film, our own perceptive apparatus enters the picture. But is there any reason to think that our eye's retina is

CHAPTER FIFTEEN

not also a physical system, and as such is subject to the linear laws of the theory? If we want to describe which signals actually reach the brain, we would once more have to relocate the boundary between the vague quantum world and the world of definite events, and we are then led to place it between the optic nerve and the brain. But even the brain is a physical system constituted of protons, neutrons, and electrons, and traversed by electrochemical reactions and the rest—processes that we have no reason to think are not governed by our formalism. And thus it makes perfect sense to relocate the boundary once again between the brain and the mind, as in the last sketch of the figure.

15.7. A Famous Historical Example of Entanglement: Schrödinger's Cat

As we mentioned in chapter 8, two fundamental works were published in the year 1935 that made clear the absolutely revolutionary implications (much more revolutionary than any other aspect of the formalism) of entanglement. In their celebrated work Einstein, Podolski, and Rosen showed how the entanglement between two microsystems, joined with a few ordinary postulates, would lead one to the conclusion that the formalism is fundamentally incomplete. Schrödinger, again with reference to entanglement (which he defined as "the characteristic trait of Quantum Mechanics, the one that enforces its entire departure from classical lines of thought"), in a long and profound article stressed the problematic situation that arises, as we have explained, when we take seriously the possible—or rather, inevitable—entanglement between a microsystem and a macrosystem. In order to impress the reader in the most forceful way, Schrödinger came up with a fictional physical process that has acquired worldwide fame. We can cite his own words:

> One can even set up quite ridiculous cases. A cat is penned up in a steel chamber, along with the following diabolical device (which must be secured against direct interference by the cat): in a Geiger counter there is a tiny bit of radioactive substance, so small, that perhaps in the course of one hour one of the atoms decays, but also, with equal probability, perhaps none; if it happens, the counter tube discharges and through a relay releases a hammer which shatters a small flask of hydrocyanic acid. If one has left this entire system to itself for an hour, one would say that the cat still lives if mean-

FROM MICROSCOPIC TO MACROSCOPIC

while no atom has decayed. The first atomic decay would have poisoned it. The Ψ function of the entire system would express this by having in it the living and the dead cat (pardon the expression) mixed or smeared out in equal parts.

We can take this argument up ourselves with reference to the experiment discussed in Section 15.2. For this purpose it will be enough to replace the measuring apparatus of Figures 15.1, 15.2, and 15.3 by the measuring apparatus represented below (Figure 15.5):

This apparatus consists of the usual box equipped with two small regions sensitive to the impact of a single photon, one of which is such that, when it is activated (by the photon following the ordinary ray), it does not induce any change to the situation inside the box, while the other, as in Schrödinger's example, if activated (by a photon following the extraordinary ray), causes the breaking of a phial containing a lethal gas.

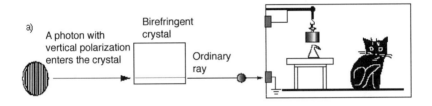

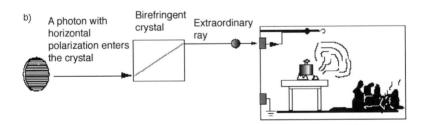

FIGURE 15.5. Preliminaries for preparing a linear combination state of "live cat" and "dead cat." If the two processes being considered develop as shown, then, by traversing a crystal with a photon in the superposition of the two states, under the assumption that Schrödinger's linear equation governs the entire process, there will automatically be produced the embarrassing physical situation described in the text and illustrated in Figure 15.6.

CHAPTER FIFTEEN

This leads us to repeat our argument: if a photon with vertical polarization is introduced into the birefringent crystal, it will propagate itself along the ordinary ray; it will with certainty strike the region at the lower part of the box, and will not bring about any unpleasant consequences for our cat. On the other hand, if the photon is horizontally polarized, it will follow the extraordinary ray and will activate the upper sensitive part of the box, triggering a process that will end with the breaking of the phial and the death of the cat.

The next step is obvious: a photon polarized at 45° (which we know very well is a linear superposition of the two previous states) is sent through a birefringent crystal. At its exit from the crystal, it cannot be thought of as localized along one of the two rays; rather, it is in a superposition of the two states, each of which corresponds to one of the two possible positions (see Figure 4.5). The linear nature of the evolution brings it about that finally the situation of Figure 15.6 is realized, wherein the most important and embarrassing element is the + between the two macroscopically (and conceptually) very different states that appear to us. This + implies that we cannot say that the cat is alive or that it is dead, and that this simultaneous presence of two incompatible potentialities can only be eliminated by carrying out an observation to determine if the cat is dead or alive. In a literal sense, if we accept that by looking in the box we will find out for sure that the cat is either dead or alive, we are led to the conclusion that our very act of conscious perception is what determines the fate (happy or sad) of the poor cat.

Schrödinger's example strikes the imagination immediately. It represents such an impressive illustration of the crucial problem of quantum theory that it is not surprising that it has become a kind of paradigmatic symbol of quantum theory. I would like to underline this fact with refer-

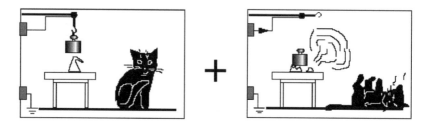

FIGURE 15.6. The final state of the diabolical machinery thought up by Schrödinger: a cat that is "potentially" alive or dead, but "actually" neither!

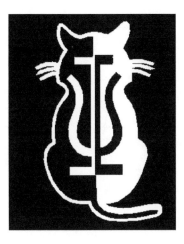

FIGURE 15.7. The logo of the ISQM conference, held every three years in Japan, and dedicated to the foundations of quantum mechanics.

ence to an interesting graphic versions of it. It represents the splendid logo of the ISQM (International Symposium of the Foundations of Quantum Mechanics in the light of New Technology), held every three years in Tokyo (Figure 15.7).

For the conclusion of this chapter, it seems fitting to include two illustrations done by John Bell, shown together with Figure 15.4 at the two conferences earlier mentioned. They are excellent illustrations of the wisdom and subtle humor of this brilliant scientist. John wanted to replace the diabolical device of the rather "macabre" original with something simpler and more agreeable. Above all, as he emphasized, in one of the two alternatives of Schrödinger's example, *even the cat does not know what is going on half the time*—i.e., if he dies. Consequently, Bell replaced the glass phial of poison with an electric switch that pours a little milk into a saucer. The cat—which is hungry—will certainly lap up the milk if it is there in the saucer. But since, as in the original case, it is not legitimate to say that the atom has or has not decayed in the hour we have left the system alone, Bell's cat ends up in the superposition of a cat with a full stomach and a cat with an empty stomach. Once again there are the potentialities of the cat being satisfied or still hungry (Figure 15.8), but there is no actuality about these properties until the exact moment when the cat's owner checks in on the cat to see what happened (Figure 15.9).

CHAPTER FIFTEEN

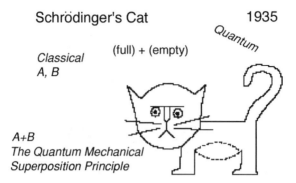

FIGURE 15.8. Bell's cat, in the superposition of being | a cat with a full stomach⟩ and | a cat with an empty stomach⟩. This is Bell's nonlethal version of Schrödinger's cat, featured in the thought experiment of the following figure.

This will conclude the chapter: what we have to see next is how the scientific community has been reacting to these difficulties, and what solutions have been proposed to overcome them.

. . . . . .

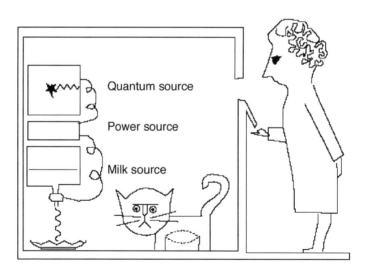

FIGURE 15.9. The experimental setup imagined by Bell as a replacement of Schrödinger's.

FROM MICROSCOPIC TO MACROSCOPIC

APPENDIX 15A: IDENTIFYING THE SUPERPOSITION

The problem to be confronted here is how to distinguish experimentally an ensemble of systems all of which are associated with the same entangled state (15.11), which we begin with by recalling it here:

$$|\text{final}\rangle = \frac{1}{\sqrt{2}}\left[|V, A_{+1}\rangle + |H, A_{-1}\rangle\right] \quad (15A.1)$$

from an ensemble E_{red} of systems, half of which are associated with state $|V, A_{+1}\rangle$ and the other half with $|H, A_{-1}\rangle$.

We consider various possible processes of measurement and we evaluate the probabilities of their outcomes according to the laws of the theory. Of course, in accordance with the position discussed in the text, in the first case the prescriptions of the standard formalism have to be applied to state (15A.1), with the postulate of the reduction of the wave packet, while in the second case we have to take into account that, in choosing at random a system, it would have a 1/2 probability of being associated with the first and the same probability of being associated with the second of the two states of the ensemble E_{red}. The calculation has then to be carried out, determining the value of the probabilities of the outcomes in the two cases and giving an equal probability to each.

Suppose we carry out a measurement that identifies the position of the pointer on the apparatus. As was shown in Section 15.5, if we use the symbols $P(A = +1)$ and $P(A = -1)$ to indicate the probabilities that the pointer will point to +1 or −1, we find they are equal:

$$P(A = +1) = 1/2, \; P(A = -1) = 1/2. \quad (15A.2)$$

both for the state (15A.1) and for the ensemble E_{red}. This type of measurement does not therefore enable us to distinguish between the two cases.

Now let us carry out a plane polarization measurement of photons in directions other than vertical. To begin with a precise example, we choose a test of 45° polarization. How should we proceed in order to evaluate the relative probabilities? The analyses of Sections 4.7 and 6.4 tell us that given one state, we ought to express it as a linear combination of normalized states that correspond to possible and mutually exclusive outcomes of the measurement in which we are interested. In our case this means that we ought to express the states of polarizations $|V\rangle$ and $|H\rangle$ in terms of the states $|45\rangle$ and $|135\rangle$, which represent, respectively, the states where the photon passes or fails the test of 45° at polarization. We already know the relations that exist between these states (recall equations [3.2], [2.7], and [2.8]):

$$|V\rangle = \frac{1}{\sqrt{2}}\left[|45\rangle + |135\rangle\right], \; |H\rangle = \frac{1}{\sqrt{2}}\left[|45\rangle - |135\rangle\right]. \quad (15A.3)$$

373

CHAPTER FIFTEEN

Substituting these formulas into the state (15A.1) we get

$$|\text{final}\rangle = \frac{1}{2}[|45, A_{+1}\rangle + |135, A_{+1}\rangle + |45, A_{-1}\rangle - |135, A_{-1}\rangle]$$

$$= \frac{1}{\sqrt{2}}|45\rangle\left\{\frac{1}{\sqrt{2}}[|A_{+1}\rangle + |A_{-1}\rangle]\right\} + \frac{1}{\sqrt{2}}|135\rangle\left\{\frac{1}{\sqrt{2}}[|A_{+1}\rangle - |A_{-1}\rangle]\right\}. \quad (15A.4)$$

Note that the states within the curly brackets are normalized states of the measuring apparatus. The equation (15A.4) tells us immediately that the photon states $|45\rangle$ and $|135\rangle$ have coefficients equal to each other and both equal to $1/\sqrt{2}$, which implies that they have a 1/2 probability that the photon overcomes or fails a test for 45.

We now can go on to the case of the ensemble E_{red} and begin by considering a system in the state $|V, A_{+1}\rangle$. This state is now written in such a way as to make clear its decomposition in terms of the states $|45\rangle$ and $|135\rangle$:

$$|V, A_{+1}\rangle = \frac{1}{\sqrt{2}}|45, A_{+1}\rangle + \frac{1}{\sqrt{2}}|135, A_{+1}\rangle, \quad (15A.5)$$

from which it follows that it has a 1/2 probability of overcoming a test of polarization at 45°. The corresponding formula for the state $|H, A_{-1}\rangle$ is then

$$|H, A_{-1}\rangle = \frac{1}{\sqrt{2}}|45, A_{-1}\rangle - \frac{1}{\sqrt{2}}|135, A_{-1}\rangle, \quad (15A.6)$$

and this too tells us that a photon like this has a 1/2 probability of passing a test for 45°. Since by choosing a system at random from the ensemble E_{red} this system is with equal probability in the state $|VA_{+1}\rangle$ or in the state $|HA_{-1}\rangle$ and for both of these occurrences it has an equal probability of passing or failing the next test of polarization, it follows that even in this case there is a 1/2 probability that the photon passes or fails a test for polarization at 45°. Once again, this type of measurement does not permit us to distinguish between the state (15A.1) and the statistical mixture E_{red}.

I will not take the time now to show how measurements that have to do exclusively with observables of the apparatus likewise do not permit the distinction. Instead, I will move on to show that measurements of correlation that combine available noncompatible observables with those that characterize either the state of the photons or of the apparatuses in (15A.3), do permit distinguishing the state (15A.1) from the statistical mixture E_{red}.

As far as regards the photons, we have already considered observables incompatible with vertical polarization, that is to say, polarization measurements at 45°. Accordingly, we must substitute the expressions (15A.3) into all the states in which we are interested. We must now consider an observable for the apparatus

FROM MICROSCOPIC TO MACROSCOPIC

that is incompatible with the observable that corresponds to the states for which the index is +1 or −1, respectively. Without getting lost in the details about how such a measurement could take place, let us imagine that there exists an observable Z of the apparatus that assumes the value X when the state of the apparatus is

$$|X\rangle = \frac{1}{\sqrt{2}}\left[|A_{+1}\rangle + |A_{-1}\rangle\right] \tag{15A.7}$$

and the value Y when the state is

$$|Y\rangle = \frac{1}{\sqrt{2}}\left[|A_{+1}\rangle - |A_{-1}\rangle\right]. \tag{15A.8}$$

The inverses of these equations are obviously

$$|A_{+1}\rangle = \frac{1}{\sqrt{2}}\left[|X\rangle + |Y\rangle\right], \quad |A_{-1}\rangle = \frac{1}{\sqrt{2}}\left[|X\rangle - |Y\rangle\right]. \tag{15A.9}$$

The reader will certainly have understood that in the same way in which a measurement of polarization at 45° is incompatible with a measurement of vertical polarization, in a completely analogous way, a measurement of the observable Z is incompatible with one intended to reveal if the pointer of the apparatus points to +1 or −1. In fact, the relationship between the states $|V\rangle,|H\rangle$, and $|45\rangle,|135\rangle$ is exactly the same as the relationship between the states $A_{+1}\rangle,|A_{-1}\rangle$ and $|X\rangle,|Y\rangle$. We can now proceed in our analysis.

In the case of state (15A.1), by expressing the states $|V\rangle,|H\rangle$ and $|A_{+1}\rangle,|A_{-1}\rangle$ in terms of states $|45\rangle,|135\rangle$ and $|X\rangle,|Y\rangle$, from (15A.3) and (15A.9) we derive

$$|\text{final}\rangle = \frac{1}{\sqrt{2}}\left[|45,X\rangle + |135,Y\rangle\right], \tag{15A.10}$$

which tells us that in a joint measurement of 45 polarization and of the observable Z of the apparatus, the outcome "the photon passes the test" is perfectly correlated with the outcome "X" and alternatively, the outcome "the photon fails the test" is perfectly correlated with the outcome "Y," each of the two pairs of outcomes occurring with an equal probability. On the contrary, the outcomes " the photon passes the test and the apparatus is found in the state Y" or "the photon fails the test and the apparatus is found in the state X" will never happen

If we move on to consider the ensemble E_{red}, we see that its members will be described either by the state

$$|V,A_{+1}\rangle = \frac{1}{2}\left[|45\rangle|X\rangle + |45,Y\rangle + |135,X\rangle + |135,Y\rangle\right], \tag{15A.11}$$

CHAPTER FIFTEEN

or alternatively by the state

$$|H, A_{-1}\rangle = \frac{1}{2}\big[|45\rangle|X\rangle - |45, Y\rangle - |135, X\rangle + |135, Y\rangle\big]. \qquad (15A.12)$$

From equations (15A.11) and (15A.12) we see that in both cases the four pairs of outcomes (45,X), (45,Y), (135,X), and (135,Y) for the joint measurements are equally probable (with a probability of 1/4). It follows that these probabilities are those that characterize the outcomes of the correlation measurements of the indicated observables. We should notice in particular that the pairs of outcomes (45,Y) and (135,X), which can never arise when the state is (15A.1), have the same probability as the other pairs in the case of E_{red}.

This concludes the demonstration that a distinction between the state (15A.1) and the mixture E_{red} is possible, but requires very difficult correlation measurements of observables that are incompatible with those that characterize the terms of (15A.1) itself.

. . . . .

CHAPTER SIXTEEN

In Search of a Coherent Framework for All Physical Processes

> An interpretation of a theory is an attempt to answer questions which that theory has left open. . . . Not every attempt to answer questions left open by the theory counts as an interpretation. If answers are given which alter the empirical predictions, we don't call that an interpretation but a rival theory. . . . Every time we discover a new tenable interpretation, we understand more, we understand the theory better. Even if we find that the interpretation is tenable only by some criteria, or if eventually it fails completely, we have gained a good deal more understanding from that. What we have found in each case, is a partial answer to the general question: how could the world possibly be the way the theory says it is?
> —Bas C. van Fraassen

THE FOREGOING CHAPTER was dedicated entirely to a clear statement of the conceptual problems encountered by the theory in the description of macroscopic systems and the processes of measurement. We spoke of the existence of scientists who still think that the problems occupying us here (and which have engaged some of the most brilliant minds of the century) are in reality pseudoproblems, or merely verbal problems requiring merely verbal solutions.[1] But I have also called the attention of the reader to the fact that such opinions as these are decisively in decline and that, on the contrary, among those scientists who are seriously concerned with these matters, a nearly unanimous consensus is arising around the fact that these are genuinely scientific problems that require, as John Bell said, "mathematical work by theoretical physicists, rather than interpretation by philosophers."

I hope that the reader has understood fully the essential aspects of the difficulties being encountered: if the theory (and actually only its linear character, i.e., the superposition principle) is taken seriously as a conceptual framework for describing reality, then it permits—or rather, requires—the occurrence of situations that are not compatible with the definiteness and precision of the macroscopic world as we perceive it. We inevitably must accept the fact that a macroscopic system can be described by a state that is the linear superposition of two states, each of which corresponds to a distinct and well-defined macroscopic situation ("the pointer on the apparatus is in this position" + "the pointer on the apparatus is in that position" or "the cat is alive" + "the cat is dead"). And yet, precisely because of the + that stands between the two terms, this situation does not permit us to assert (unless we work out new, more appropriate hypotheses about the interpretation of the theory) that one or the other of the situations in question is in fact the one that characterizes the individual system we happen to be investigating.

As we have seen, the overwhelming majority of the experts in the field are convinced that this problem has to be faced, and, in fact, during the sixty years in which it has been a subject of debate, many interesting suggestions for solving it have been proposed. In this chapter my plan is to survey the most relevant of these suggestions, and to point out what seem to me to be their most salient features, their advantages and disadvantages.

But instead of following a chronological order of presentation (which might seem to be the most natural way of going about it), it will actually make more sense to order these proposals on the basis of the relevant physical aspects that characterize them.

16.1. Strategies for Identifying Possible Solutions

One particularly appropriate method for understanding the essential elements of the various solutions to the so-called problem of the objectification of the macroscopic properties of physical systems consists in considering the essential elements of the formalism. Quantum mechanics, like every physical theory beginning with the Galilean revolution, accounts for physical processes by means of a mathematical formalism that instructs us precisely in (1) how the states of the physical systems we are interested in should be described, (2) what entities are going to be associated with measurable physical quantities, (3) how to move from the knowledge of a system's state at a certain instant of time to a knowledge of the state at later times, and finally (4) how to obtain from the knowledge of the state information about the outcomes of future processes of measurement. Within the standard quantum framework these four aspects are represented, respectively, by (1) the state vector, (2) the observables of the system, (3) the evolution equation, and (4) the probabilistic rules that govern the outcomes.

A first distinction among the proposed solutions concerns the position taken up by each one on how to characterize formally the states of individual physical systems. The question can be summed up in the following question: Is quantum mechanics complete or does it have to be completed? In other words, we are asking: Does the state vector represent by itself the most accurate possible specification of an individual physical system? If the answer to this question is in the affirmative, the position is the orthodox one. If, on the other hand, the answer is in the negative, then we have taken the road of the *hidden variable theories* we analyzed above, in Chapter 9. In the section to follow, this kind of theory will be investigated again, this time with specific reference to the measurement problem and to macroscopic systems, and we shall see that, although this perspective does indeed allow us a way out of the impasse of the superposition of macroscopically distinguishable states, there is also a price to be paid for it.

But let us examine the next alternative. We assume that the standard theory is complete, which is to say that the states of individual physical systems are exhaustively described by the state vector. We can begin by considering a system that at a certain instant is characterized by a state vector that is an embarrassing superposition of two states corresponding to macroscopically diverse and mutually irreconcilable situations—let us say, for example, the dead cat and the live cat. As I have already pointed out, by being fundamentally probabilistic, the theory requires the study of a statistical collection to verify its predictions. In Section 15.5 and in Appendix 15A we showed how, in practice, the distinction between a collection that is "quantum mechanically homogeneous" (where all the members of the collection are in the same embarrassing state we are considering), and a collection that is "nonhomogeneous" (where some of the members are in the state "live cat" and some in the state "dead cat") requires extremely complicated correlation measurements involving all the constituents of the process under consideration. It is obvious, then, that when we limit (whether in principle or for practical motives) the class of

CHAPTER SIXTEEN

quantities considered as observables permitted for the system, we might conclude that since we have no experimental means for showing the superposition, it would then be legitimate to identify the embarrassing collection of so many cats "neither alive nor dead" with the reasonable collection "some cats alive and some cats dead." These observations suggest that appropriate assumptions about the observables of the system, in particular a *limitation of the physical quantities which can actually be measured*, would be able to remove us from the interpretative difficulties we have encountered. This line of thought will be analyzed and further articulated in section 16.3.

Until now we have tacitly assumed a somewhat "traditional" position about reality. A whole family of possible solutions to the problem that torments us tries to go around it by assuming very "strange" positions of the physical world: positions that configure that world with more articulation than we are accustomed to. Once again, a quantum mechanically "homogeneous" situation, that is, one characterized by a unique and very precise state vector, is assumed capable of being associated with a large variety of physical situations. However, this time the multiplicity does not refer to different members within a collection of physical systems of our universe, but to a system of hypothetical universes different from each other. This is how we arrive at the *many worlds* or *many minds* interpretations. The latter kind, instead of multiplying realities, multiply the perceptual modalities of the subject.

Finally, as mentioned previously, another essential element of the formalism to which we can resort in order to go around the obstacles is the evolution equation of the theory. In fact, as we observed before, quantum mechanics already in its orthodox formulation assumes that the linear evolution does not have general validity: when a process of measurement occurs it is necessary to suspend Schrödinger's law of linear evolution and to have recourse to a new dynamic principle, described by the reduction of the packet. There emerges then a fundamental dualism of physical processes. The real problem becomes that of identifying exactly when, that is, for which specific systems and in which precise situations, one or the other principle of evolution should be followed. We will study some interesting proposals for locating this line of demarcation, and we shall see that they require taking an extremely precise philosophical position about the *special role of the conscious observer* in determining the state of the entire universe.

Keeping our reference to the dynamic laws of the theory we can follow a line of thought that has been recently proposed. This refutes the ambiguous dualism of the case just considered and requires a modification of Schrödinger's own equation, which is substituted by an equation that satisfies two apparently opposing needs, that is, it does not enter into a conflict with any of the thoroughly verified predictions of the standard theory for microsystems, and at the same time *spontaneously* induces a *dynamic reduction* of the embarrassing superpositions of macroscopically distinguishable states to one of the nonambiguous terms, and does this in an extremely brief time. In the terms used by van Fraassen quoted at the head of the chapter, this theory appears to "rival" standard quantum mechanics rather than offer a new "interpretation."

We can now proceed to an analysis of the various possibilities just catalogued.

16.2. Incompleteness of the Standard Theory: Hidden Variable Theories

The hidden variable theories formed the object of an especially detailed analysis in chapter 9. It would help to recall here two essential points of that analysis:

1. Such theories claim to assign to any observable a precise value, determined by the hidden variables themselves. It is therefore obvious in particular that the observable "position of the pointer on the apparatus" will always have a definite value for each individual system, and that this will depend on the value of the hidden variables that characterize it.

2. Even so, we cannot conceal the fact, discussed in detail in section 9.6 and in Appendix 9A, that every conceivable theory which constitutes a completion of quantum mechanics in a deterministic sense cannot avoid having the majority of the observables turn out to be contextual. This means that the value the theory allows us to consider as "possessed" by the observables is not unambiguously determined by the hidden variables themselves, but is dependent on the whole context, and, in particular, is nonlocally dependent on this context. As I have already emphasized, the conceptual status of a contextual property is certainly much less "objective" than that of a noncontextual one. So then the problem arises of identifying the noncontextual variables and, in my view, the only logically consistent approach to be taken toward such theories is to say that the only properties of which such theories can speak, and to which they can attribute an objective

CHAPTER SIXTEEN

value, are just the values assumed by this kind of variable—i.e., the noncontextual ones.

To understand the meaning of this assertion fully, it will be helpful to refer to the pilot wave theory. As we pointed out repeatedly, this theory describes the physical situation in terms of the wave function and the hidden variables that are the positions of all the particles of the system. Let us study from this perspective a measuring process such as that of Figure 15.3. In Bohm's theory, the final state vector of the system is precisely that of equation (15.7), that is, the superposition of states corresponding to the pointer in position −1 and to the pointer in position +1. But the hidden variables of the system, that is, the positions of all the particles that come into play, are perfectly definite. This means, literally, that the pointer is in fact *either* in position −1 *or* in position +1, and that we ought not to consider the reduction process. Obviously, the fact that the wave function is different from zero even in the "empty" region where the pointer is not located will have dynamic consequences. But it can easily be shown that in the case of a macroscopic system, the effect of this part of the wave function upon the future evolution of the system can be disregarded to a very high degree of approximation. Therefore, if someone wants to, he can include a kind of formal wave-packet reduction process within the theory, but this is not by any means necessary. Even if someone were to take into account in an absolutely rigorous way the very slightest effects of the *empty* wave function, he would not end up with any contradiction and no difficulty would arise with respect to the pointer: it would always actually be found in a very precise position, in accordance with our sensory experience.

It follows from this that in the context of this theory, the assertion "the cat is in fact *either* living *or* dead" is perfectly legitimate. In fact, the positions of the cat's own particles are certainly different in the two cases, and since they are the perfectly definite variables of the theory, even though the wave function includes the component "the cat is dead (alive)," in fact "the cat *is* alive (dead)." In Figure 16.1 is a summary representation, with reference to this famous example, of the position one has to assume if one adopts the point of view of the hidden variable theories.

Of course, we have not encountered any problem in the above analysis just because the relevant variables for the process in question were precisely the positions that are the only noncontextual, and hence objective, variables: the position of the pointer on the apparatus corresponds to a property objectively possessed (that is, independently of our choice to go

A FRAMEWORK FOR PHYSICAL PROCESSES

$$|\Psi\rangle = (1\sqrt{2}) \{| \text{ Live cat }\rangle + | \text{ Dead cat }\rangle\}$$

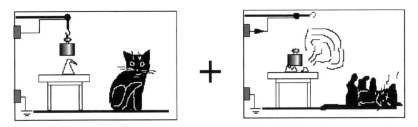

But if one knows $|\Psi\rangle$ & the hidden variables λ

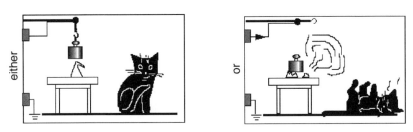

Figure 16.1. What the Schrödinger's cat problem looks like in the context of a hidden variable theory. Although the wave function is a superposition of the states |live cat⟩ and |dead cat⟩, the specification of the value assumed by the hidden variables λ permits the assertion that the cat is in fact alive (or, alternatively, dead).

"see" if it is there), the particles of the cat's tail are in the positions that correspond, as time passes, to the fact that it is "wagging" (if the cat is alive) or "not wagging" (if it is dead). But the contextuality remains.

In conclusion, it seems to me that the de Broglie–Bohm theory offers a coherent and extremely interesting solution to the problem of the theory of measurement, provided we accept without reservation that the theory can only and exclusively make assertions relative to (present and future) positions of all the particles of the universe. To say that a system has a certain energy, a certain velocity, etc., is understood simply as affirming that if the energy is measured (or if any physical process that depends on the value of the energy takes place), then the future distribution of the particles (for example, the position of the pointer on the gauge) will be the one unambiguously predicted by the theory. Any further connotation that one wants to give to assertions about contextual quantities would have a status that could certainly not be defined as "objective."

CHAPTER SIXTEEN

16.3. LIMITING THE OBSERVABLES OF MACROSYSTEMS

As explained in section 15.6 and demonstrated in Appendix 15A, any procedure that permits the experimental distinction between the state (15.11) and the statistical mixture E_{red} (corresponding to a collection of which half of the systems are associated with the state $|V,A_{+1}\rangle$ and half with the state $|H,A_{-1}\rangle$) requires measurements of correlations involving observables of the apparatus that are incompatible with the observable that corresponds to the position (+1 or −1) of the pointer.

The first idea[2] advanced was that, because of the macroscopic nature of the measurement apparatus, it presents a so-to-speak "classical" behavior, so that it would simply be impossible to carry out a measurement of any observable that would be incompatible with the observable "position of the pointer." It is clear that such an attitude solves the problem, since it would always be possible to replace a collection associated with the state vector (15.11) with one associated with an appropriate statistical mixture, whose members would all have precise macroscopic properties. Interpreted in this way, the theory would require that all the macroscopic characteristics that are actually accessible in the case of an individual physical system would end up being objectively possessed. But—alas—this escape route is not practicable for the simple reason (which we will not demonstrate but will only state) that the hypothesis advanced comes into a contradiction with the hypothesis that the apparatus is able to "measure," or give us "information" in a reliable way, about the properties of the microsystem it is devised to ascertain.

Since the idea of limiting "in principle" the actual measurable quantities is not practicable, it is natural to wonder if it cannot be substituted by a "factual" impossibility, which would then provide us with the escape we need. The idea is fairly simple and can be expressed in elementary formal terms.

We consider again the state (15.11), but we understand, as emphasized previously, that the pointer of the apparatus can never be considered to be perfectly isolated from its environment. This means that the state $|V,A_{+1}\rangle$ will be correlated with the air molecules, with the photons that fall on the pointer, or even with the neutrinos that strike it, bringing into play the state $|V,A_{+1}, E+\rangle$, in which the symbol E+ is an abbreviation for the general state of the environment that is correlated with the position +1 of the pointer. Analogously, the state $|H,A_{-1}\rangle$ brings to the state $|H,A_{-1},E-\rangle$. The actual state that concerns us immediately after the process of measurement will not be the state on which we have based our argument, that is, (15.11), but rather will be (still assuming the unlimited validity of quantum evolution)

$$|\text{final}\rangle = \frac{1}{\sqrt{2}}\left[|V, A_{+1}, E +\rangle + |H, A_{-1}, E -\rangle\right] \qquad (16.1)$$

Now, the situations of the entire environment summed up in the symbols E+ and E are mutually exclusive and incompatible, and if it is not possible to carry out correlation measurements among all the constituents of the environment, the system, and the apparatus, we know that there is no practical possibility of distinguishing the situation corresponding to the state (16.1) from the situation corresponding to the statistical mixture E_{red} of a collection in which half of the members of the (system + apparatus + environment) are in the state $|V, A_{+1}, E+\rangle$ and the other half in the state $|H, A_{-1}, E-\rangle$. The conclusion is clear: it is perfectly legitimate to assert that, in fact, we will have one of the two alternatives now considered (which, we should note, correspond to situations perfectly consistent with our perception that the pointer is really either +1 or −1), in so far as *from the practical point of view* there is no way of showing that things are not like this. I wanted to underline the crucial point of the argument by showing in italics the fact that limitation of measurability for macrosystems or for the still more complex environment is not going to be assumed here as a postulate, but accepted as an unavoidable practical fact.

Bell discussed this problem many times. He repeatedly explained that the approximation of substituting a collection whose members are all described by state (16.1) with the collection E_{red} "is manifestly an approximation. The approximation is good enough in practice, and perhaps it will always be good enough in practice"—i.e., enough so as not to be falsified in a laboratory. Actually, with his usual subtle humor and critical spirit, Bell suggested the use of the acronym FAPP (for all practical purposes) to show the necessity of limiting our statements to conclusions that have only a practical validity, in order to avoid logically embarrassing situations. He also proposed that every theorem of quantum mechanics should end with the letters QED-FAPP (quod erat demonstrandum ["what was to be proved"]—for all practical purposes).

As in the previous section, I have summed up in Figure 16.2 the meaning of the position I have just outlined.

Let us now consider seriously the position of those who, having taken account of the considerations just developed, assert that the theory does

CHAPTER SIXTEEN

The real state of affairs is the one pictured here:

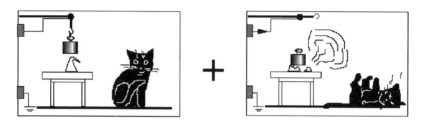

But, owing to insurmountable difficulties of a practical nature, the above situation cannot be distinguished from this one:

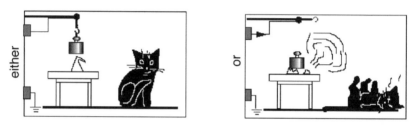

Figure 16.2. The escape route from the difficulties of the formalism that corresponds to taking account of the practical limitations on the measurability of physical quantities at the macroscopic level. The theory asserts that in reality the situation is really the one corresponding to the superposition, which is incomprehensible (on the basis of our sensory experience), but because of practical difficulties it is legitimate to introduce an approximation which allows us to say that one or the other of the reasonable situations represented in the lower panel of the illustration really occurs.

not present any interpretative problems. At a certain level it is legitimate (not in principle, but for all practical purposes) to say that the embarrassing situation (16.1) never happens, but that we are always confronted with one of the reasonable states of the mixture E_{red} that correspond to macroscopic situations in accordance with our perceptions. From this perspective, the position under investigation could be considered reasonable and acceptable. The situation would be, at first sight, analogous to that of those who would maintain that, despite the fact that mechanical laws are completely reversible, it is legitimate to assume FAPP the validity of the irreversible laws of thermodynamics. In fact, we all know that these laws are approximations that increase in accuracy with an increase of the number of constituents involved in the system in question (typically a gas), and we also know that in order to demonstrate explicitly that they do not represent "true" natural laws but only represent an approximation, we would need to have a Maxwell's demon at our disposal, who could easily produce a violation of the laws.

If things were really like that, that is to say, if the relationship between the linear evolution of the theory and the postulate of the reduction of the wave packet corresponded, *mutatis mutandis*, to the relationship between the reversible laws of classical mechanics and the irreversible laws of thermodynamics, we might say we were satisfied. But this is not the case.

We begin by taking into consideration the classical mechanics of a perfect and isolated gas. Well, there is an important theorem of the great Poincaré stating that, provided we wait a sufficiently long time (a period now known as "Poincaré's recurrence time"), the state of the system would return as closely as possible to its present state. Let us consider the true import of this theorem for a real gas. At the beginning, the gas is confined to a corner of a container. Then the gas expands. What do the irreversible laws of thermodynamics tell us? That the gas will occupy the entire volume accessible to it, and that it will remain in this situation of thermodynamic equilibrium practically forever. Poincaré's theorem assures us that this statement is false because if we wait a sufficiently long time (billions and billions and billions of years) the state of the system will return as "close" as possible to the initial state, which means that the gas will spontaneously compress. After this long wait we might legitimately conclude that we have obtained experimental proof that thermodynamics has been violated and represents only an approximation to the correct theory (the perfectly reversible mechanics). But all this does not really amount to much. In fact, we can ask ourselves, does the discovery (requiring an endless stretch of time) that the laws of thermodynamics are only approximations of the correct dynamic laws falsify the assertion that now (and for intervals of time on a cosmic scale) our gas will actually expand and, once having reached its thermodynamic equilibrium, will remain in that state except for very small fluctuations? Certainly not!

Now let us look at the (only apparently) analogous problem of the relationship between the "true" quantum linear evolution and the approximate description that makes recourse to the reduction of the wave packet. In this case, too, unimaginable experiments or interminable waiting periods are needed to demonstrate that in fact, the assertion that now the pointer shows +1 or −1 constitutes an approximation to the true situation. But we need to ask ourselves: if we wait long enough, or if we imagine ourselves capable of performing a marvel like that of Maxwell's demon, would the demonstration that we would then obtain—that the reduction hypothesis is equivalent to giving an approximate description of the real situation—falsify the assertion that now, and for a long stretch of time, the system that we have before us or the constituents of the collection that we are studying are really either in the state $|V, A_{+1}, E+\rangle$ or in the state $|H, A_{-1}, E-\rangle$? The answer would clearly be in the affirmative. The assertion that *now* our systems are in the reasonable states just described is demonstrated to be absolutely false by an experiment that would take place within an undetermined time and that certainly surpasses by many orders of magnitude the expected duration of life on earth.

But there is more. When it comes to classical mechanics or thermodynamics, the states of physical systems and their evolution are perfectly reasonable and compatible with our sensory experience. In the case of quantum theory, the state (16.1) is absolutely deprived of any physical meaning in so far as it does not correspond to any of our perceptions (we perceive something as being here or there), while the approximate states of the mixture are perfectly meaningful and consistent with our experience. I do not believe that there has ever occurred, in the history of science, a situation according to which a theory that is supposed to have universal validity brings results to which no meaning can be attributed, while having recourse to an approximation brings about an acceptable situation.

Despite these reservations, we cannot get away from acknowledging that for all those who do not think that science can lead us to any form of comprehension of the real but that its aim finally only amounts to making predictions of future events or to the construction of useful devices (that is to say, for all those who assume an instrumentalist position in regard to scientific knowledge)—for persons such as these, the solution outlined in this section might be acceptable.

16.4. THE MULTIPLICATION OF WORLDS AND THE FOLIATION OF MINDS

In the preceding section our point of departure was the consideration of a single physical system in an embarrassing superposition of macroscopically and perceptually diverse states. A state of this kind means that, if there exists, on some level, a nonambiguous way to determine "which macroscopic state the system is in" the outcome will be that, in half the cases, it will be in one term of the superposition, and, in the other half of the cases, in the other. We have seen how, if for all practical purposes it is impossible to perform measurements of observables that are incompatible with measurements of the pointer's position, we would be allowed to say that our system (or all the systems of the collection we are using in order to verify the statistical predictions of the theory) is, in fact, and with equal probability, *either* in the situation that actually has the property "the pointer shows +1" or "the cat is alive" *or*, alternatively, in the situation with the property "the pointer shows –1" or "the cat is dead."

The positions we intend to analyze in this section come from a completely different perspective. In 1957, H. Everett III published a work which he prefaced by expressing very clearly his dissatisfaction with the occurrence of superpositions of macroscopically distinguishable states within the theories. He was also dissatisfied with the fact that the standard formalism required two principles of evolution (i.e., Schrödinger's linear evolution and the reduction of the wave packet) and that it was incapable of specifying when one or the other of these two incompatible dynamic processes ought to be applied. For Everett the solution clearly required eliminating the reduction process from the formalism: Schrödinger's equation has an absolutely universal validity. Of course, since we are interested not only in the description of the physical process but also in our own perceptions, it is indispensable to include the observer himself in the physical description. The resulting state, once again, is governed by the linear evolution laws of the theory.

To understand fully Everett's hypothesis, it will make sense to recall the von Neumann chain discussed in section 15.4, and to explicate, with reference to our state (16.1), the situation that follows upon the interaction of the systems we are examining with ever more complicated systems. At a certain moment we will find ourselves facing a situation described by the formalism as follows:

$|\text{Final state}\rangle = (1/\sqrt{2})$ [|V, the pointer shows +1, the computer stores +1 in its memory, ..., the observer sees the pointer at +1⟩ + |H, the pointer shows –1, the computer stores –1 in its memory, ..., the observer sees the pointer at –1⟩].

It should be noted here, as already emphasized in chapter 15, that for each of the superposed states there is a perfect coherence among the "situations" of all the coinvolved systems. Nevertheless, the state reported above is still more embarrassing than the ones already discussed in this chapter. If, with Everett, we exclude the reduction of the wave packet, the actual state of affairs corresponds to the fact that, in the first term of the sum, the observer has one perception, and in the second, another, incompatible with the first. What are we to make of this peculiar situation? Since we have definite perceptions we ought in some way to suppress the term of the state that corresponds to the perception we are not having. But this would amount to a reduction of the state vector, a procedure absolutely prohibited by Everett's assumption that Schrödinger's evolution has an unlimited validity.

There is only one way out of the difficulty, strange though it is. We imagine that the various parts of the wave function which, as repeatedly emphasized, correspond to macroscopically and perceptually incompatible situations but which must all be taken into consideration, refer in fact to different worlds among which there is no relation at all. In other words, in Everett's interpretation, which has come to be known as the *many-worlds interpretation*, all possible outcomes according to the formalism actually occur, but take place in a multiplicity of worlds that do not communicate with each other. Thus, in the case at hand, there will be a world in which "the photon" really "has vertical polarization, the pointer shows +1, . . . , I perceive that the pointer shows +1," and an alternative world in which "the photon" really "has horizontal polarization, the pointer show −1, . . . , and *a replica of me* perceives that the pointer shows −1." Note that in each world there is a perfect consistency; for example, there can never be a world in which the pointer shows −1 and the replica of me perceives a +1! Once more, Figure 16.3 offers a schematic representation of the situation that corresponds to this interpretation.

The argument just developed is valid for every kind of measurement and/or process that leads to the superposition of states corresponding to macroscopically diverse situations. All measurement outcomes actually occur, all outcomes included in the state are realized in a collection of distinct worlds. We note that, as in the preceding section, in this perspective too, no process of reduction is ever verified and Schrödinger's equation governs the entire physical process.

The real meaning of the interpretation is not easy to grasp, but it can be intuited by placing its perspective side by side with the one we all unconsciously adopt, and which makes the superposition of distinct macroscopic states so embarrassing. In a certain sense, without becoming aware of it, we have grown accustomed to thinking of the observer as a spectator of events taking place on the world stage, as if he were external

A FRAMEWORK FOR PHYSICAL PROCESSES

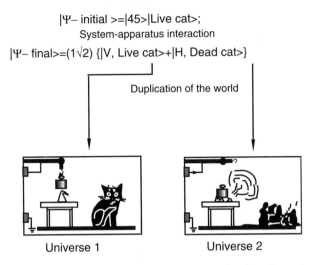

Figure 16.3. The many-worlds interpretation of Schrödinger's cat. The state vector that initially describes a living cat, as a consequence of the system-apparatus interaction, contains two terms that correspond to the two alternatives (happy) and (fatal) for the cat. But the two terms in fact refer to two distinct worlds that have no connection with each other. There is a universe in which the cat is dead, and another universe in which his replica survives.

to the whole process. If one takes such a position, he faces a scenario that does not permit him legitimately to assert that "the cat is alive" or alternatively that "the cat is dead." But the interpretation we are considering requires, so to speak, a radical "coinvolvement" of the observer in the process: he himself cannot withdraw from the ironclad, universal laws that govern all events and therefore the one reality is represented by a wave function that *contains all possibilities*. Among these there also exist those with observers having incompatible perceptions. In a certain sense it could be said that, while we had been led to face the problem of the superposition because the theory speaks of two situations but I perceive only one, and this fact made it necessary to suppress one of the two alternatives, in the new perspective, the "I" has consciousness of only one possibility because the "I" that is making an assertion about that world is the "I" associated with that outcome. But there exist other "I"s with completely different perceptions.

The idea, as difficult as it seems, has a certain inner logic to it and a fascination. Perhaps some of the fascination derives from the many exam-

CHAPTER SIXTEEN

ples in literature of visions of this kind. One of the most lovely of these is found in Jorge Luis Borges' "Garden of Forking Paths." In the story the author describes a situation that (for anyone interested in the foundations of quantum mechanics, at least) cannot help but recall the solution just proposed.

In Borges' story one of the protagonists, Stephen Albert, speaks of the genial Ts'ui Pen,

> Governor of his native province, learned in astronomy, astrology, and in the tireless interpretation of the canonical books, chess player, famous poet and calligrapher—he abandoned all this in order to compose a book and a maze. . . . When he died, his heirs found nothing save chaotic manuscripts. His family . . . wished to condemn them to the fire; but his executor—a Taoist or Buddhist monk—insisted on their publication.

The other protagonist, Yu Tsun, a relation of Ts'ui Pen, replies: "Their publication was senseless. The book is an indeterminate heap of contradictory drafts. I examined it once: in the third chapter the hero dies, in the fourth he is alive. As for the other undertaking of Ts'ui Pen, his labyrinth . . ." Then Stephen Albert explains to Yu Tsun how he arrived at the conclusion that the book and the labyrinth were one and the same:

> to no one did it occur that the book and the maze were one and the same thing . . . . two circumstances gave me the correct solution of the problem. One: the curious legend that Ts'ui Pen had planned to create a labyrinth which would be strictly infinite. The other: a fragment of a letter I discovered. "I leave to the various futures (not to all) my garden of forking paths. . . . The phrase "not to all" suggested to me the forking in time, not in space. A broad re-reading of all the work confirmed the theory. In all fictional works, each time a man is confronted with several alternatives, he chooses one and eliminates the others; in the fiction of Ts'ui Pen, he chooses—simultaneously—all of them. He creates, in this way, diverse futures, diverse times which themselves also proliferate and fork. Here, then, is an explanation of the novel's contradictions. Fang, let us say, has a secret; a stranger calls at his door; Fang resolves to kill him. Naturally, there are several possible outcomes: Fang can kill the intruder, the intruder can kill Fang, they both can escape, they both can die, and so forth. In the work of Ts'ui Pen, all possible outcomes occur; each one is the point of departure for other forkings.

and the dialogue continues to its climax, which is what especially interests us: Stephen Albert says:

> The *Garden of Forking Paths* is an incomplete, but not false image of the universe as Ts'ui Pen conceived it. In contrast to Newton or Schopenhauer, your ancestor did not believe in a uniform, absolute time. He believed in an infinite series of times, in a growing, dizzying net of divergent, convergent, and parallel times. This network of times which approached one another, forked, broke off, or were unaware of one another for centuries embraces *all* possibilities of time. We do not exist in the majority of these times; in some you exist, and not I; in others I and not you; in others, both of us. In the present one, which a favorable fate has granted me, you have arrived at my house; in another, while crossing the garden, you found me dead; in still another, I utter these same words, but I am a mistake, a ghost.

Yu Tsun replies: "In every one . . . I am grateful to you and revere you for your recreation of the garden of Ts'ui Pen." And Stephen: "Not in all of them. Time is constantly bifurcating toward countless futures. In one of these, I am your enemy."

It is unnecessary to emphasize how some passages of this splendid story anticipate the positions that must be assumed if we want to take the many-worlds interpretation seriously. It is enough to point to the "bifurcation in time", the fact that normally, "each time a man is confronted with several alternatives, he chooses one and eliminates the others"; but here, "he chooses—simultaneously—all of them," and the peculiar result that "this network of times embraces *all* possibilities." Finally, to relate the idea one more time to the peculiar destiny of the famous cat, featured in the analyses of the last chapter: "In the present one, which a favorable fate has granted me, you have arrived at my house; in another, while crossing the garden, you found me dead."[3]

This literary diversion leads me to mention an aspect of this interpretation that some consider very positive. I am referring to the so-called anthropic principle.[4] As all those who have heard about it know very well, it is based on the observation that the existence of life in our universe appears to be absolutely exceptional, and really, as has been emphasized, has a practically zero probability of occurrence.[5] In fact, it is sufficient to change some constants of nature by an almost infinitesimal quantity, and life would not be able to occur (of course, in doing this the very important fact is overlooked, in my view, that we are referring exclusively to life as we know it, as based on the chemistry we know about, and so on). But leaving aside my own personal reservations about this kind of argument, I will go on to explain how it can be said that the many-worlds interpretation reinforces, and is reinforced by, the anthropic principle as such. Since in this world view any event with a probability *not identical with zero* does in fact happen, we are able to think that reality is such that there exists a possibility that the wave function of the universe contains a term in which, in fact, life occurs. The fact that this probability is absolutely irrelevant does not matter, since, of course, we exist in that part of the wave function and we know nothing about the other ones. As Borges wrote, "In the majority of these times" (which are nothing other than all the other branches of the countless evolving ramifications) "we do not even exist." Another aspect of this interpretation that some writers think interesting comes from the fact that, if we consider only a single one of the many multiple bifurcations that have taken place, everything is consistent, and, one must add, all past history in this branch makes perfect sense. This would reflect our sensation that past history is certain. On the other hand, even in the branch where I happen to be right now (along with my readers), processes that can have diverse macroscopic outcomes are on the verge of being verified. In connection with them, a further multiplication of worlds will take place and consequently our future will be ontologically uncertain. To illustrate this fact clearly, it should be enough to think of a case where we ourselves are in the box where Schrödinger's famous cat is, just before the impact of the photon: in the diverse futures that will soon be disclosed, there is one in which we are still here to share our ideas, and another in which we have all passed on to a better life.

The most serious criticism that has been advanced of the many-worlds interpretation is based on the application of the principle known as Occam's razor: Entities should not be multiplied beyond what is necessary (*entia non sunt multiplicanda praeter necessitatem*). And in this interpretation, entities are indeed multiplied to infinity.

This last observation presents us with the opportunity to mention an interpretation that in a certain sense recalls this one but also differs from it in some essential aspects. It has been elaborated by David Z. Albert and Barry Loewer in order to reply to some of the criticisms directed against the many-worlds interpretation we have not had the opportunity to discuss. It is usually referred to as the *many-minds interpretation*. In outline, the idea can be expressed as follows: instead of an infinity of worlds and therefore an infinity of doubles of every conscious observer, we are to think of only one world and a single example of every conscious observer, but each observer is equipped with an infinity of minds (or a mind structured into an infinity of sheets or "pages" [*folia*]), each one of which perceives one of the diverse outcomes of every process in which perceptually distinct outcomes are possible. With reference to this scenario, it could be said that, whereas in the many-worlds interpretation all possible events take place, in the many-minds interpretation, all possible perceptions take place. Of course, to secure an intersubjective agreement, it must be admitted that the various sheets of the minds of all the percipient beings are synchronized in such a way that two persons present at the same physical process associated with perceptually diverse states (for example, at the superposition of the state "the photon counter has made a click" and the state "the same photon counter has made no sound"), then the "mind" of the one person of the pair who has consciousness of having heard the sound would have to be synchronized (that is, be in the same term of the superposition) with the mind of the other observer. This way it can be guaranteed that if, for example, one of the subjects asks the other, "Did you hear anything?," the mind of the other who answers, "Yes," would be perceived by the same mind of the first observer that had the same sensation, and analogously for the alternative answer to the question (which also takes place and is perceived, but by a different sheet).

I do not intend to take any more time to describe this approach. A brilliant book, published not long ago, has been dedicated to it, entitled *Quantum Mechanics and Experience,* by my friend and colleague David Z. Albert. As in the previous cases, Figure 16.4 shows a summary of the situation that holds with the many-minds interpretation.

16.5. QUANTUM HISTORIES

In this section I shall briefly describe a recent proposal that has encountered a certain amount of success in the United States. In a certain sense, it straddles the two interpretations I have just discussed. I have not mentioned it until now for the reason that its defenders have assumed rather diverse positions about the true meaning they think it has. I will begin with a summary exposition of the formal rules on which it is based, and then will go on to explain how it can be considered, in a very precise way, as an articulation of the position treated in section 3 of the present chapter that depended on practical limitations on the measurability of observables of macrosystems, and then, alternatively, how it can be understood as a variant of the many-worlds theories.

The rules of the game are the following: we think of a system (for example, a photon) initially in a very precise polarization state (let's say,

CHAPTER SIXTEEN

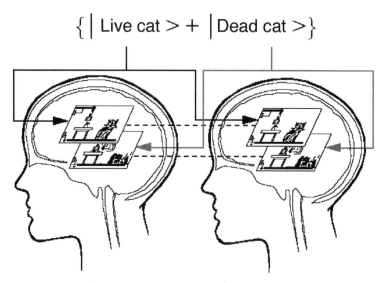

Figure 16.4. In the many-minds interpretation, only one world is assumed to exist, but each term of the wave function that corresponds to a perceptually different situation is correlated to a different "perceptual stratum" of everyone's mind. Of course, to guarantee intrasubjective agreement the different "perceptual levels" of the different subjects have to be perfectly synchronized as shown by the dotted lines.

vertical). Next a "quantum history" is formulated, that makes certain assertions about this system with regard to its polarization for a sufficient succession of instants. For the sake of simplicity we suppose that the history consists in the assertion that the photon at noon (the initial instant) has vertical polarization, at 12:30 has 45° polarization, and at 1 P.M. has polarization of 60° from the vertical. We call this story (Story 1). Of course, the story considers the system in question as if it were undisturbed, that is, that it is never being subjected to any process of measurement. Now a very precise probability is attributed to this story, and it is precisely the probability that quantum mechanics would attribute to the situation in which the photon has been subject at the times indicated to polarization tests for the directions indicated, and that it has passed all these tests. The procedure is now generalized to any stories whatsoever, in which as many times as desired are taken into account, with as many assertions about any properties whatsoever. In accordance with this, the schema also attributes a very precise probability to any story we can pos-

A FRAMEWORK FOR PHYSICAL PROCESSES

sibly conceive, representing the probability that such a story could actually be verified.

At first view, the idea appears quite original and appropriate. In fact, there never are any measurements: in particular, if a situation were considered of the type that presents itself in a process of measurement, the entire system (system + measuring apparatus) would have to be treated and an appropriate story would have to be "told" which would automatically be associated with a very precise probability. This represents the probability that the story is true, that is, it corresponds to what actually happens at the times considered and with reference to the properties taken into consideration. Nevertheless a difficulty immediately arises: it is contradictory to think of "all the conceivable stories," because that would not respect the unalterable rules that the collection of relative probabilities has to obey.

In order to grasp this point we can begin by considering a story we shall call (Story 2) that coincides with (Story 1) at noon and 1 P.M., but differs at 12:30 by asserting that at that instant the photon was polarized at 135°. Since the probabilities of the two stories are those that the standard theory would attribute to a process in which the particle at 12:30 is subjected to a measurement of polarization at 45° and passes it (Story 1) or fails it (Story 2)—a photon either passes or fails with certainty a measurement for polarization, and to fail a test for 45° is to pass a test in the orthogonal direction, i.e., 135°—the sum of the probabilities of the two stories coincides, as it logically should, with the probability that the schema attributes to the story in which nothing at all is said about the polarization of the photon at 12:30. But suppose now we consider many other stories, identical to the preceding insofar as concerns the assertions about 12 noon and 1 P.M. but which assert that at 12:30 the photon has a polarization plane in any arbitrary direction n (Story n). It is obvious that if all the stories of this type are considered and their relative probabilities are summed up, the resulting sum will grow with the number of stories taken into account, and it can be shown that such a sum increases without limit. Clearly, when a collection of events is considered (in our case, n stories), and every event has a probability, it must still be granted that the sum of the probabilities does not exceed 1: there must be something wrong with the interpretation.

The element that brings in the inconsistency is easy to identify, and at the same time shows the escape route. If we do not suitably limit the members of the collection of stories that can be considered at one time, stories will be included such as the ones considered just now, which the theory tells us are inconsistent because they involve incompatible properties. At 12:30 a photon can "have" 45° polarization or it can "have" 135°, but if one of these cases is granted, then it cannot be said at the same time to "have" vertical polarization. It is therefore necessary to make a radical revision of the proposal: all the stories are possible, but they must be regrouped into families of compatible stories (technically, "decoherent"). Once one family is chosen, within it all the probability axioms are respected, and the relative probabilities for the various members can be considered and taken seriously. But it is absolutely forbidden to include in a family a member of an incompatible family, without violating the axioms of the theory of probability.

These premises having been laid down, I can now briefly show my reasons for saying that this position can be assimilated either into the interpretation discussed in section 3 above or into the "many-worlds" interpretation discussed in section 4. In fact, without slowing down the discussion with technical analyses foreign to the plan of this book, I will limit myself to mentioning that in the case where the assertions being made in two distinct stories are compatible in the precise sense of standard quantum theory, then these two stories are in the same family. But now, suppose that in practice at the macroscopic level all the measurements are compatible, it follows that all the stories on the macroscopic systems are decoherent, that is to say, they can be considered together. It is also understandable how the adoption of the quantum histories point of view can lead to the many-worlds theory. In fact, when we take account of the fact that, strictly, incompatible stories present themselves anyway, it is natural to escape the logical difficulties that emerge from considering all possible stories by associating every family of stories mutually consistent with each other with a different universe. In this case the interpretation that associates the probabilities indicated with the various stories is perfectly consistent for the world to which it refers. No relationship can be drawn between different families that are incompatible with each other, just as no relationship exists among the multiple worlds of the many-worlds interpretation.

We have now gone through an analysis of those proposals that simply deny, on various principles, that the reduction of the wave function happens, and that attempt in spite of this fact to account for the real or apparent process of macro-objectivization of the properties of the systems of our daily experience, a process that has to take place if any consistency is required between the implications of the theory and our definite perceptions. We can now proceed to an analysis of proposals that do accept that the reduction process occurs. Of course, the crucial problem for them will be, as I have repeatedly mentioned, that of identifying their precise modalities, to make clear when we are dealing with a quantum system and when with a classical system; when we are dealing with a deterministic evolution and when with a genuinely stochastic evolution; when with a reversible process and when with an irreversible process.

16.6. Objectification as a Dynamic Process

As I have already repeated numerous times in the course of this book, the orthodox view overcomes the problem of the objectification of macroscopic properties simply by admitting that there exist two classes of systems and physical processes, governed by different sets of laws, that is, the linear evolution and the collapse (or reduction) of the wave packet, respectively. This position represents a precise philosophic option that was repeatedly stated by Bohr from 1927 onwards: "it is decisive to recognize that, however far the phenomena transcend the scope of classical physical explanation, the account of all evidence must be expressed in classical terms." Classical conceptualizations, elaborated on the basis of our sensible experience with the systems of everyday life, are thus configured as logical prerequisites for the elaboration of a formalism intended to describe a reality that transcends this type of experience. Bohr continually insisted on the fact that the greater part of the world ought to be considered "external" with respect to the "quantum systems" we are concerned with. He always laid stress on the fact that in the real world, there are definite events instead of vague potentialities and therefore, at that level, ordinary language and accustomed logic were appropriate. However strange the processes we must confront in our exploration of the world, they must always lead to assertions expressible in ordinary language and in consistency with traditional logic.

And yet we have shown, as far back as chapter 6, how Einstein's subtle attacks compelled Bohr to adjust his position: to fight back against the implications of the *Gedankenexperimente* presented by Einstein at the Solvay Conferences, Bohr saw himself constrained to include parts of the measuring apparatus in the "quantum world," and was clearly embarrassed about defining, in a nonambiguous way, where to locate the boundary between these two aspects of reality, irreconcilable by his own admission.

When von Neumann elaborated a rigorous formulation of the theory, he was naturally led to reconsider the problem of the effect of a process of measurement on a microsystem, and he studied it deeply. This brought him to some very interesting conclusions; above all, the entire process was described in a very precise way by the postulate of the reduction of the wave packet. Secondarily, the fundamental characteristics of the process were put fully into the light with the introduction of the "chain" that bore his name (treated in chapter 15). But, at the same time, this process made even more evident the fact that, at least on a certain level, the chain (that is, the law of linear evolution) has to be broken in order to retrieve the agreement with the data of our sense experience. Von Neumann himself suggested an escape route that was taken up again by Wigner, which we will analyze in the following section. Before we take up that discussion, I think it is an opportune moment to point out that if linear evolution is accepted for some systems but denied for others, a dualistic position will be adopted toward physical systems: fundamentally different systems are then in existence that obey mutually incompatible laws.

16.7. Reduction on the Part of the Conscious Observer, and Wigner's Friend

Inspired by the observations and suggestions of von Neumann, Wigner faced with great resolve the problem of identifying the line of demarcation between the two classes of processes mentioned above, processes that require completely different kinds of treatment. This acute thinker was very clear on the point that the theory did not contain any formal element that in any way defined this boundary. And so he imagined a von Neumann chain that contains in fact two observers. The state he came up with can be written explicitly, including a specification of the various stages of his argument:

> *First stage* — initial state:
> $|\Psi, 0\rangle = |45\rangle|\text{detector ready to register}\rangle|\text{Wigner's friend, waiting}\rangle|\text{Wigner}\rangle$.
>
> *Second stage* — the photon interacts with the detector which, as usual, is supposed to carry out a measurement of vertical or horizontal polarization:
> $|\Psi, 1\rangle = (1/\sqrt{2})\{|V, \text{detector has clicked}\rangle + |H, \text{detector has not clicked}\rangle\}$
> $|\text{Wigner's friend, waiting}\rangle|\text{Wigner}\rangle$.
>
> *Third stage* — the sound wave (if the detector clicks) reaches the auditory organ of Wigner's friend:

|Ψ, 2⟩ = (1/√2){|V, detector has clicked, friend perceives the click⟩
+ |H, detector has not clicked, friend has no audible perceptions⟩}|Wigner⟩.

With reference to the last stage, Wigner poses the question: is it reasonable to assume that things are as they are implied by the vector we are looking at here: in particular, that my friend cannot be thought of as having a definite perception about the fact of having heard or not having heard a click? And now Wigner goes on to consider a later stage in the chain, when he "tries to determine" his friend's state, subjecting him to a process of measurement. The obvious way to do this is to ask him if he has perceived a click or not. The argument now proceeds in a clear, linear fashion:

- Wigner knows with certainty that he will receive a single and very exact answer. His friend is reduced to a precise state of consciousness by Wigner's attempt to ascertain in which one of the two terms of the superposition his friend is. At the stage we are considering, Wigner has no doubts that the reduction has in fact taken place.
- Is it in any way acceptable to think that it was the act of Wigner that produced the reduction? No, for two fundamental reasons. On the one hand, to assume such a perspective would constrain Wigner into a position of extreme solipsism: he could only speak about his own perceptions. On the other hand, Wigner himself refutes such a conclusion: he does not think it legitimate to consider himself fundamentally different from his friend when it comes to reacting to stimuli.
- The inevitable conclusion is that the reduction has to take place before his question: his friend, like him, will certainly have definite perceptions.

The question that Wigner asks at this point is: is there a characteristic element in the chain that permits me to distinguish a stage where it is reasonable to place the collapse? Does a level exist that can be identified as qualitatively different from the preceding ones in the series of processes that have taken place? The reader will be able to intuit where Wigner's argument is inevitably leading him: on the entire chain, the only point that can be considered peculiar is the one that involves the passage from the purely physical processes to their perception. Therefore, instead of trying to locate the separation (as Bohr tried) between "small" and "large," Wigner chose to locate it between "material world" and "mind." And in this way emerges the position (which seems idealist to me but certainly has a certain reasonableness to it) of the *reduction on the part of the consciousness*: all the systems of the universe obey the linear laws of quantum

CHAPTER SIXTEEN

evolution, with the exception of the acts of perception on the part of a conscious observer. After having chosen this path, Wigner dared to journey much farther. He observed that, having accepted that physical reality and the act of conscious perception are placed on different levels, then, whereas all the scientific theories of the past attributed to the act of consciousness an exclusively passive role (a perceiving being becomes conscious of the state of the physical world around him), quantum mechanics, supplemented by the interpretation of the reduction on the part of the consciousness, attributes an absolutely peculiar role to the observer, and describes in the simple language of the reduction of the wave packet a process hitherto unsuspected, that is to say, the occurrence of a reaction of the conscious mind upon the physical world that leads to effects of great importance: it is the act of becoming conscious that makes reality pass from the limbo of potentialities to the clarity of actualities.

As with the other positions examined in this chapter, I have chosen to illustrate this argument schematically in Figure 16.5.

The reader should now understand perfectly how this firm position, combined with the assertions of some of the adherents of orthodoxy[6] already mentioned, inspired various thinkers to assert that in fact quantum

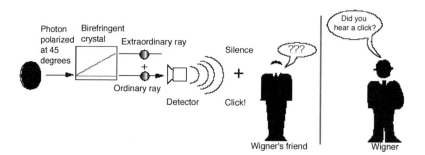

Figure 16.5. The argument that brings Wigner to assume the position that the act of perception on the part of a conscious observer induces the collapse of the wave packet. According to Wigner, nothing within the chain that precedes the instant when the sound wave reaches his friend is excluded from the material reality governed by the laws of the theory. On the other hand, Wigner does not think his friend lacks definite perception before the moment when he asks him. The conclusion is clear: the rupture of the theory's linear laws of evolution happens at the level of conscious perception; the theory describes all physical processes but must be modified to take account of mental processes.

mechanics has led to an "undoing of the Copernican revolution" by placing man once more, and in a much more radical sense, at the center of the universe.

Some comments are in order. The position has a certain inner coherence to it, but it implies some very peculiar positions. For example, it makes absolutely senseless all the accurate assertions of the cosmologists about the state of the universe millions and millions of years ago. The universe, until a conscious observer appeared in it, was a reign of infinite potentiality; at the first act of conscious perception it was instantaneously endowed with a precise actuality. This seems to attribute to man an absolutely disproportionate role.

However, it must be emphasized that this is not the really weak point of the interpretation we are now considering. As Bell pointed out so acutely, it is infected by a serious flaw in its origin. The solution proposed was dictated by the desire to remove the ambiguous separation between quantum and classical that characterizes the orthodox interpretation, and to locate precisely the boundary between the two. However, we must ask ourselves: does the criterion adopted effectively respond to the need that motivated it? Does perhaps our own present knowledge allow us to specify in a precise way what is meant by conscious beings? Perhaps we can say that, once again, the claim to have found a consistent solution is based, in fact, on returning, once again, to the identification of a boundary that turns out, like the one Bohr tried to define, to be essentially ambiguous. Bell's comment on this point is illuminating: "Was the wave function of the world waiting to jump [the jump which led to macroscopic definiteness] for thousands of years until a single-celled living creature appeared? Or did it have to wait a little longer, for some better qualified system . . . with a PhD?"

We can leave this unsettling proposal with this question unanswered, and go on to examine our final alternative. This one assumes that processes of reduction do happen, and that therefore macroscopic properties are actualized, but it attempts to treat all systems (micro, macro, and conscious) on the basis of a universally valid dynamic principle, and within a formalism that assumes the completeness of the description of physical systems in terms of state vectors. Unlike the other approaches, this one does not represent a new attempt to interpret the theory but takes shape as a rival theory in so far as it changes (although in a way which is very difficult to test) the empirical predictions of the theory, and therefore, in principle, allows for experiments that can distinguish between it and the standard theory.

CHAPTER SIXTEEN

16.8. The Dynamic Reduction Program

As already anticipated, the latest attempt to overcome the difficulties of the theory in treating macrosystems and the process of measurement assumes that the quantum description is complete (and consequently that the states of the individual physical systems are exhaustively described by the state vector), but considers the possibility that Schrödinger's equation has to be modified. Naturally, two kinds of problems arise. Above all, we have to imagine what types of modifications need to be introduced in order to reach the primary objective, namely, a unified description of all physical processes that does not, however, allow any embarrassing superpositions of macroscopically distinguishable states. The second problem facing us, as already mentioned, is that of making these modifications satisfy two requirements that are prima facie contradictory[7]: that is, they must lead to no measurable disagreement with the predictions of the standard theory for microsystems (predictions confirmed beyond any doubt by an enormous number of experimental data), and yet at the same time must be capable of introducing an extremely important and almost instantaneous modification of the evolution of macrosystems, in such a way as to lead, in extremely brief periods of time, to the dynamic suppression of the superpositions of distinct macroscopic states—such suppression coming about as a consequence of the unique dynamic principle that governs all physical processes.

A suitable way to get at the problem would be to ask ourselves what objectives we want to reach and what would be the characteristics of such objectives, and then to confront these characteristics with Schrödinger's equation, trying to imagine what modifications need to be introduced. At the macroscopic level, the dynamic will in essence have to reproduce the reduction process. But then we must ask ourselves: what are the most relevant characteristics that differentiate this process from the linear evolution of the standard theory? The answer is clear: while the quantum evolution of the state vector is perfectly deterministic and linear, the reduction of the wave packet is a fundamentally stochastic and nonlinear process. In fact, the outcomes of measurement are genuinely random, and their probabilities of occurrence are governed by the square of the modulus of the wave function, so that, as discussed in chapter 4, the probability associated with the sum of the two states does not coincide with the sum of the probabilities associated with the terms of the superposition. Actually, it is just this difference that gives rise to interference phenomena.

This is how we get our first indication of how to proceed: we try to modify the evolution equation by including terms that describe possible stochastic and nonlinear processes that influence the wave function itself. These processes will also have to make some properties objective (recall that because we are in a genuinely quantum context it is impossible to make all conceivable properties objective), according to modalities that are ineffective at the microscopic level but are very effective at the macroscopic level. A simple observation provides us with a new indication. We ask ourselves: what are the properties of a macrosystem, the objective absence of which are the most embarrassing from the point of view of interpretation? Everything points equally to the answer: it is position. This point was already expressed very lucidly by Einstein: "A macro-body must always have a quasi-sharply defined position in the objective description of reality." And moreover a fundamental characteristic of our perceptions is that they correspond to objects that have position and precisely defined spatial extension. The obvious conclusion is that the dynamic will have to tend toward making objective the positions of the particles and consequently the objects constituted by them.

It is worthwhile emphasizing that, in the same way as when we consider a hidden variables theory, taking account of the inevitable contextuality of the greater part of the observables requires that we make a choice about which things are not contextual, and thus acquire a very particular level of "objectivity," so in a genuinely quantum perspective, since not all the observables can be definite insofar as incompatible properties always exist, a choice has to be made about which of these are going to be privileged.

We are now ready to present the line of thinking that in 1985 led the present author and his colleagues Alberto Rimini and Tullio Weber to propose a phenomenological model (1986), now universally known in the literature as the GRW theory (derived from the three authors' initials), a model that incorporates the ideas I have just delineated in a precise mathematical fashion. I will describe this model in its simplest version. In the years immediately following its formulation it gave rise to various formal developments to which not only the original three authors but also Philip Pearle, Fabio Benatti, and Renata Grassi have contributed. These further researches have brought about a formal schema that not only preserves all the advantages of the original model but is in addition also very elegant from the mathematical point of view, making systematic use of the theory of dynamical stochastic equations. But the essential and crucial elements of the theory can be understood more easily and directly with reference to the original model.

The basic idea of the proposal can be summed up in very simple terms: we assume that all the elementary constituents of the physical world endowed with mass (by this is meant all the protons, neutrons, electrons, and so forth), apart from obeying Schrödinger's linear dynamic that is appropriate for the problem in question (and therefore takes account of the forces that act upon them), also at certain times undergo—at random and with an average frequency—spontaneous processes of localization in space. Before going into detail about the specific modality of these processes, I would like to stress that they are to be understood as fundamental natural processes that owe nothing to interactions with other physical systems or to deliberate actions on the part of conscious observers. On the contrary, the idea is that the space-time in which physical processes develop exhibits some fundamentally stochastic, random aspects, which induce precisely the spontaneous localizations of the microscopic constituents of the universe.

Let us try to be more specific. Certainly some questions arise, which we can represent by the interrogatives HOW, WHEN, and WHERE?

1. How? What are the precise modalities of the localization process?
2. When? How frequently do they occur?
3. Where? Since, as already mentioned, these processes lead to localizations of the microconstituents of any system whatsoever, which, according to the standard formalism, are in general associated with wave functions appreciably extended through space and which, correspondingly, cannot be thought of as being in one precise region or another, we must ask ourselves: in what region, among the ones "potentially" possible, will the systems end up being confined?

We can begin the answer to these questions with reference to a single particle, let us say a proton, and we suppose for simplicity's sake that it moves along the x axis. We also assume that the wave function $\Psi(x)$ that describes it is appreciably different from zero in a fairly extensive interval. Now we consider a function $L_{x^*}(x)$, called the localization function, which is going to play a fundamental role in the scheme. This function is different from zero only in an interval of amplitude δ centered around the value x^* and assumes the constant value $1/\sqrt{\delta}$ in this interval.[8] This function is represented in Figure 16.6.

With these premises, we can now formulate precise answers to the questions formulated above.

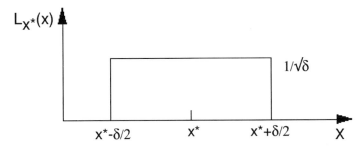

Figure 16.6. Graphic representation of the localization function $L_{x^*}(x)$ around the point x^*. The function is different from zero only in an interval of amplitude δ around the center x^* and within that interval it takes the constant value $1/\sqrt{\delta}$.

1. HOW? If at a certain instant our proton undergoes a localization around point x^* its wave function changes according to the following prescription:

$$\Psi(x) \to \Psi_{x^*}(x) = N\Phi_{x^*}(x),$$
$$\Phi_{x^*}(x) = L_{x^*(x)}\Psi(x). \qquad (16.2)$$

Taking account of the characteristics of the function $L_{x^*}(x)$ illustrated above, the prescription (16.2) can be expressed most simply by saying that the effect of the localization is to make the original function $\Psi(x)$ equal to zero outside the interval $(x^* - \delta/2, x^* + \delta/2)$ and to leave it unchanged within the interval itself, except for multiplying it by a factor N that guarantees that the area subtended by the square of the function $\Psi x^*(x)$ is equal to 1, an operation that makes certain that if a measurement is carried out, the particle will be found with certainty in the interval within which it has been localized. The process is illustrated in Figure 16.7.

2. WHEN? For every single microconstituent of any physical system whatsoever, the probability of undergoing a process of spontaneous localization is constant in time and is characterized by an average frequency λ, such that the probability that a process like this occurs in a short temporal interval Δt is equal to $\lambda \Delta t$.

3. WHERE? The localization can take place around any point x^* such that the particle, according to the standard formalism, has a nonzero probability of being found in the interval δ associated with such a point if a measurement for position is carried out on it. However, the theory we are analyzing assigns a precise probability of a localization being verified

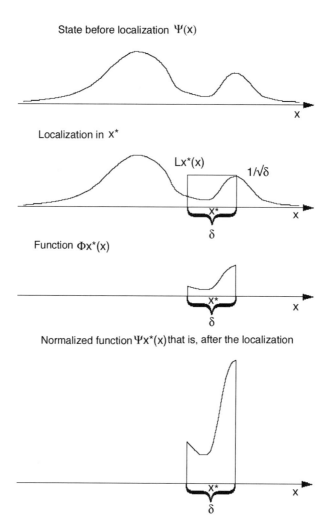

Figure 16.7. Representation of a spontaneous localization process over a system whose wave function before the process is that represented in the first graph. If the localization takes place around point x*, this brings about the function represented in the third graph. Afterward, this function is normalized in such a way that the area subtended by its square is equal to 1.

around one point rather than another. More precisely, the probability density that a localization taking place at x* is given by the area subtended by the function $|\Phi_{x^*}(x)|^2$. This means simply that the spontaneous localizations occur preferably where the standard theory asserts there is a greater probability of finding the particle in a position measurement. But, we must keep in mind, here no observer carries out any measurement: nature itself (Einstein's God?) chooses to induce such a process according to random choices but with precise probabilities. It is worthwhile pointing out that the formal rules are such that, if at a certain instant a localization happens, the sum of the probabilities that this is happening at any one of the points of the real axis is equal to 1 (in fact it is precisely this that requires the localization function to have the term $1/\sqrt{\delta}$ in front of it).

Note that the processes of localization (which occur along with the usual evolution of the standard theory) represent a stochastic and nonlinear modification of the theory. The stochasticity derives from the fact that both the instants as well as the precise positions where the localization process occurs are governed by genuinely probabilistic laws. Further, the fact that the probability of a localization at a certain point rather than at another depends on the square of the relevant portion of the wave function and not on the function itself implies that the process is not linear in the state vector.

To illustrate how the mechanism now presented tends to making positions objective, we consider the case that we have so often already encountered in the course of this book, in which a particle is in the superposition of two localized and spatially separated states (we can think, for example, of a wave function of a particle at the exit of a Stern-Gerlach apparatus):

$$\Psi(x) = \frac{1}{\sqrt{2}}[\Psi_L(x) + \Psi_R(x)]. \quad (16.3)$$

In the above equation the two states $\Psi_L(x)$ and $\Psi_R(x)$ are well localized (with respect to the distance δ that characterizes the theory) in two regions around the points L and R, whose spatial separation is much greater than δ. Suppose now that the particle in question undergoes a localization process. By the rules stated, it is obvious that there is an appreciable probability that the localization is happening only around point L or around point R. In the specific example being considered, these two probabilities are practically equal, and equal to 1/2. The reader will certainly grasp the fact that the probability that a localization is happening

CHAPTER SIXTEEN

elsewhere, for example in the region that stands between L and R, is practically nil, since the wave function $|\Phi_{x^*}(x)|^2$ that follows such a process subtends an infinitesimal area.

The conclusion is now clear: if a particle in a state such as the one being considered undergoes a localization process, it becomes, so to speak, constrained to choose whether to stay around L or around R. In other words, the random and nonlinear dynamics leads to the suppression of one of the two terms of the superposition (16.3) (needless to say, the resulting function is normalized so that the probability that, once it has been localized, it will be found around L—if this is the outcome that has taken place between the two possible ones—is equal to 1). The process just described is illustrated in Figure 16.8.

We can now go on to illustrate the characteristic of the theory that is most important for our purposes, namely, that which induces the macro-objectification of the properties for a macroscopic system. We consider the embarrassing situation that follows when the standard formalism is applied to a situation of the type of measurement process that leads to a superposition of states of a macroscopic indicator, a state such as the one in equation (15.7). We focus our attention on the state of the pointer. The pointer is an object containing a very high number of particles and of course is a practically rigid system. This requires that in the states we have synthetically characterized as $|A_{-1}\rangle$ and $A_{+1}\rangle$, the particles of the pointer will actually be all round point S or all around point D, respectively. In other words, if, by making it explicit that the system depends on the spatial variables of the N constituents, we express as $\Psi_L(x_1, x_2, \ldots, x_K, \ldots, x_N)$ the wave function that corresponds to the state $|A_{-1}\rangle$ of the pointer and as $\Psi_R(x_1, x_2, \ldots, x_K, \ldots, x_N)$ the wave function that corresponds to the state $|A_{+1}\rangle$, then what will happen is that $\Psi_L(x_1, x_2, \ldots, x_K, \ldots, x_N)$ is different from zero only when all the position variables $x_1, x_2, \ldots, x_K, \ldots, x_N$ assume values near to S, and analogously, $\Psi_R(x_1, x_2, \ldots, x_K, \ldots, x_N)$ is different from zero only for the variables $x_1, x_2, \ldots, x_K, \ldots, x_N$, all near R. We can write the functions explicitly at the instant that interests us:

$$\Psi_{(x_1,\ldots,x_K,\ldots,x_N)} = \frac{1}{\sqrt{2}}[\Psi_{L\,(x_1,\ldots,x_K,\ldots,x_N)} + \Psi_{R\,(x_1,\ldots,x_K,\ldots,x_N)}]. \quad (16.4)$$

We suppose now that one of the elementary constituents of the pointer, let us say the kth particle, undergoes a spontaneous localization. We know that this can occur only in regions of the variable x_k where the wave function is nonzero, that is to say, when x_k assumes a value x_L^* that

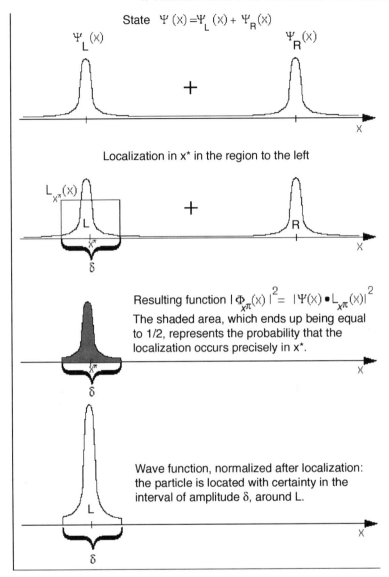

Figure 16.8. The effect of a spontaneous localization on an elementary constituent associated with a wave function that does not permit us to say, as we know very well, that the particle is in L or in R or in both or in neither. The potentiality of "being somewhere" is actualized in the dynamic process of spontaneous localization.

CHAPTER SIXTEEN

is "around L" or alternatively, a value x_R^* that is "around R," and that the two cases present themselves with equal probability. If a localization has taken place, supposing for example that the eventuality x_L^* has occurred, then the function (16.4) has to be multiplied by the function $L_{x_L^*}(x_k)$, which is different from zero only in the interval δ around x_L^*. But now by multiplying $\Psi_R(x_1, \ldots, x_K, \ldots, x_N)$ by this function we get 0, insofar as the indicated function is different from zero only for x_K near to L. The conclusion is clear: a localization process for the kth particle of the pointer can occur only at one of the two positions that are possible for it in correspondence with the two states of the superposition, and if the localization happens in one of the two positions, it leads to the suppression of the other term of the superposition. In other words, localizing one of the microscopic constituents of the pointer is equivalent to localizing the entire pointer in one of the two positions in which it "potentially" can be found.

The situation just now described is schematically represented in Figure 16.9. In this figure the wave functions $\Psi_L(x_1, \ldots, x_K, \ldots, x_N)$ and $\Psi_R(x_1, \ldots, x_K, \ldots, x_N)$ are states written like products of wave functions for every constituent that has a subscript L or R to call the reader's attention to the fact that in the states being studied the constituents are all in one of the two regions.

The mechanism just described shows how an amplification of spontaneous localizations occurs and makes clear how the parameters λ and δ can be chosen so as to realize the purposes for which the theory was designed. Let us recall that such parameters govern the localizations of microscopic constituents. Supppose the following choices are made:

$$\lambda = 10^{-16} \text{ sec}^{-1}, \quad \delta = 10^{-5} \text{ cm}, \qquad (16.5)$$

which were those proposed by GRW in their original work. The choice of λ implies that a microsystem, let's say a proton, undergoes on the average a spontaneous localization every 10^{16} seconds, which is every 100 million years. This means that a proton can stay in the embarrassing state of not being either here or there for an incredibly long time. In this model, as in the standard model, microsystems truly enjoy the *cloudiness of waves*, as Bell put it so appropriately. But now we are interested in what happens to the pointer of the apparatus. It contains about an Avogadro's number of elementary constituents, that is, 10^{24} particles. Even if all of these have only a probability equal to 10^{16} of undergoing a process in one second, because $10^{16} \times 10^{24} = 10^7$, it follows that every second 10^7 particles are localized. In other words, at least one of the particles of the

A FRAMEWORK FOR PHYSICAL PROCESSES

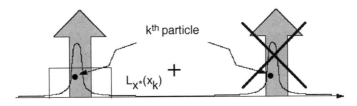

Figure 16.9. The dynamic reduction model's amplification mechanism. The figure shows how a spontaneous localization of a single elementary constituent of the pointer leads to the localization of the entire pointer and to the suppression of one of the two terms of the embarrassing superposition of macroscopically distinct states.

pointer will almost certainly undergo a localization in a ten-millionth of a second. And since, as we have seen, every localization of a constituent requires the localization of the entire pointer, this implies that, on the basis of the unique dynamical principle that governs all physical systems, the theory cannot tolerate a superposition of the kind of (16.4) to last longer than a ten-millionth of a second. In other words, the macroscopic pointer must decide immediately if it is pointing to −1 or +1.

As we have seen for all the other cases analyzed in this chapter, Figure 16.10 is meant to represent how the dynamical reduction model solves the Schrödinger's cat problem. When we realize that the two situations—live cat and dead cat—certainly correspond to different spatial distributions of a macroscopic number of microscopic constituents of the cat, we can understand perfectly why Bell wanted to sum up the fundamental characteristic of the theories presented in this section by saying: "Any

413

CHAPTER SIXTEEN

$|\Psi\text{- Initial} \rangle = |45\rangle |\text{Live Cat}\rangle;$

$|\Psi\text{- Final}\rangle = (1\sqrt{2})\{|V, \text{Live Cat}\rangle + |H, \text{Dead Cat}\rangle\}$

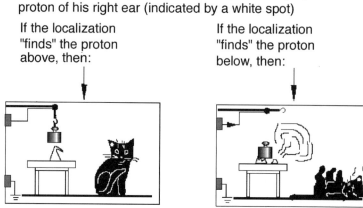

Figure 16.10. A view of the destiny of Schrödinger's celebrated cat that would follow from the adoption of the dynamical reduction model.

embarrassing macroscopic ambiguity in the usual theory is only momentary in the GRW theory. The cat is not both dead and alive for more than a split second."

To conclude this section, it seems fitting to recall briefly the conceptually relevant characteristics of the dynamical reduction theories. They do not add variables to the formalism, that is, they assume that the state vector represents the most accurate possible description of an individual physical system. In this sense, from the point of view of their formal structure, they are genuinely quantum theories in the traditional sense of the word. Nevertheless, they introduce dynamical modifications of the evolution equation that governs all physical systems. The fundamental characteristic of the new, universal law of evolution is the following. When it is applied to the description of a microsystem it implies predictions that coincide with those of standard quantum mechanics with such a level of precision that any discrimination between the two theories is impossible in the present state of our technology. If the theories are applied to the description of a measurement process, then the unique dynamical principle that stands at the base of the theory implies precisely the effects that the standard formalism incorporates in the principle of the reduction of the wave packet. In the theory, not only is this principle not in contradiction with the dynamical equation for microsystems (as it is in the standard theory), it is actually the rigorous consequence of that equation. Finally, if the theory is applied to the description of macrosystems, it brings predictions that are in agreement with those of classical physics (as is also the case with standard quantum mechanics), but—and here lies all the difference—it does this without presenting any embarrassing superpositions of macroscopically diverse states. Furthermore, the theory contains parameters, that is to say, the frequency of localizations and their spatial accuracy, which precisely define and permit a quantitative evaluation of just when the quantum mechanical treatment is appropriate, and when, and for what systems, the quantum mechanical treatment has to be replaced by the reduction of the wave packet, and so on. For every specific problem and every specific physical process the theory gives absolutely precise answers to questions of this kind: is this system to be treated as a quantum system or as a classical system, is this process reversible or irreversible, and so on. In brief, the theory defines in quantitatively precise terms where the boundary is to be located between the cloudy world of potentialities and the definiteness of actualities: the crucial problem, that is, to which the standard formalism has no answer. To cite Bell once again: "I do think however that [these theories] have a certain kind of goodness . . . in the sense that they are honest attempts to replace the wooly words by real mathematical equations—equations which you don't have to talk away—equations which you simply calculate with and take the results seriously."

In the next chapter, we will analyze in more detail some of the implications of the theories presented in this last section, in order to show more clearly their peculiar characteristics and to illustrate the vision of the world they make possible.

CHAPTER SEVENTEEN

Spontaneous Localization, Properties, and Perceptions

> For myself, I see the GRW model as a very nice illustration of how quantum mechanics, to become rational, requires only a change which is very small (on some measures!). And I am particularly struck by the fact that the model is as Lorentz invariant as it could be in the nonrelativistic version. It takes away the grounds of my fear that any exact formulation of quantum mechanics must conflict with fundamental Lorentz invariance.
> —*John Stewart Bell*

THIS CHAPTER represents the sole exception to the commitment I took up in the preface, to present all the points of view on the difficulties of the formalism without privileging any single perspective, and leaving it up to the reader to choose freely the most congenial. By looking in more detail at the program of dynamical reduction, I will inevitably be placing more emphasis on it than on the other interpretations. All the same, I believe my readers will understand my desire to dedicate one part of this book (albeit a very limited part) to a fuller elaboration of an area my colleagues and I have been scientifically investigating for a number of years. While it is true that various books already exist which are entirely dedicated to the points of view I have analyzed in the previous chapter[1]—points of view that still capture the attention of the scientific community active in the field—the GRW theory, despite its being explained and commented upon in most textbooks on the foundations of quantum mechanics that have been published since 1986, and despite its being treated in many published articles, has not until the present work been presented in a systematic and comprehensive way, except in specialized review articles appearing in journals not normally available to the nonspecialist. In addi-

SPONTANEOUS LOCALIZATION

tion, since many aspects of the theory require a radical change of perspective from all the positions just reviewed, a more thoroughgoing analysis is needed in order to evaluate its implications adequately, and in order to grasp fully the vision of the world it offers.

17.1. Macro-objectification: What Properties Are Involved?

As I have repeatedly stated in this book, the conceptually most relevant point that the quantum theory forces us to face concerns the legitimacy of considering some physical properties as objectively possessed[2] by individual physical systems. The answer given by the standard theory is well known: one should limit oneself to considering as possessed properties only those properties for which the outcome of a measurement can be predicted with certainty, all other properties being left, as it were, suspended in the limbo of potentiality. Furthermore, as long as we assume that the theory has its linear character, this conclusion is valid both for microscopic systems and for macroscopic ones, and it is exactly with reference to these latter systems that the thorny problem of macro-objectification arises.

In order to overcome these difficulties within the theoretical framework of dynamical reduction theories—a framework that assumes the completeness of the description of the states of physical systems in terms of the wave function—a choice is required about which variables to make objective. As we have seen, the GRW theory—on the basis of valid argumentation—suggests that the appropriate choice would be to attribute a privileged role to the position variables[3] of all the elementary constituents of the universe. This model is constructed in such a way that the tendency toward objectivization it incorporates is practically without effect at the microscopic level, so as not to contradict the results of all experiments carried out over the last three-quarters of a century. These experiments have revealed interference effects closely bound up with the delocalization associated with a spatially extended wave function. Nevertheless, the localizations are amplified in proportion to the number of constituents of any physical system, leading to the objectification of the positions of macro-objects.

One may immediately ask, "Granted that this model assigns an objective status to the positions of macrosystems; are these the only properties required for such systems? Or are there others as well?" The question is

CHAPTER SEVENTEEN

important, if we remember, for example, that the analogous choice of privileging positions in the framework of the de Broglie–Bohm theory leads us to the conclusion that almost no other property is objective in the full sense of the word, thanks to the peculiar aspects connected with the phenomenon of contextuality.

We can now consider the implications of the theory. For example, consider the case of two states of a macroscopic system that at a certain instant perfectly coincide in position, but have macroscopically diverse velocities. Of course, at the instant under consideration (and even for the interval of time during which many of the microconstituents will be undergoing a process of spontaneous localization—i.e., a few millionths of a second), the superposition of the two states can occur and persist. But, as soon as Schrödinger's equation (which governs the system in any case) leads to the spatial separation of the states of the superposition as a consequence of the diverse velocities of the two states (we suppose that one of them corresponds to a stationary sphere, the other to a sphere in the same location but moving at the rate of one centimeter per second), any process whatsoever of localization for a microconstituent constrains the macroscopic system to choose between being in the position that results from the state with zero initial velocity and being in the one that results from the state with initial velocity of a centimeter per second.

The example just now discussed (with analogous considerations being valid for all the situations we can consider as macroscopically diverse) shows clearly how, if even in an indirect way (that is to say, as a consequence of suppressing superpositions of diversely localized states), the dynamic of this model conspires to make macroscopic properties other than position objectively possessed as well—velocity, for example. This fact remarkably clarifies the physical meaning of the theory and permits an evaluation of the relevant physical differences (other than the obvious formal ones) between it and the pilot wave theory. For this latter theory, the positions are objective anyway, whether at the macroscopic or microscopic level, but the other variables have a somewhat peculiar state as a consequence of contextuality. In dynamical reduction models, which tend to make only the positions objective, these positions cannot be considered objective when it comes to the microsystems, whereas all the macroscopic properties of macrosystems are dynamically objectivized as a kind of secondary effect of the objectification of their microconstituents' positions.

17.2. A Critical Evaluation of the Dynamical Reduction Program

The GRW theory certainly has some limitations, which have brought some criticisms. Above all, it takes shape as a purely phenomenological model, that requires the consideration of processes that can appear arbitrary, invented solely for the purpose of getting the desired result. Further, the model contains two phenomenological parameters, the frequency and accuracy of the reductions, which, if the theory is going to be taken seriously as a fundamental theory for all physical processes, would come to acquire the conceptual status of new constants of nature, analogous with what the speed of light and Planck's constant meant in connection with the two most recent scientific revolutions of relativity and quantum mechanics, respectively. It would seem rather excessive to introduce two new physical constants.[4]

We should also point out that the theory is fundamentally irreversible. As I shall discuss a little later on, this is a feature that can be considered an advantage or disadvantage of the formalism, depending on a particular author's point of view. Finally—and this is really the crucial problem—despite the repeated and concentrated efforts of various researchers over the last half decade or so, it has not been possible to elaborate a completely satisfactory relativistic generalization of the theory, and this is a problem that simply has to be solved before the new proposal can be taken seriously.

In the light of these recent attempts, I do not think it proper, in the present circumstances, to attempt to endow this theory with the status of a true scientific revolution. Its real merit, in my view, is that it constitutes an explicit example of how new lines of thought can be followed in order to escape the conceptual impasse encountered by the standard theory in its treatment of macrosystems. Thanks to the new theory, it has been possible to identify some of the characteristics that any future theory of the same type would have, and we get a clearer picture of the price that has to be paid for these characteristics.

An appropriate way of evaluating the theory would appear to involve asking ourselves what the founders of quantum mechanics presumably would have thought about the theory—those, at least, who were never quite satisfied with the orthodox position. Bell himself undertook such an exercise when he spoke at the meeting held at the Imperial College of London to celebrate the centenary of Erwin Schrödinger's birthday. On

CHAPTER SEVENTEEN

that occasion, Bell dedicated his remarks to an analysis of the GRW theory and its implications for an EPR situation. In the conclusion of his paper, Bell (1987b) asked himself what Schrödinger would have thought of the theory:

> I think that Schrödinger could hardly have found very compelling the GRW theory as expounded here—with the arbitrariness of the jump function and the elusiveness of the new physical constants. But he might have seen in it a hint of something good to come. He would have liked, I think, that the theory is completely determined by the equations, which do not have to be talked away from time to time. He would have liked the complete absence of particles from the theory, and yet the emergence of "particle tracks" and more generally of the "particularity" of the world, on the macroscopic level. He might not have liked the GRW jumps [the localizations], but he would have disliked them less than the old quantum jumps of his time. And he would not have been at all disturbed by their indeterminism [which, as the reader should have grasped, is even greater than in the standard theory]. For as early as 1922, following his teacher Exner, he was expecting the fundamental laws to be statistical in character: "[O]nce we have discarded our rooted predilection for absolute causality we shall succeed in overcoming the difficulties."

And at this point Bell concluded his talk with the passage quoted at the beginning of the present chapter.

Another method of evaluating the most profound meaning of these researches could proceed from an attempt to repeat Bell's exercise, using Einstein in place of Schrödinger. I have decided to undertake such an analysis myself, no matter how irreverent it might appear, since it presents me, among other things, with the opportunity to be more precise than I was able to be in chapter 6, in my description of the "fourth phase" of Einstein's thought. As I mentioned before, in 1949, in honor of Einstein's seventieth birthday, P. A. Schilpp published a collection of writings by some of the greatest scientists of the time under the title *Albert Einstein, Philosopher-Scientist*. Many protagonists in the debate on quantum mechanics contributed to the work, and of course, many of them, while expressing their profound appreciation for Einstein's contributions to the birth of the theory, were not silent about their differences of opinion from him when it came to the orthodox interpretation. Einstein decided to conclude the book with a remarkable work of his own entitled "Reply to Criticisms." This piece was specially composed to combat the objections that had been made to his positions. Reading some passages from this work shows us how much Einstein's position had changed over the years. He took up a typically microscopic process: the decay of a radioactive system. As is well known, quantum mechanics describes the situation by means of a wave function that is the superposition of two states, that of the nondecayed system and that of the products of the decayed system (as in the elementary process that determines the fate of Schrödinger's cat). The coefficients of the states in the superposition vary with time, for which at every instant they furnish, with their squares, the probability that, if a measurement were carried out, the system would be found either to be nondecayed or as the products of decay. Einstein raises the problem of the objective instant at which the unstable system decays. He observes that, obviously, the quantum formalism does not allow any assertion about the instant of disintegration. This, Einstein says, could disturb me, but I already know the answer that the orthodox theory gives me: this question starts and finishes in the unwarranted assumption that there objectively exists (that is, independently of any observation whatsoever) something that can be identified with the precise instant of disintegration. In fact, the formalism has taught us that no experiment can possibly be made to determine the instant of disintegration for an isolated system, and that such a determination requires an uncontrollable disturbance that prohibits us from obtaining the desired information. At this point Einstein emphasizes that what in fact disturbs him is not the immeasurability of the instant of disintegration, but rather that the theory does not even allow us to pose the question whether such a moment exists.

But, leaving aside his dissatisfaction on this point, he goes on to show how he had clearly identified the true problem with the theory. He said,

> As far as I know, it was Erwin Schrödinger who first called attention to a modification of this consideration, which shows an interpretation of this type to be impracticable. Rather than considering a system which comprises only a radioactive atom, one considers a system which includes also the means for ascertaining the radioactive transformation. Let this latter include a registration-strip, moved by a clockwork, upon which a mark is made by tripping the counter. In this consideration the location of the mark on the strip plays the role played in the original consideration by the time of the disintegration.

(The reader will surely note the strict analogy with the process of measurement involving Schrödinger's cat: here, the states of the cat being alive or dead are replaced by "there is a mark on the strip" and "there is no mark on the strip".)

CHAPTER SEVENTEEN

At this point, Einstein expounds his own dogmatic point of view with great clarity: "The location of the mark on the registration strip is a fact which belongs entirely within the sphere of macroscopic concepts." And, if one were to adopt the orthodox position, one would be forced to conclude that

> the location of the mark on the strip is nothing which belongs to the system per se, but that the existence of that location is essentially dependent upon the carrying out of an observation made on the registration strip. Such an interpretation is certainly by no means absurd from a purely logical standpoint; yet there is hardly likely to be anyone who would be inclined to consider it seriously. For, in the macroscopic sphere it simply is considered certain that one must adhere to the program of a realistic description in space and time; whereas in the sphere of microscopic situations one is more readily inclined to give up, or at least to modify, this program.

Later in the essay Einstein returns to the point in order to articulate his own position more clearly: "The 'real' in physics is to be taken as a type of program, to which we are, however, not forced to cling a priori. . . . However, no one is likely to be inclined to attempt to give up this program within the realm of the 'macroscopic' (location of the mark on the strip of paper 'real')." And now he draws the conclusion that is so important for our immediate interests: "but the 'macroscopic' and the 'microscopic' are so inter-related that it appears impracticable to give up this program in the 'microscopic' alone."

If we recall that the realism requirement, as specified above, makes reference in a specific way to the possibility of a causal, spatiotemporal description of physical processes, a possibility that is precluded by the linearity of the standard theory and by the nature of pure potentiality that this assigns to physical properties, then the preceding phrase can be read as a clarification that Einstein did not consider viable (nor anyone else until 1985) the line of research that has now been shown to be practicable through the elaboration of the program of dynamical reduction. The GRW theory, as Bell says, "allows electrons [in general, microsystems] to enjoy the cloudiness of waves," which means that it accepts the fact that the behavior of systems like this cannot be squeezed into conceptual schemes that have been elaborated on the basis of the objects of our everyday experience, but at the same times permits us to recover a fully realistic conception, in Einstein's precise sense, at the macroscopic level, allowing tables and chairs, and ourselves, and black marks on photographs, to be rather definitely in one place rather than another, and to be described in classical terms.

This section seems now to have reached its goal. Without implying that this grants them the conceptual status of "true theories," I have put on exhibit theories of hidden variables that are consistent and predictively equivalent to quantum mechanics (in particular, the pilot wave theory), and this represents a fact of extreme conceptual and epistemological importance, insofar as it shows that a deterministic completion of this theory is possible, as well as a transformation of the probabilities of various physical processes from the nonepistemic to the epistemic. In just the same way, the models of dynamical reduction constitute an explicit demonstration that theories can be conceived that are different from but predictively equivalent to quantum mechanics on the microscopic level, but which also furnish a solution of the problem of macro-objectification that is coherent and based on dynamical principles of universal validity. Such models, like the pilot wave theory, are formulated in a precise mathematical way, and thus allow us to assess the price that must be paid for obtaining the results they can obtain. Like all hidden variable theories, the de Broglie–Bohm theory requires accepting the contextual nature of the greater part of the observables. The GRW theory requires recognizing that the dynamical equation, on which rests the entire theoretical framework that has proven so successful in the most varied fields, must be modified in a stochastic and nonlinear direction. As Bell stated, both these schemes represent two "exact" theories, compatible with all the predictions of quantum mechanics. It is appropriate now to stress once again that by the term "exact" this thinker intended to describe theories "which neither need nor are embarrassed by an observer."

17.3. AN ILLUMINATING HISTORICAL PARALLEL

In this section, to assist the reader in understanding the meaning of the GRW theory more deeply, we can take time out to present an historical parallel proposed by Bas van Fraassen, one of the great American philosophers of science, on the occasion of the conference entitled "The Interpretation of Quantum Theory: Where do we Stand?" organized at Columbia University, New York, in April, 1992, by the Institute of the *Enciclopedia Italiana* and the Italian Academy for Advanced Studies

CHAPTER SEVENTEEN

in America. Bas van Fraassen, from the Department of Philosophy at Princeton University, had been invited to open this important conference. He proposed to set up a seminar that presented some fascinating parallels between recent proposals for overcoming the difficulties of quantum mechanics and key philosophical positions in the history of human thought. After having pointed out the obvious parallelism between the many-worlds theories and the possible worlds of Leibniz, and with Epicurus's (much earlier) plurality of worlds, he declared explicitly his desire to concentrate on two specific cases: the GRW theory and the pilot wave theory. With reference to the first he stated:

> The world picture of Lucretius' *De Rerum Natura* is indeterministic, due to the atomic swerve. In this development of ancient atomism, a strictly deterministic theory was modified by introducing a very small bit of indeterminism—the very rare and slight, but otherwise unpredictable swerve in atomic motion. The reasons were not dissimilar from ones we hear for indeterminism today.
>
> According to Lucretius, the atoms have been "falling straight down" from all eternity—that is their natural motion. If there is no exception to this, then of course their spatial configuration is conserved in all but the "up/down" direction. In that case, a world such as we know does not evolve. By postulating that there is a random perturbation of this natural motion, we allow for the possibility of collisions, which result in further divergences from the "straight down" motions, still further collisions, and whatever interactions may happen when atoms are in contact. Too high a probability of swerves, or swerves which are exceedingly large or fast, and the world becomes too chaotic, of course. But somewhere between the extremes of total conservation of the "horizontal" configuration and total chaos lies the actual evolution of the universe.
>
> This was a solution to a problem for the earlier deterministic atomism. Apparently that had not included the assumption that all atoms are merely "falling straight down." The determinist can allow for the evolution of the actual universe by postulating sufficiently interesting laws of motion, plus an "initial" configuration (at some past time) which happened to be just right (however chaotic it looked at the time) to evolve—via those laws of motion—into the atoms' present configuration. It appears that Democritus, and possibly Epicurus, conceived of the world this way. But obviously such a conception immediately confronts the challenge: tell us what those interesting laws of motion are, that can transform chaos or very simply structure into such highly organized, complex structure! However vague the answer, it had better make it plausible that the indicated laws can lead to such an evolution of the universe.
>
> Resorting to indeterminism, this challenge is to some extent defused. The

SPONTANEOUS LOCALIZATION

laws of motion can be asserted to be exceedingly simple, but to govern only the "normal" case.

By hypothesis, the indeterministic divergences from normality cannot be described except in a very general way. All the indeterminist can be expected to do is to show that he has allowed for a certain *possible* evolution. The less he says about the swerve of the initial configuration the more possibilities he leaves open, so that expectation is easily fulfilled. On the other hand, the law of motion he does give is one of pristine clarity and simplicity.

Today we don't let people off so easily of course. Now that indeterminism takes a respectable form only in statistical guise, we expect a demonstration that things will behave as they do at least with high probability. The challenges an indeterministic theory must confront are not much less stringent than those faced by their deterministic rivals. Apart from that , however, we do have today a striking parallel to Lucretius' theoretical swerve.

If Epicurus's atoms did nothing but fall straight down, there would be no collisions or interactions of any sort at all. If quantum mechanical systems do nothing but obey Schroedinger's equation, it has been argued, then there are no measurement outcomes, or indeed any macroscopically definite events at all. Cats do not die, neither do they stay alive—the most ordinary facts in the world around us are unaccountable.

Enter the new Lucretius: Professor Ghirardi. Postulate a very slight "swerve" with exceedingly low probability—a swerve now being a departure from the evolution of state described by Schroedinger's equation. In consequence, macroscopically definite events become possible. The Born probabilities, calculated on the basis of the state-vector, become probabilities, in effect, of events due to the swerves that perturb the natural or normal evolution of the quantum-mechanical state. Unlike Lucretius, of course, the GRW model is designed so as to make the probabilities exactly calculable and in accordance with what they should be, if our world is to look the way it does in fact.

I do not wish to add anything to this statement, which in a most effective way manages to sum things up correctly and to provide a certain conceptual and historical reinforcement to the model now under consideration.

17.4. Spontaneous Reduction as a Rival of the Standard Theory

As just stated, models of dynamical reduction bring about precise mathematical modifications of the standard theory and as such, at least in principle, imply precise physical consequences that contradict the standard theory. In this respect such theories constitute not merely reinterpretations of quantum mechanics with the same empirical content but genuinely rival theories capable of being falsified in a laboratory. Consequently we would particularly be interested in finding the *experimenta crucis*, the "crucial" experiments that would permit us to distinguish between the two theories. This problem has been fully debated, and in the recent literature various physical processes have been analyzed in detail that could bring about meaningful tests of the theory. Let us take a look at a few of these, not only for the reasons just mentioned, but also because they will shed some light on some aspects of the GRW theory that may have escaped notice, owing to the concise nature of the presentation in the foregoing chapter.

To begin with, one effect that deserves our utmost attention is the fundamentally irreversible character of the new dynamics. This requires (contrary to what happens in every classical theory as well as in standard quantum theory) that the energy is not a conserved quantity. The reader can easily understand the reasons for this effect by recalling that every increment in the definition of the position of a system (such as that localizations induce) must induce a spread of the momentum. In particular, the spontaneous localizations of the theory introduce high components of velocity for all the particles of the universe. But more elevated velocities imply a greater kinetic energy, a quantity "equivalent" to the temperature of the system: the localizations tend to "heat up" the universe. This effect can be explicitly evaluated. It will interest the reader to know that the total increase of temperature caused by the countless localizations of all the particles of the universe from the big bang until now amounts to about one-hundredth of the temperature of the black-body background radiation of the universe.

Some comments are in order. Above all: since the effect in question depends on the parameters we are considering, it makes clear that however "fantastic" the model may appear, it is not so arbitrary as some may think: the change of a factor of one hundred in the parameters is enough to imply consequences that would be in conflict with well-established facts. Secondly, the considerations just developed pose a question for us: Can we take seriously a theory that includes among its principles the fundamental irreversibility of the laws of nature? The truth is that different scientists have given absolutely different answers to this question. In many of the seminars I have presented I have encountered negative reactions motivated precisely by this aspect of the formalism. But on at least

SPONTANEOUS LOCALIZATION

as many other occasions, I have experienced confirmations that this is considered a remarkable asset of the theory. In fact, as everyone knows, the problem of the "vector of time"—the fact that it goes in a precise direction—is one of the great problems in the epistemological debate about science. In various works he has published, and in particular in his last book *Time and Chance*, my friend and colleague David Albert has identified the irreversibility of the GRW model as one of its greatest assets. It would provide a solid logical foundation for the theorems of statistical mechanics.

Before concluding this section, allow me to mention that there appear to be other natural candidates for bringing into the open possible discrepancies with respect to the predictions of the standard theory, namely, effects like superconductivity, usually referred to as "macroscopic quantum effects." In processes like these macroscopic quantities come into play (such as the current that passes through a superconductor) but the processes in question require a quantum treatment to be described correctly. As far as concerns the specific problem of superconductivity, this can be analyzed in great detail in the framework of dynamical reduction models. Various authors, among them M. R. Gallis and G. N. Fleming, A. I. M. Rae and A. Rimini, have shown that, in effect, the resistance of a superconductor is slightly different according to this theory from what would be predicted by the standard theory. Alas, the effect is many orders of magnitude smaller than can be shown using today's technology.

The last effect that must be mentioned consists in the fact that a process of localization of a constituent, for example an atom, can lead to the excitation or dissociation of that same atom. It would therefore seem that a field is opened here for possible crucial tests of the theory. However, this is not the case, and the analysis of this effect once more brings into the light how the physical implications of the dynamical reduction models are truly irrelevant for microsystems. For the case in question, the processes of localization imply that for one mole of hydrogen (that is, for a total of 10^{24} atoms) spontaneous localizations could excite atoms at the rate of one per second, an effect obviously too small to be shown, since it is remarkably smaller than the excitations or dissociations caused by various natural processes, such as the effect of thermal motion (even in the neighborhood of absolute zero), the interaction with particles of cosmic radiation, the background radiation, and even the fluctuations of the gravitational field caused by quantum gravity.

The hopes of reaching a crucial test for the model in a reasonable time are extremely dim. In such conditions, to assume the point of view that it requires is fundamentally a choice dictated primarily by the position one is inclined to assume with respect to science. This is not intended to be taken, nor has to be taken, as a criticism of the theory. Indeed, in the preceding chapters we have considered in great detail certain theoretical models, such as the pilot wave theory, which require a priori the same empirical content as quantum mechanics. The choice between the two alternatives, in such cases, will never be decided on an experimental basis, but this does not make any less interesting an analysis of the different conceptual implications of the positions. In the present case, something more is at work as well: the GRW theory is scientifically falsifiable with respect to standard quantum theory. The fact that current technology still does not permit a discrimination between the two is in a certain sense accidental, and should not cause surprise. After all is said and done, the really powerful point of the theory is that it differs from the standard theory in essentially one respect, that is, in the fact that it does not allow superpositions of macroscopically distinguishable states to persist. However, we also know very well that to make experimentally evident the coherence of two macroscopic states (that is, to distinguish the pure state from the statistical mixture), is an achievement almost despaired of from the practical point of view.

17.5. THE ROLE OF THE OBSERVER AND THE DEFINITENESS OF THE OBSERVER'S PERCEPTIONS

In this section we will show how the GRW theory is capable of explaining our definite perceptions. I would like to begin by stating openly that for obvious reasons I feel a little anxious about confronting a theme of such importance. On the one hand, our knowledge about the processes of perception is still very vague, and there is yet a long way to go before we can show precisely just what specific physical-chemical processes need to be accounted for. Secondly, to undertake the risk that such a task presents for a theoretical model is justified only if there are valid reasons for considering that model to be well established and of general validity, and this is certainly taking a chance with reference to the theory we are now treating. On the other hand, some recent arguments in the scientific literature, particularly those appearing in a remarkably successful book, make it impossible for us to shirk the task, as the reader will easily understand from the following.

In 1988 David Albert and Lev Vaidman published a paper containing

SPONTANEOUS LOCALIZATION

serious reservations about the GRW model. The argument has been rehearsed several times by Albert and constitutes one of the most important points made in Chapter 5 of his *Quantum Mechanics and Experience*, which we have already cited a number of times. It will make the most sense now to pass immediately to the argument. The authors Albert and Vaidman consider the specific experimental situation illustrated in Figure 3.8 and repeated in Figure 17.1: a neutral atom with spin 1/2 in the state |x up⟩ is sent into a Stern-Gerlach apparatus for which the magnetic field is inhomogeneous in the direction of the z axis. As we know well, at the exit from the magnetic region the state of the atom will be a superposition of the states corresponding to a trajectory deflected upward and another one deflected downward. At a certain distance from the magnet is placed a fluorescent screen with a peculiar characteristic—a characteris-

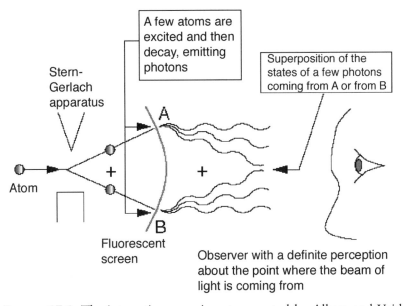

FIGURE 17.1. The interesting experiment suggested by Albert and Vaidman, supposed to represent a serious obstacle to the program of dynamical reduction. A nucleus with spin 1/2 is placed in a superposition of two states corresponding to propagation toward two points of a fluorescent screen. At impact, the atom excites a limited number of screen atoms, which subsequently emit a little stream of photons that propagate themselves toward a conscious observer. Because of the extreme sensitivity of the visual mechanism, the observer has a perception, and this will of course be a definite one: he will see a spot on the screen at A *or* at B.

429

tic that has been very well studied and yet is not difficult to realize in practice—if an atom hits it at a point within a certain region, a relatively small number of its atoms (let's say, around ten) are excited to a level with a relatively brief half-life, by means of which almost immediately they all decay, each one emitting a photon. Of course, since the state of the incident atom is a superposition of the state deflected upward (which hits the screen near point A) and the state deflected downward (causing it to strike the screen near point B), the final situation is the superposition of a state in which ten photons emerge from point A and a state in which ten emerge from point B. Up until this point there is nothing new with respect to what we know will take place in such a situation.

At this point Albert and Vaidman introduce the truly original element of their argument. They take full advantage of the well-known fact that the threshhold of visual perception is extremely low: it has been demonstrated that only seven or more photons have to strike the observer's retina to produce a nonambiguous perception. Now before going further I would like to point out that, in my opinion, this observation represents the main point of interest in the work we are now considering: the two authors have appropriately identified a case in which a genuinely microscopic situation (a few photons proceeding from a certain region) is able, thanks to the extremely sensitive nature of our perceptual apparatus, to determine a precise perception on the part of a conscious observer. But let us proceed with their argument. In front of the screen is situated a man who observes it as shown in Figure 17.1. Inevitably two facts must be recognized:

- Before the observer looks at the screen there is no doubt that, according to the GRW theory, there is an effective superposition of two situations: photons emerging from A and photons emerging from B, each bringing two different perceptions. In fact, the number of systems involved (the original atom, the atoms of the screen, and the photons) is so low that the probability that a process of localization would bring about the suppression of one of the terms of the superposition can be considered, speaking strictly, to be null.
- Without any doubt we can expect that the conscious observer does not end up in a state of mental confusion: he will certainly perceive either a light coming from A or one coming from B.

The conclusion drawn by the authors from these facts is: "The GRW theory implies that no measurement is absolutely over, no measurement absolutely requires an outcome, until there is a sentient observer who is actually aware of that outcome."

SPONTANEOUS LOCALIZATION

I would immediately like to call the attention of the reader to the subtle deception into which these authors have drawn us. Apart from the considerations I will soon be developing, the statement just quoted is misleading. Indeed, as the authors know very well, and as the reader will surely have anticipated, it is completely incorrect to say that the presence of an observer is required in order for the process to have a definite outcome—to say that would mean attempting to assimilate dynamical reduction models into Wigner's idealistic model. There is no doubt that if instead of a conscious observer two Geiger counters were placed beyond the screen, for which the stream of photons would initiate an avalanche of a very large number of particles, the theory implies, on the basis of universal laws, that only one of the counters will be activated. And therefore it is not at all true to say that the theory requires an act of conscious perception in order to lead to definite outcomes.

But these observations do not exhaust the debate. Independently of the fact that it is extremely easy to imagine situations that represent simple variations on the one considered, and that conduce to the dynamic suppression of the superposition without any need for recourse to an act of conscious perception, the acute observation that in practice our sense apparatus is so constituted as to be able to bring about definite and distinct perceptions as a consequence of stimuli that differ only for the states of a microscopic number of constituents, poses a real problem and constitutes an interesting challenge for the GRW theory. I only need to say at this point that I have in fact taken up this challenge, availing myself of the valuable collaboration of Professor Antonino Borsellino, a brilliant theoretical physicist and one of the first Italians to devote himself seriously to biophysics, and with the help as well of two young researchers, Franca Aicardi and Renata Grassi. In 1991 we studied Albert and Vaidman's criticism in detail and undertook a very general analysis of how it would be possible to model the process of perception within the theoretical scheme adopted. In order to avoid making even the least untrustworthy assumptions, we attempted to simplify the problem to the maximum extent by limiting our considerations to the transmission of an electric signal along an axon that connects the retina of an observer to the lateral geniculate body, and from there to the higher visual cortex, where electrical stimuli are transformed into perceptions. As is well known, the axon is coated with a layer of myelin and is interrupted at very short intervals by so-called Ranvier's nodes. The function of these nodes is precisely that of allowing sodium and potassium ions to be filtered from the inside to the outside of the axon, bringing about a polarization and de-

CHAPTER SEVENTEEN

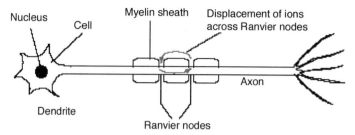

FIGURE 17.2. Schematic representation of nerve transmission along an axon. The axon itself is surrounded by a myelin sheath (with a thickness of a hundred-thousandth of a centimeter) and sodium and potassium ions come in and go out along the Ranvier nodes, leading to the transmission of an electric signal along the axon itself.

polarization of its segments, which determine the propagation of the electrical signal that induces perception (see Figure 17.2). I recall the heated discussions with Borsellino who insisted on including ever more extensive parts of the cerebral cortex in the process, while I decided firmly not to have recourse to any hypotheses not rigorously proved concerning so complex a process, so long as such hypotheses were not absolutely necessary for bringing about a suppression of the "superposition of perceptions." I recall the pleasant surprise we had in discovering that, on the basis of well-established facts regarding the transmission of signals in the brain, we could reach the conclusion that the process involves the displacement—to the amount required by the model's parameters (the thickness of the myelin sheath is precisely of the order of 10^5 cm, the characteristic distance of localization)—of a sufficient number of constituent elements (the protons and neutrons of the sodium and potassium ions) as to imply that in a time briefer than is characteristic for a conscious perception (which is on the order of one-hundredth of a second [10^{-2} sec]), the universal dynamic that stands at the basis of the model itself was leading to the suppression of one of the terms of the superposition.

I do not wish to be misunderstood. I am certainly not suggesting that we must give a purely physicalistic explanation for the incomprehensible process of conscious perception. I am only showing that if we concentrate our attention on the purely physical characteristics of the process that we know have to take place in the mind in order for a conscious perception to emerge, then the model, as it has been conceived, assures us that the potential superposition of physical signals in the brain that will correspond to distinct perceptions is dynamically suppressed in a time shorter than what is necessary for precise perceptions to emerge.

I have presented these considerations with a certain reluctance insofar as I do not like to venture into fields in which too little is known for the making of reliable assertions, and also because I do not approve of the recent mode of using the peculiar formal structure of quantum mechanics to make logical "quantum leaps," and ascend to fantastic conclusions about the human psyche. But such considerations as these have to be presented in order to show that the argument (of Albert and Vaidman) just set forth is completely irrelevant for the GRW model. On the contrary, if we wanted to take it very seriously, we would have to conclude that it constitutes a remarkably strong point of the theory: in a certain sense it can be asserted that, on the one hand, the theory tends to make objective the average distribution of the particles of the universe, and on the other hand, the process of conscious perception requires a precise distribution of particles in the brain itself, which, in turn, is rendered objective by the basic dynamic. The problem of psycho-physical parallelism here finds a solution within the model that respects to a remarkable degree what is usually adopted in a genuinely classical scheme.

17.6. Dynamical Reduction and Relativistic Requirements

A problem that cannot be ignored is whether models of the kind we are speaking of can be generalized so as to render them compatible with the requirements of the theory of relativity. In fact this is a problem that comes to the fore with reference to standard quantum theory and to theories of hidden variables. Let me attempt to sketch the problem in general.

Over the years since its first formulation, Schrödinger's equation (as mentioned in section 15.1) has undergone various generalizations in a relativistic sense.[5] Some of these generalizations are the Klein-Gordon equation, Dirac's equation, and Weyl's equation (appropriate, respectively, for [a] the description of particles of spin zero with mass—or mesons, [b] particles of spin 1/2 with mass—or electrons, protons, and neutrons, and [c] particles of spin 1/2 without mass—or neutrinos). All of these present serious mathematical problems, and have led to modern quantum field theories which describe the evolution of elementary systems in terms of linear equations that satisfy perfectly the requirements of the theory of relativity. But none of them is capable of giving reasonable indications about the path to be followed when it comes to generalizing the process of the reduction of the wave packet in a relativistic sense.

The situation could be summarized as follows:

1. The genuinely quantum relativistic theories that deny any reduction of the wave packet do not lead to conflicts between the relativistic requirements and the measurement process, since they assume that any process whatsoever is governed by the same linear laws that represent the relativistic generalization of Schrödinger's equation. Of course, as we have seen, such theories have to accept processes like the proliferation of universes or the foliation of minds, or have to assign a very peculiar role to the act of consciousness on the part of an observer. But these processes are located, so to speak, beyond the space-time continuum to which the undeniable requirements of relativity theory necessarily refer.

2. The genuinely quantum theories (using the expression to indicate that they are based on relativistic equations for the state vector that conserve the fundamentally linear character of microsystem evolution) that contemplate—in the manner of Bohr—the occurrence of a nonlinear process of reduction in connection with the interactions between microsystems and macrosystems must confront the problem we have referred to above—the problem, that is, of how to take account in a consistent way of the reduction process (other than being consistent with respect to relativity requirements in the sense discussed in the preceding chapter). Interesting critical studies by Y. Aharanov and D. Albert and more recently by R. Grassi and the present author have identified some of these characteristics and have pointed out possible ways out. I would like to emphasize that the problem does not derive from the fact that the process of reduction permits, for example, superluminal effects that would enter into an incurable conflict with the basic requirements of relativity theory. This point was exhaustively discussed in chapter 11, where it was shown that events of the kind mentioned do not occur. But certainly the fundamental nonlocality that characterizes quantum mechanical processes produces a certain tension between these theories and the theory of relativity, a tension that in a certain way is more difficult to overcome when we take into account the instantaneous objectification process of the potentialities "contained" in the wave function.

3. The hidden variable theories must also face up to the problem of their compatibility with relativistic requirements. Once again, physical effects of the kind that can be used to induce a direct conflict with rela-

tivity are not in question, in so far as these theories, being hypothetically equivalent to the standard theory, do not permit superluminal effects, as it does. Nevertheless, as has been pointed out by John Bell and rigorously demonstrated by the present author and Renata Grassi, these theories present some characteristics that are completely peculiar from the point of view of a relativistic concept of physical processes. On the one hand, it is possible and relatively easy to work out theories which represent with respect to modern field theories the relativistic analogues of the de Broglie–Bohm theory. Bohm himself, together with Hiley, has in fact exhibited a perfectly consistent model equivalent to the relativistic theory of the free meson field. Of course, the model requires that suitable field observables play the role that positions play in the original theory. But the peculiarity of this relativistic version of a hidden variable theory can be grasped by looking at the way it has to be formulated. For this purpose, we place ourselves in a precise system of reference that then comes to assume a privileged role. In this system, the relativistic field theory in which we are interested is formulated (and, in accordance with the requirements of Lorentz invariance, it has exactly the same form in any other inertial reference) and suitable prescriptions are offered for defining a hidden variable theory that would reproduce the desired theory in this system. Since the physics of the genuine quantum theory is the same in all systems of reference, it is clear that the empirical implications of the hidden variable model constructed in this way would be the same for all observers: implications for any other observer are obtained by translating the description of the observer of the privileged system into that other observer's own language. But the theory is not, to use technical jargon, covariant, that is, the theory assumes a different formal structure according to the system of reference being considered. As Bell observed, the position is analogous to that of relativity before Einstein's theory: there is a privileged system of reference—and thus an absolute meaning can be given to simultaneity—but the contraction of distances and the expansion of time conspire to deceive moving observers, leading them to the conclusion that light has the same speed with respect to everyone—in such a way that they will be able to say they are at rest. Bell's conclusion appears to me to be quite profound:

> For me, this is an incredible position to take—I think it is quite logically consistent, but when one sees the power of the hypothesis of Lorentz invariance in modern physics, I think you just can't believe in it. . . . So that for me is a big defect. However, what I want to insist is that this theory agrees with experi-

ment—when you are dismissing this theory on the grounds of non-Lorentz invariance, you are requiring more than agreement with experiment. And I think it's very reasonable to require more than the agreement with experiment. I think that theoretical physics owes much to insisting on more than agreement with experiment.

In the context of dynamical reduction theories, the problem of compatibility with relativity requirements becomes instead a precise technical problem. The situation is curiously the opposite with respect to the one of hidden variable theories. In the present case, it has been demonstrated that, contrary to what happens with any deterministic completion of quantum mechanics, such theories do not have any characteristic that would prevent a "genuinely" relativistic generalization, that is, one that would not require the assumption of a privileged system of reference. But these theories in turn encounter mathematical difficulties that to this date are insurmountable, despite the combined efforts of various scientists (including the present author) over the last five years or so; such difficulties are due to the apparent impossibility of incorporating genuinely stochastic processes into a spatiotemporal framework. Technically what happens is that the nonconservation of energy (which is also characteristic of nonrelativistic models) turns out to be catastrophic at the relativistic level. Even if there is no theorem to demonstrate it, the stochastic and nonlinear terms of the theory seem inevitably to induce an infinite increase of energy per unit of volume and per unit of time, and that is certainly unacceptable.

As far as regards the problem we have treated in this section, the situation is somewhat peculiar for all the currently available theories that attempt to give an objective description of the world (at the microscopic level, or even just at the macroscopic level). On the one hand, the hidden variable theories admit relativistic generalizations that nevertheless require the assumption of a privileged system of reference; on the other hand, the dynamical reduction models appear to allow at the purely formal level the elaboration of genuinely relativistic generalizations, but in fact, nobody has succeeded as yet in providing one without unacceptable divergences. It is not by chance that Bell chose to conclude one of his last works by referring to the pilot wave theory and the GRW theory as follows: "The big question, in my opinion, is which, if either, of these two precise pictures can be redeveloped in a Lorentz invariant way [i.e., as satisfying relativity requirements]?"

CHAPTER EIGHTEEN

Macrorealism and Noninvasive Measurements

> One might imagine that there are corrections to Schrödinger's equation which are totally negligible at the level of one, two, or even one hundred particles but play a major role when the number of particles involved becomes macroscopic (say, of the order of Avogadro's number).
> —*Antony J. Leggett*

THE PASSIONATE DEBATE about the new theory with which this entire book is concerned, and the analysis developed in the last few chapters on the conceptual problems presented by that theory, should have made very clear to the reader that the most important thing being asked of the theory is whether it is really possible for macroscopically distinguishable superpositions of states to occur in nature? It is obvious that if it were possible to answer such a question in a laboratory, we would be face to face with a discovery of huge importance for science, and more generally, for human knowledge itself. In fact, if the answer were positive, as a consequence of the fact that our perceptions are always definite, nothing would be left to do but attribute a special status to the perceptive process, by adopting Wigner-like positions or "many minds" interpretations; on the other hand, if the answer were negative, we would have to admit that some fundamental mechanism that violates the superposition principle is effectively operating in nature, independently of any role played by consciousness, and this fact would then open the way to a series of investigations intended to give precise answers to the questions HOW?, WHEN?, and WHERE?, such as we have developed in connection with dynamical reduction models.

Recently, an interesting line of research in this direction has been launched by Antony J. Leggett, who has attempted to identify processes that involve macroscopically distinguishable superpositions of states, and

CHAPTER EIGHTEEN

for which experiments can be carried out that allow macroscopic coherence to be tested. Leggett combines two assumptions: one of these is equivalent to asserting that such superpositions cannot take place (which we will call *M.R.* for "macrorealism"), while the other assumes the possibility of carrying out a kind of measurement which can be called noninvasive (we will call it *N.I.M.* for "noninvasive measurement"); his idea is that these two assumptions, when combined, permit the derivation of a sort of Bell's inequality of a temporal kind, a violation of which would make evident the existence of a superposition of macroscopically distinguishable states. Even if the conclusion Leggett wanted to draw from the experiment needs to be modified somewhat, the result would be enormously important for the reasons we have already shown.

18.1. THE MACROREALISM HYPOTHESIS

Leggett's macrorealism assumption M.R. can be expressed best by using his own words:

> **Assumption** *M.R.* If every time a macroscopic system is observed, it is inevitably found in one of two (or more) macroscopically distinguishable states, then it will be possible to attribute to it, almost at any instant—that is, even when it is not being subjected to observation—the property of really being in precisely one of those states.

The hypothesis incorporates in the most natural possible way the idea that, at least at the macroscopic level, macroscopic properties that appear to characterize a system every time we observe it ought to be possessed by the system even in the absence of our process of observation. As Einstein would say, the moon is definitely there, even when nobody is looking at it.

18.2. NONINVASIVE MEASUREMENTS

Leggett's second hypothesis requires a more detailed analysis. A measurement of an observable of a physical system is said to be noninvasive when it is irrelevant for the subsequent evolution of the system whether the measurement happens or not. More precisely, noninvasivity calls various elements into play: the state of the system at the moment of mea-

NONINVASIVE MEASUREMENTS

surement, the outcome given by the measurement, and the kind of evolution we are interested in.

We can begin with a banal observation: in both classical physics and quantum mechanics it is easy to imagine measurements that, when they provide a certain outcome, are noninvasive for particular states of the system being measured. The prototype of such measures would be so-called measures with negative outcomes. We can directly analyze them in a quantum context by making reference to a completely banal example. Let us say we have a system that can be prepared in one of two states that leads it at successive moments to propagate along two different trajectories. We do not know if it will follow trajectory U (up) or trajectory D (down; see Figure 18.1) because we do not know the details of how it was prepared. But we do know with certainty that it actually is in either one or the other state in question.

Suppose now that we place along one of the two paths that it can pass along (let's say, D) a perfectly efficient detector, and suppose that the detector does not register the passage of the system. After enough time has passed for the system to have reached the detector, we can say that our measurement has as its outcome that it lets us know that the system was in fact propagating along path U, and it can also be said that such a measurement did not disturb the later evolution of the system in any way. We note that the measurement is noninvasive only for the state that corresponds to pathway U, or, equivalently, only when it gives a negative re-

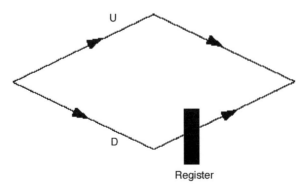

FIGURE 18.1. A particle is being prepared without the knowledge of the person running the experiment either in the state corresponding to path U or in the state corresponding to path D. If a detector is inserted on path D and does not click, it is legitimate to conclude that the particle is passing along path U and that the detector did not change its evolution in any way.

sult. Moreover, the noninvasivity pertains in this instance to the specific property which interests us, that is, the modality of propagation of the system in space.

With reference to the hypothetical macroscopic system considered in the preceding section, we can now formulate the *N.I.M.* requirement:

> ***Assumption*** *N.I.M.*: For a macrosystem, granted the M.R. hypothesis, it is assumed possible in principle to carry out a measurement that could tell us which macroscopic state that system is in, among the various states possible for it, without influencing in any way its later behavior, at least as far as concerns the evolution between the macrostates in question.

Some observations are in order. As we shall see, it is not necessary to be able to carry out measurements that are noninvasive independently of their outcome. For the argument that I will soon develop, it will suffice to have apparatuses that are not invasive when they give a certain outcome (obviously we will have to use similar apparatuses, although they can be otherwise completely different, for each result that interests us). It must be emphasized that in the *N.I.M.* assumption, the noninvasivity of the measurement (when the apparatus gives a certain response) is referred exclusively to its outcome and not to the state of the system (this observation must emphasize that the requirement is much more stringent than that of the example we have just discussed above).

We are now in a position to state the general argument: using conjointly the two assumptions *M.R.* and *N.I.M.*, it can be demonstrated that they imply a relation about the probabilities of obtaining certain outcomes at various times chosen for the macroscopic observables that interest us—a relation that is the exact analogue of Bell's inequality. In other words, the conjunction of *M.R.* and *N.I.M.* implies in general a violation of the predictions about the outcomes that must occur if the superposition principle is valid for the macroscopic system and for the states in question. It is therefore possible, if the preceding conditions can be satisfied, to carry out a test of macrocoherence.

18.3. Leggett's Argument

Given the premises I have just set forth, Leggett argues as follows. The situation we are confronted with is radically different from that of Bell's traditional inequality, despite the analogies. In fact, in the standard case of an EPR experiment, the hypotheses that are made in order to derive the desired inequality are those of microrealism (that is, that there are elements of reality that are possessed by the two microsystems prior to the process of measurement) and of B.L., of locality in Bell's sense. Now, these two requirements are logically independent one from the other. In the present case, however, Leggett argues, the two hypotheses M.R. and N.I.M. are logically dependent in the sense that N.I.M. is logically implied by M.R. In fact, says Leggett, what meaning would there be in asserting that the macrosystem is always in one of two macroscopically distinguishable states and at the same time saying that a noninvasive measurement for one of these states could change the later evolution of the system?

The observation seems pertinent but in reality it is not conclusive. The reader can easily convince himself of this by considering the quantum experiment of a particle that can be in the superposition of two different trajectories and can give rise to a process of interference, by adopting the perspective of the pilot wave theory. We try to represent this process in schematic terms in Figure 18.2. In it are shown two possible trajectories

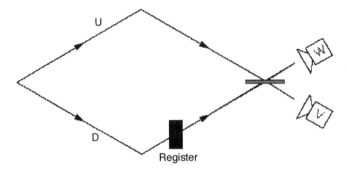

FIGURE 18.2. A particle in the superposition of two paths, in the absence of a detector along D can only activate detector V when the device is set to have constructive interference in direction V and destructive interference in direction W. Nevertheless, the insertion of the detector D, even if no particle is registered, can bring about the activation of W. If this standard quantum experiment is considered in the perspective of the pilot wave theory which assures us that in fact the particle is being propagated along a precise path, and since we know that the theory reproduces the same outcomes as quantum mechanics, we are led to conclude that the evolution of the particle even though it passes by path U has been influenced by the insertion of the detector along path D: the operation is therefore invasive.

of the type used in neutron interferometry and it is supposed that the system is in the superposition of two states corresponding to the two trajectories, and that things are regulated in such a way that, in the absence of a detector along D, there would be constructive interference in V and destructive interference in W.

In order to understand our argument the following must be kept in mind:

1. If the particle has been prepared in the state that corresponds to the fact that it will travel with certainty along the upper path (U), then placing a detector on the other path will not have any effect upon the particle's evolution. In other words, the measurement that corresponds to asking ourselves, "Is the particle traveling on the lower path?," when the answer is the negative, is an absolutely noninvasive measurement, so long as the particle is associated with a wave function different from zero only on the upper path.

2. Let us now consider a particle in the superposition of the two states that correspond to the two trajectories. According to the de Broglie–Bohm theory, it is perfectly legitimate and appropriate to say that, actually and objectively, the particle passes along either the upper trajectory or the lower one.

3. We recall that the pilot wave theory nevertheless yields predictions that are in perfect accordance with quantum theory, and that, according to this theory, even if a screen placed on the lower path does not reveal the particle, this still leads to the reduction of the wave packet, a process that by destroying the interference changes the evolution of the system relative to the position variable that interests us. In particular, when the screen is inserted in the path and does not register anything, its presence alone causes the fact that the counter W can register (with probability 1/2) the arrival of a particle.

What, then, is the moral of this story? Clearly, it is that the *M.R.* and *N.I.M.* requirements are logically independent. In fact, the pilot wave theory constitutes an explicit example of a perfectly coherent theory, in which, even though the assumption is valid that the particle with certainty passes by one only of the trajectories (the analogue of the *M.R.* requirement), and even though it is true that, when the particle is associated with a wave function that is nonzero only on the upper path the insertion of a screen on the lower path is absolutely noninvasive, nevertheless, the same operation *is* invasive (i.e., it changes the subsequent evolution) when the wave function is nonzero on both the paths, even if the

NONINVASIVE MEASUREMENTS

particle in fact passes by the other path. In short, it is possible to construct in explicit fashion a perfectly coherent theory that satisfies *M.R.* but does not satisfy *N.I.M.*, and therefore it is not correct to say that *M.R.* implies *N.I.M.*

Despite this necessary qualification, Leggett's argument still keeps all its fundamental conceptual importance. In fact, the peculiarity of it just now brought to the fore, that the attempt to identify which of the two possible paths the particle is following, and that *not* finding it along one of them brings about some effects for the particle itself within a theoretical scheme that assures us that in fact it always takes only one of the two paths, constitutes a fact that is in no way reconcilable with a classical conception of the process. In other words, this inevitable invasivity of a process that is noninvasive for "particular quantum states" is nothing other than a reflection of the nonclassical nature of the process itself. In the pilot wave theory it is a reflection of the fact that the quantum mechanical potential is determined as well by the "empty" part of the wave (another way of saying that the coherence between the two terms of the superposition has precise physical consequences). All the same, the reader will certainly understand that even if we cannot always speak of superpositions in such a general context as the one in which we have put ourselves (in fact, we have not precisely defined what theory we are using, apart from insisting that it be empirically equivalent to quantum mechanics), the inevitable invasivity of noninvasive measurements is nothing other than the counterpart of quantum coherence within this very general framework. For this reason, I think we can say that what Leggett is proposing, even if it cannot be directly structured into a test for macrorealism, is nevertheless a test for proving the existence of macroscopic systems that exhibit fundamentally quantum behavior. And that is what is really interesting.

It is worth mentioning that Leggett's proposal is not purely speculative, and there are in fact well-founded hopes that the experiment in question could be carried out within a few years' time.

18.4. How to Carry Out an Experimental Test of Macrocoherence

Leggett himself, in collaboration with Anupam Garg, observed that modern technology puts at our disposal certain instruments that are especially suited for an experiment of the kind just discussed. The system to be measured is made up of a superconductor ring, interrupted by a Josephson junction. It is not important for the reader to understand the technical aspects of the system in question; the only really interesting thing to know is that it can be easily put into a state that is the superposition of two states corresponding to the fact that two "macroscopic" currents (about 50 nanoamperes) circulate in opposite directions within the ring. The intensity of the two currents varies with time: at a certain instant only the current running to the right is present, and then it diminishes, as a component going to the left emerges, increasing to a maximum value (at the instant when the other current disappears) and then the process repeats itself from the beginning. The system's evolution perfectly regulates the two terms of the superposition and therefore we are dealing with a state that exhibits characteristics of coherence between two "macroscopically" different states.

There are essentially two problems still remaining. The first is that of guaranteeing a sufficient isolation of the system from the surroundings in such a way as to prevent, through the unavoidable entanglement of the system with the environment, a loss of coherence between the two terms of the superposition, so that it would become impossible to see it in the laboratory.

This point has been analyzed in great detail by C. D. Tesche, who demonstrated that, by working at low enough temperatures, coherence can be maintained for a period of time long enough for a meaningful test to take place.

The second problem is that of constructing an experimental setup that would allow the carrying out of perfectly noninvasive measurements when the system is actually in one of two "macroscopically diverse" states. The main idea is once again to use a superconductor ring with a Josephson junction that is traversed by direct instead of alternating current. This second superconductor can be coupled to the system to be measured in such a way that if, for example, the current is running to the right, nothing happens to the measurement apparatus and it does not react on the system, but if the current is to the left, then the measuring apparatus is saturated and reacts abruptly upon the measured system. The second ring can then be assimilated to a measuring apparatus that is able to carry out noninvasive measurements when the state of the system is the one corresponding to the "right current." Of course, another noninvasive measuring apparatus would have to be constructed for a "left current," but that is not a problem. The apparatus is essentially the same with a reverse orientation.

The practical possibility of realizing Leggett's test entirely depends on the possibility of making this second direct current superconductor capable of performing noninvasive measurements. It seems that this possibility is within the reach of modern technology. In various laboratories, such as those of the "National Institute of Nuclear Physics" (*Istituto Nazionale di Fisica Nucleare*) in Rome, a serious program of experiments is being carried out with the intention of overcoming some of the technical difficulties that still impede the successful construction of the right instrument. Reasonable estimates are that within a few years we will have what we need. But it seems superfluous to emphasize once again the conceptual importance of this project. It would represent a first step toward the exploration of that unknown territory that lies between genuinely microscopic systems and the macroscopic systems of our experience. Such systems have been aptly referred to as the "laboratory cousins of Schrödinger's cat."

The analysis now developed has led to the identification of the peculiarity of certain processes of measurement, that is to say, processes that by having a negative outcome would seem at first sight not to bring any disturbance to the measured system, but on the contrary are remarkably invasive for it. Obviously, as discussed in relation to the pilot wave theory, processes of this type can very well take place at the microscopic level as well. It seems fitting to close this chapter by describing one last incredible exploit that is made possible thanks to the quantum nature of microscopic systems.

18.5. How to Observe a System without Interacting with It

Suppose we have to solve the following problem: we have a number of plates, some of them simply glass plates and some photographic plates. They cannot be distinguished by touch, and thus the only way to distinguish them is by shining a light on them and observing the result. But—and here is the point of interest—the photographic plates are made so that when they are exposed to even a single photon they suddenly react and become completely darkened and incapable of any further exposure. Our goal is to succeed in identifying some photographic plates that have not yet been darkened, so that we could use them in some way—to see, for example, if a photon has hit them. It will be perfectly obvious to the reader that the problem as so formulated does not admit of a solution within classical physics.

CHAPTER EIGHTEEN

But let us see if we can perform this extraordinary feat by using quantum mechanics and the invasivity of measurement with a negative outcome.

To solve our dilemma we make use of a Mach-Zehnder interferometer, like the one we considered in section 4.4, and as shown in Figure 4.8. The only changes we need to make to the setup are to insert one plate on the upper path (U) that is completely the same as one of the glass plates in our supply of plates, and then to allow for the insertion along the other, lower path (D) of another plate taken from our supply. The situation is illustrated in Figure 18.3. Let us analyze the process.

A photon is sent into the apparatus. Suppose we now insert a glass plate along path D. Now the process will be the same as that discussed in section 4.4, and we have two perfectly equal situations on the two branches, because the plates are identical. This means that there is constructive interference along the path leading to counter V and destructive interference along the path leading to W. In other words, every time a simple glass plate is inserted in the region of the camera and a photon is sent, then the counter V is activated with certainty.

What would have happened if we had put in a plate coated with the photographic emulsion? In half the cases it will intercept the photon, and

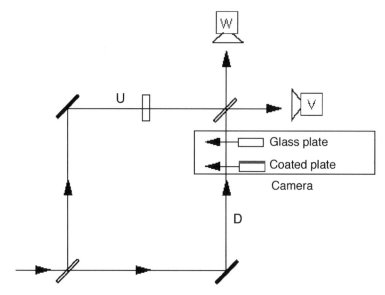

FIGURE 18.3. The incredible experiment that permits the identification of a coated glass without in any way interacting with it (which would have the effect of making it useless). As explained in the text, the performance of the feat analyzed here would be impossible within any classical scheme.

there will be no registration either by the counter V or by W, but the plate will be black and will have to be discarded, since it has become unusable. But now let us consider the other half of the times. With these, a process of measurement with a negative outcome has occurred, meaning that the plate has registered no photon on path D. Now, as we know very well, a reduction of the wave packet happens and the photon is collapsed onto path U. This meets the semitransparent mirror placed in front of the two counters V and W, and thanks to the properties of this mirror, the photon will, with equal probability, continue its path to counter V or become diverted toward counter W.

Let us summarize the situation again for the case when a photographic plate has been inserted. In one-half of the cases, it will be darkened (and then discarded), and in one-half of the cases in which it has not been darkened, the photon as it exits will activate counter V as in the case when a glass plate was inserted, and therefore we will not know if the plate in question is purely of glass or whether it has photographic emulsion and is (consequently) now unusable. In the rest of the cases (one-fourth of the total) the counter W will go off, and this indicates two things with certainty: (1) above all, that what we have inserted cannot be a glass plate because in this case we know that only counter V can go off, and (2) on the other hand, the same plate cannot have intercepted the photon (that is, it has not been darkened) because in that case it would not have been able to emerge and go on to activate counter W.

In conclusion, by taking advantage of the invasivity for the upper branch (U) of the measurement with negative outcome along branch D (note that in fact the evolution of the photon along U is really changed by the insertion of the photographic plate, since otherwise it would not be able to activate counter W), we succeed in identifying one part (in fact, one out of four) of the photographic plates without darkening them, and we have them available for use in our future research.

The ingenious experiment just now described (or rather a more refined variant thereof) was suggested by A. C. Elitzur and L. Vaidman in a recent work and seems to have found some interesting applications in delicate quantum experiments.

With this curious feat that once again illustrates the mysteries of the quantum world, we have reached the end of our discussion. All that remains for the concluding chapter is to sketch out the larger problems and some of the interesting lines of research that will most likely be pursued in coming years by scientists active in the field.

CHAPTER NINETEEN

Conclusions

> My own inclinations are to try to hang on to both—quantum realism and the spirit of the relativistic space-time view. But to do so will require a fundamental change in our present way of representing physical reality . . . a profound change of viewpoint, which makes it hard to speculate on the specific nature of the change. Moreover, it will undoubtedly look crazy!
> —*Roger Penrose*

OUR FASCINATING JOURNEY, undertaken to grasp the secrets of nature, has given us the "new eyes" that, as Thomas Kuhn would say, every scientific revolution provides (in our case the quantum revolution) to those who are interested in plumbing its mysteries. It has been articulated into five phases which may not have escaped the attentive reader's notice, even though these phases are not identified clearly by divisions of the text. We began by illustrating the difficulties that the scientific community had to face at the beginning of the twentieth century, and then by showing how a restricted group of scientists, thanks to their genial intuitions and fortunate hypotheses, were able to pass through moments of great frustration to enthusiastic hopes, from profound discomfort to an almost ecstatic wonder, lifting the veil that hid the surprising realities of the world. Next, we explored the debate that was ignited over the interpretation of the theory, a theory that has acquired a precise structure and a formal appearance of great elegance, but which poses serious problems for anyone who looks with a critical eye on the conception of scientific knowledge that it seems to imply, especially if one refuses to assume a purely instrumentalist position. The giants of this titanic struggle were two of the most important figures of the century: Niels Bohr and Albert Einstein, whose dialogue and disagreement were fully treated. Chapters 7 through 9 then showed how the confrontation of these two exceptional figures—a confrontation that also involved the other celebrated protagonists in this un-

CONCLUSIONS

believable intellectual adventure, from Heisenberg to Schrödinger, from Born, Pauli, Jordan, and von Neumann to de Broglie and Bohm—led to an increasingly sharper focus on the true problems of the theory, compelling many thinkers on a quest for discovery, leading them to take ever more precise and defined positions. The achievements analyzed in this part of the book furnish interesting material for reflection for anyone interested in the evolution of scientific thought. I believe it can be said that the uncritical acceptance of the Copenhagen ideology represented a brake on the development of new ideas, on the elaboration of alternative models, and above all on the very comprehension of the revolutionary aspects of reality that were emerging. It became very difficult for dissenting voices to be taken seriously, and all too easy for those who shared the "victorious" position to take things that really required further serious analysis to be accepted for unequivocally stable truths and principles. The third part of the book then showed how it needed a genius like Einstein and a profoundly motivated,[1] lucid thinker like Bell to shake the hold of the "orthodox" clan. In an article from the Abstracts of the 1994 Conference "Advances in Quantum Phenomena,"[2] authors Alain Aspect and Roger Grangier concluded as follows, after reviewing the various experiments demonstrating the violation of Bell's inequality: "We hope that even those who do not have an immediate interest in problems of this kind will be convinced that Einstein pointed out one of the most extraordinary properties of Quantum Mechanics. We must thank J. Bell for having provided us with the possibility of experimentally evidencing this extraordinary characteristic of nature." If the uncritical acceptance of the orthodox position had continued to dominate the scientific scene without challenge, it would not have been possible to approach the topics that have led to new and unsuspected developments of great conceptual importance. Moreover, even when turning to some applications of the theory, we showed in the fourth part of the book (that is, in chapters 12 and 13) how it was these very discussions, considered by many to be "philosophical" and "sterile," that led to some eminently practical results and seem to have opened up perspectives that up until a few years ago nobody could suspect, even in such technologically advanced fields as computers.

We thereby reached the fifth and last part of the book, which raises and leaves unanswered many problems of a technical, conceptual, and epistemological nature. The arguments treated in this part seem to me to be the most stimulating of all, because, as I have explained at length, the debate on these issues is anything but settled and provides most interesting stimuli for a critical reflection on science. But to go further: we have also seen

449

CHAPTER NINETEEN

how this very debate, which once again it would be extremely easy to judge as Pauli did (and as many physicists still do) to be as sterile as the most abstract medieval theological debates, has led, on the one side, to the elaboration of theories that are potential rivals of quantum mechanics and can therefore suggest where violations of it may some day be found, and on the other side, to the identification of possible experimental tests that make use of modern technology and can be crucial for determining the theory's validity.

What are the prospects, then, for further research in this field? As Roger Penrose fittingly emphasized in the statement placed at the beginning of this chapter, the identification of a fully satisfying way to solve the problems that still afflict the splendid theory to which this book is dedicated would probably require so radical a change in our point of view that it is difficult and risky at the present time even to imagine what it would be like. Any affirmation of possible future developments could only have an extremely limited value.

Nevertheless, I do not think it would be right to bring this book to a close and take leave of the reader who has patiently followed me over so long a distance, without expressing a few personal opinions, which I humbly offer for consideration, asking the reader not to magnify them beyond what they are, namely, the pure and simple reflections of someone who has devoted a great deal of time and energy to the themes treated in this book.

It is recognized that in recent years an interest has increased in positions like that of the many-worlds theory, and that personalities of high prestige like Stephen Hawking, Murray Gell-Mann and many others have adopted that point of view. Similarly, a certain fascination has been exercised upon some of those who are interested in these themes by the positions that in one way or another call into question the role of the conscious observer, positions ranging from those of von Neumann and Wigner to the very different ones I have classified as the "many-minds" interpretations of the formalism. I think the renewed popularity of these positions is to be attributed in an appreciable degree to the enormous interest for some time now, on the part of both the scientific community and the public in general, in all discussions that in one way or another are connected with the processes of perception—in short, the search to "grasp" the human mind. One revealing example of this should suffice. In recent years there has been a multiplication of books written at various levels and with various degrees of rigor that attempt to meet this exciting problem whether the human mind is just like a computer or

CONCLUSIONS

whether the plane upon which the mind resides far transcends that of any "physical" instrument; to put it more simply, the question whether the difference between the brain and a very refined computer is at bottom a quantitative or qualitative difference. That all this is happening ought not to surprise us if we take into account, on one side, the absolutely extraordinary progress that has taken place in the fabrication of ever more powerful and rapid calculators that appear to exhibit or imitate typically human behaviors such as apprehension, the capacity to adapt to new situations, and so forth, and which are already decisively superior in various fields to human beings (almost everyone agrees that Kasparov's victory at chess over a supercomputer was probably the last time that will happen in human history, and that the next generation of computers will be unbeatable). On the other side, recent years have also witnessed remarkable progress in scientific understanding of the superior functions of the human being. Understandably, the human race is particularly attentive to all these events that cause us to face once again the basic themes that quite rightly have occupied the center of attention for the great thinkers of all times, but for the first time there is "a certain impression" that science can answer these problems with its own method, which has proven so effective in the understanding of natural processes. Personally, I do not think that this impression has as yet been justified by our new knowledge. In fact, I do not mention these matters in order to enter into problems for which my opinion has no more value than any other human being who thinks seriously about such unfathomable topics. I merely wish to point out that the widespread interest in certain interpretations of quantum mechanics that in some way call into question the conscious perceiver is a reflection of a larger interest in the whole thematic of cognitive processes.

With such premises set down, I will risk stating a few opinions on the theme that interests me, namely, the most promising aspects of the approaches analyzed in the last chapters. At the risk of seeming pedantic—or of raising the suspicion that I do not speak freely out of fear of displeasing some readers—I would like to make it clear that my fundamental position is that of someone who is convinced that there is still a long way to go and that it is not really possible even to imagine what the outcome of researches will be in this field, which (as I believe this book has amply demonstrated) is the object of a lively new interest. My own position is a rather open one, and can be characterized not so much by a precise expectation that things will turn out as I would like them to (I think that all readers perfectly understand that my sympathies lie with the theories that permit the most comprehensive recovery possible of an objective vision of reality), as by an extreme intellectual curiosity. If the reader will allow one more quotation from John Bell, I think it appropriate to recall a statement from one of his last works, showing how he, too, despite the accusations of ideological prejudice that were often thrown at him, was extremely open to the future developments of science: "Suppose, for example, that quantum mechanics were found to resist precise formulation. Suppose that when formulation beyond FAPP [= for all practical purposes] is attempted, we find an unmovable finger obstinately pointing outside the subject . . . to the Mind of the Observer, to God, or even only Gravitation? Would that not be very interesting?"

So here then, in a synthesis, are my ideas. I do not very much like the "many-worlds" interpretations. They seem redundant and definitively sterile. As regards all the interpretations that make an appeal to the conscious observer I would like to be very explicit. The problem of the emergence of consciousness in the physical world appears to me to be an absolutely central problem that has passionately engaged human interest in the past and will do so again: it deserves the utmost attention. I would not in any way deny that perhaps, in the future, discoveries will be made that find close connections between this problem and the problem of the quantum theory of measurement, namely, the reduction (or collapse) of the wave packet. But I do not think that the solution of the problems with the theory will come from the solution of the problems of conscious perception. I think that the problems of quantum mechanics must be faced and resolved in exclusively physical terms, and independently of—I dare to say, with priority of—the problems of perception. It may be that, once we have elaborated a solution in exclusively physical terms of this eminently physical problem we will find ourselves in a better position to deal with other more difficult and more profound problems. I hold that the solution of the problem of macro-objectivization represents an indispensable prerequisite for going any further along the long road of scientific discovery, and I also think that, as difficult as it may be, it is still a much less difficult problem than understanding the processes of consciousness.

Among the proposals we have analyzed, I would place great importance on the pilot wave theory, even if the difficulties in principle that it appears to involve, and which make a genuinely relativistic generalization of it almost impossible, are certainly cause for discomfort. I would like to mention that recently D. Duerr, S. Goldstein, and N. Zanghi have taken important steps toward getting over this seemingly insurmountable obstacle.

CONCLUSIONS

We then come to the theories of dynamical reduction. As I have repeatedly underlined, their more serious limitations derive from their phenomenological character. I am certain that these theories cannot be configured into definitive models; as I have already made clear, at most I hope that by having shown the practicability of a solution that had not been tried before, they can give useful indications for the elaboration of a truly satisfying theory that would include their positive aspects. As I have explained, the attempts to generalize these theories in a relativistic sense have been fruitless and have shown difficulties that do not seem easy to overcome. It is my impression that the ideas on which the models are based would require a truly qualitative leap before they can be incorporated into a genuinely relativistic framework: a transformation into theories much more general than those we have been considering.

In this connection I cannot fail to mention two facts of some importance. Some modern relativistic theories have recourse to new mathematical formalisms that attempt to incorporate aspects of the physical world that have only recently been identified. A typical example would be the theories that take the universe to be, so to speak, a constellation of microscopic black holes. Now, as the nonspecialist public well knows, everything that interacts with a black hole cannot reemerge, and if this perspective is assumed in a quantum context, one is forced to admit that parts of the wave function become, so to say, "sucked back" into a black hole. This inevitably brings about a loss of coherence among the "surviving" parts of the wave function. Some physicists, especially the group centered around John Ellis of CERN's theoretical division, have studied this type of model and maintain that at the nonrelativistic limit they lead to equations very close to the GRW model.

My last observation concerns the role of gravity. For many years, various thinkers have suggested that the problem of the reduction of the wave packet has some direct connection with the problem of gravity. Roger Penrose in particular seems rather convinced that the solutions of the problems of quantum theory and the solutions of the problems of quantum gravity will make progress side by side. It is well known that this last problem is extremely important as well and is far from receiving any satisfactory solution.

With reference to our theme I have already related how recently my colleagues Rimini and Grassi and myself presented a variant of the GRW model which eliminates one of the two phenomenological constants that are characteristic of it by substituting it with Newton's constant of gravitational attraction, and connecting the reduction process with gravita-

453

tional effects. But the model remains fundamentally phenomenological and nonrelativistic. Penrose's researches are obviously much more ambitious[3]: he hopes to obtain something more fundamental and definitive. Nevertheless, at the present time, although he has shown on many occasions how he hopes to get over the difficulties he meets, he does not have at his disposal a precisely formulated mathematical model that would bring about the suppression of macroscopically distinguishable states.

My final observation derives specifically from my own education and personal situation, that of a physicist who believes in his profession. As the reader will certainly have gathered, I nourish a lively interest in formally logical and conceptual analyses. Fortunately, however, I have not forgotten Galileo's great lesson, and I believe that the direct confrontation of ideas with experimental facts and the speculative elaboration of new ideas by the identification of new experimental data represent the high road of scientific knowledge. In this sense even limited models like the phenomenological GRW theory that have an empirical content different from the standard theory would be able to suggest lines of experimental research; and even if such experimentation does not confirm the models, nevertheless, by bringing new horizons to research, it can lead to discoveries of processes that are in disaccord with standard theory. Analogously, experimental research in fields not sufficiently explored, such as typically the difficult realm of the *mesoscopic*, can bring new aspects of reality into the light.

My conclusion should now be obvious: with the same enthusiasm that led me to share with a wider public my wonder and amazement at the richness of reality that quantum mechanics has revealed to us, in the same way do I watch the future with lively interest; not my own future, of course, which is reaching its sunset, but that of the new generations of scientists who will surely find ever new stimuli, ever new reasons for traveling along the fateful path of research, and who will always find waiting the amazing mysteries that nature so jealously holds in store.

Notes

CHAPTER ONE

1. Rutherford discovered that some alpha particles, when made to bombard a very thin film of gold, were deflected at angles greater than 90 degrees with respect to their direction of incidence. He commented as follows: "It was quite the most incredible event that has ever happened to me in my life. It was almost as incredible as if you fired a 15 inch shell at a piece of tissue paper and it came back and hit you." (see J. Bernstein, [1991], p. 28) The only explanation possible was to assume that practically all the mass in an atom was concentrated in the nucleus, surrounded by opposing charges in a virtually empty space. The alpha particles that were most highly deflected were those that sprang back from an almost frontal collision with the nucleus.

2. Obviously, there is an important difference in the case of gravity, for, while according to Newton's universal laws of attraction, all masses are attracted to one another, two charges of the same sign, and thus two electrons, would repel each other. But this difference has a practical rather than conceptual importance in the analysis to be developed here.

3. This argument can be more easily understood and made more accessible to the modern reader through a comparison with synchrotron light machines. In these machines, because the electrons move in an orbit and do not propagate themselves in a straight line, they have acceleration, causing the "light" to be produced (x-rays or gamma rays in this case). If this resultant loss of energy were not restored at every revolution by the accelerating components of the machine, the orbits of the electrons would decrease until they fell against the chamber walls. The beam of light would disappear and the machine would cease to function.

4. It is worth mentioning that it is really diffraction which permits two persons to communicate by walkie-talkie over a hill standing between them. The waves transmitted by one apparatus diffract at the rim of the hill and reach the other unit.

5. Recalling the example given in the previous note, while it is perfectly possible for a radio wave to "go around" a hill, something of the kind could not happen in the case of visible light. As long as the sun is blocked by the hill, we are simply in the shade. The wavelength of the light is so small with respect to the dimensions of macroscopic objects that it doesn't allow for any appreciable effects of diffraction, and thus cannot "go around" the obstacles. The situation can be explained as follows: under the conditions indicated, for all practical purposes, it is possible to describe the process by resorting to an approximation (geometrical optics), which assumes the rectilinear propagation of light rays, rather than in terms of the precise equations which govern it (wave optics).

6. Clearly, the holes must have dimensions comparable to the wavelength, so that the diffracted waves overlap in the spatial region between the holes and the screen. If the openings had dimensions large enough to permit a description of the

process in terms of geometrical optics, the waves to the right of the two openings would not "superpose" or "interfere" with each other.

7. 1905 was an extraordinary year for the twenty-seven-year-old Einstein. In addition to his work on the photoelectric effect, he wrote two others in which he formulated the theory of relativity: one on Brownian movement, and another in which he proposed a new method for determining the dimensions of molecules. While mentioning these works it is interesting to observe that, except for the years 1911 and 1915, Einstein was nominated for the Nobel prize every year between 1910 and 1922, the year he finally won it. The three fundamental acquisitions of this fertile year 1905 that were given over and over again as justification for receiving the awards were the photoelectric effect, the treatment of Brownian motion, and, of course, the theory of relativity.

8. J. Bernstein (1991), p. 31.

9. His friend Weyl would later say that Schrödinger "completed his great work during a late erotic explosion in his life."

10. As soon as it was discovered that some substances were emitting "something" capable of imprinting, for instance, a photographic film, the problem immediately arose of how to identify whether these radioactive products were particles or electromagnetic rays. This illustrates very effectively how, in those days, these two types of phenomena were considered absolutely distinct and incompatible. It was only in the course of the experiments, in order to reply to this precise and important question, that J. J. Thompson identified the particle nature of some of these radiations (the so-called beta rays) which would afterwards be recognized as electrons.

CHAPTER TWO

1. By "amplitude of the field" is meant the maximum value it can attain.

2. There is a very simple reason why such lenses reduce glare. When light falls on a shiny metal surface, for instance, the light becomes polarized in the process of being reflected. As we shall soon see, when a polarized light plane passes through a polarized lens, if the typical plane of polarization of the lens does not coincide with that of the radiation which hits it, the lens reduces the intensity of the light that passes through it. The lens has a very precise polarization plane, while the polarization plane of the light varies in accordance with its orientation and other characteristics of the reflective surface, so that on average there will be a perceptible decrease in the reflected rays.

3. You can check the quality of your polarized lenses by seeing if the maximum intensity of the transmitted light practically corresponds with the incident light, that is to say, if there would be no difference when you take away the second lens, and on the other hand, if no light passes through them at all, with the correct rotation.

4. We should recall that the luminous intensity is proportional to the square of the field.

5. We can assist our imaginations here with a mechanical analogy. Instead of a ray of light, think of a string. Let's say we make the string vibrate by pulling on its end. The plane in which the string vibrates corresponds to the plane of the electrical field, which is to say, the plane of light polarization. Normally, we can

make the string vibrate in any plane we choose. But now we have to simulate the polarized lens. To keep up the analogy, this would correspond to two parallel plates, placed on either side of the string, and oriented in a certain way (this would correspond to the orientation of the plane of polarization of the filter). Now it is clear that if we try, for example, to make the string oscillate in the vertical plane, but the plates are horizontally positioned, the vibration that reaches the "polarizing lens" will not be able to go beyond it. On the other hand, if the two plates are positioned vertically, they will not affect the vibration and the oscillation of the string, and the ray of light can propagate itself without any disturbance. In most cases, the two planes will neither completely coincide nor be perpendicular to the string's plane of oscillation, so that in such conditions the vibration that "goes beyond the lens" will be reduced, and its oscillations will develop along the plane defined by the two metal "guides".

6. It should be noted that the sum of the intensities of the two rays is equal to that of the incident beam.

Chapter Three

1. As I will explain shortly, the relationship between energy in a certain volume and the position of the photons that carry it must be understood in a statistical sense. Consequently, the assertion just made would be more correctly stated as follows: "in such a way that, *on the average*, one photon per second enters the filter."

2. In fact, Bohr formulated his rules of quantization by assuming that in its movement around the nucleus, an electron cannot possess arbitrary values of angular momentum, but only certain exact values. Consequently, only certain precise orbits are accessible to electrons, and atomic states are quantized.

3. This assertion requires some qualification. It could be considered meaningless, since, if it were true that the angular momentum oriented itself so as to form only certain angles with a certain direction, it would form other angles with other directions, such angles varying arbitrarily when these directions were changed. In fact, the statement about quantization refers to a direction in space that is physically identifiable, and to a situation in which there is a connection between the angular momentum itself and a system or field that characterizes this direction, in such a way that, for example, diverse orientations correspond to diverse energies for the system. A similar state of affairs often presents itself (and should be clearly recognized) when one intends to measure a component of angular momentum along a certain axis. The assertion here made is therefore meant as follows: whenever the component of angular momentum is measured along a certain axis, then only one of the values (to be described later) will be obtained as the outcome of the measurement.

4. Even though, as just pointed out, the spin angular momentum (like any angular momentum) can never be perfectly aligned with a preset direction, in the following discussion I will always refer (whether in terminology or imagery) to the situation that corresponds to the two possible values of the component of spin just mentioned, by saying that the spin points "upward" or "downward" with respect to the same direction.

5. While the orientation of the needle always corresponds with that of the spin,

the directions of rotation can be concordant or discordant. In particular, a neutron with "upward" spin corresponds to a needle with north pole pointed "downward." For the sake of simplicity, we can assume that the two directions of turning—the clockwise action of the spin and the south-to-north magnetic action—are "concordant," which is what occurs, for example, in the case of an antineutron.

6. According to classical mechanics, if a magnetic needle is oriented in such a way as to form an arbitrary angle with the vertical, it will maintain that angle by the gyroscopic effect, because every force which tends to line up with the vertical simply produces a precession of the needle itself around this direction, in the same way as the force of gravity on a spinning top does not succeed in deviating its axis, but only in lending a precessionary movement to the top.

7. In this connection it is interesting to observe that although there are appropriate criteria that can be used for judging a numerical succession as random, the fact that these criteria are satisfied does not in itself guarantee the randomness of the succession. One good example should suffice for all instances: the numerical value of the famous π, which represents the ratio between the radius of a circle and one-half the circumference, and which has been figured out (thanks to the recent developments of algorithms) to more than a billion places, passes all the tests for randomness. For example, all the numbers between 0 and 9 appear with the same frequency, the correlation between successive numbers or numbers separated by a fixed number of places characterizes a completely random succession, and so on. On the other hand, it is clear that the sequence itself is anything but random, that the next number worked out will be perfectly determined, and that the calculation can be followed (and in fact has been followed) using a precise algorithm that arrives at a nonambiguous conclusion!

8. In fact, as we will see below, it is quite possible to formulate theories (the so-called hidden variable theories) that are rigorously deterministic and perfectly equivalent, from the point of view of their predictability, to quantum mechanics, but we will also see that such theories have their own peculiarities.

9. In fact, the general formalism can be more fully articulated, but for our purposes, need not be analyzed in exacting detail. It can provide the probability of outcome of any measure of polarization whatsoever, even circular and elliptical polarization.

10. In the more general meaning of the term, complexity calls into question the idea that in every case the study of complex systems can lead us back to the study of their constituents. It therefore follows that some of the assertions made, for example, in chapter one, concerning the possibility of deriving thermodynamic processes from classical mechanics, need to be taken with a certain measure of caution. The most characteristic example appears in the emergence at the macroscopic level of effects such as the transition of phases (e.g., from liquid to gas) or the spontaneous "breaking of symmetry," which represent "violations" of microscopic laws. There is no need at present to take time to discuss this interesting problem. With reference to the considerations we are here developing it is nevertheless worthwhile to mention that such kinds of "violation" are difficult to predict, precisely because of an especially marked instability with regard to initial conditions.

11. It is curious and interesting to observe that the structure of quantum me-

chanics, which, as we argue, is a fundamentally stochastic theory, implies, according to recent studies, that it reduces or even removes the extreme sensitivity to initial conditions that characterizes classical systems. At the quantum level, then, there would not exist an equivalent to the deterministic chaos so important on the classical level. The framework now being developed, in a paradoxical way, overturns the conception that prevailed up to a few years ago: classical mechanics, which is the prototypical deterministic system, can lead to stochastic behavior, through the emergence of deterministic chaos. On the other hand, quantum mechanics is intrinsically probabilistic, and thanks to its more stable character, is more predictable than classical mechanics. These observations do not change the conclusions of my text in the least, as every attentive reader will understand; I wanted to mention this not only because it is interesting in itself, but also because these problems represent one of the more important themes of current research.

12. We have added to the picture a drawing of the experimental apparatus in the schema used by Bohr in his debate with Einstein. It helps to keep this picture in mind, to understand some of the points of this debate.

13. To be rigorously exact, there is another complication to keep in mind, and that is, that the wave function at a certain point and at a certain instant becomes a complex number instead of assuming real values, as does the electrical field. This difference, which has important implications, both in principle and in physical nature, is nevertheless not relevant to our argument. To be precise, rather than write $[\Psi(x,t)]^2$, we would use the correct notation $|\Psi(x,t)|^2$, which in the case of complex numbers makes us substitute the "square of the modulus" (a real and positive quantity) for the square. But someone without familiarity with complex numbers can happily continue to read this last expression as a square.

14. It can be said that Heisenberg's analysis is to the new theory what Einstein's analysis was to relativity theory. Beginning with the hypothesis that light is propagated at a finite velocity, Einstein studied from that perspective how two observers could synchronize their watches, and thus arrived at the revolutionary innovations of relativity theory. Beginning with the hypothesis that every process has a double, wave/particle nature, Heisenberg studied from that perspective the possibility of determining the values of two specific magnitudes, and was led to the conclusion that there is a conceptual limit to the precision with which both could be known.

15. In fact, because of the phenomenon of diffraction of the wave function, after the passage of the opening, there would be spread in direction x, but, if the second measurement we are about to consider happens immediately after the first, the beam would not have time to expand appreciably.

CHAPTER FOUR

1. The reason for this is rather complicated, and is connected with certain conditions relating to the continuity of the electromagnetic field on the mirror surface.

2. Of course, the gravitational field also varies along the two inclined side pathways, but this will be exactly the same for both trajectories, and thus the propagation along both branches is influenced to an absolutely equal degree.

NOTES TO CHAPTER FIVE

3. In fact, as Einstein taught us, general relativity implies that even photons would be deflected by a gravitational field, but the effect in our case would be so small that it is correct to ignore it.

4. In this connection it is worth mentioning that even in classical physics we do not, in general, have exact solutions of equations of motion for systems of more than two bodies, even in cases where their interactions are governed by such a simple force as gravity.

5. Note that a state of this kind can always be prepared, given the validity of the superposition principle.

6. The theme will be resumed in chapter 6, where we discuss the effect of measurements on systems.

7. We remind the reader of the problem of the standardization of the state vector discussed in chapters 2 and 3. Such standardization is necessary in order to guarantee that the probabilities of obtaining any of the allowed results add up to 1, which clearly must take place.

8. And it may be good to recall here the discussion at the beginning of the third section in this chapter, where I pointed to the important difference between assertions relative to the measurement outcomes (separate registration by a detector) and assertions relative to situations preceding the measurement itself (as to "be" along one of the trajectories). The truth of the former does not imply the truth of the latter.

9. Einstein's fame reached an unprecedented level just when it became possible to measure gravitational effects on light rays.

10. We cannot take the time to treat this delicate issue at length, but I would at least like to mention that if it were legitimate to "retrodict" a past situation by means of a subsequent measurement, there would be an immediate violation of the uncertainty principle in situations like that of Figure 3.10. Indeed, when the diffracted photon strikes the screen, we can think of retrodicting that at its exit from the opening this photon would have the speed that would bring it to the point of impact. In this way I would be able to know both the velocity and position at the time of its passing. This would contradict Heisenberg's principle. But it is easy to show that the assumption made (i.e., that at the instant of passing through the opening, each photon has a precise velocity) would be contradictory and incompatible with the outcomes of the possible physical experiments to follow. These latter are explained perfectly when it is assumed that the only thing we can know about the system at the moment of its passage is its wave function, which will be the same for all the photons.

CHAPTER FIVE

1. The reader may recall the quotation of Lorentz at the head of the chapter.

2. It should also be pointed out that the wave function of a system of many particles (say, n particles) becomes a function of the positions of all the particles, that is to say, has the form $\Psi(x_1, y_1, z_1, x_2, y_2, z_2, \ldots, x_n, y_n, z_n)$, and thus does not "live" in the ordinary three-dimensional space but in $3n$- dimensional space (the so-called configuration space) which is very difficult to visualize.

Chapter Six

1. The position of this great scientist toward the new theory was to remain fundamentally critical, but even so underwent a series of important adjustments that are often ignored both in technical books and in popularizations. We shall have the opportunity to follow the evolution of his thought in some detail, and trace the impact it had on the most important developments.

2. Of course, to be meaningful, this definition would require the exact specification of what is intended with the expression, "laws of physics." It is not the appropriate context for entering into this delicate problem. For the present it should suffice to understand the expression as meaning "the scientific laws currently known and considered valid for the description of natural processes."

3. On the surface, this fact may seem paradoxical, if one considers the motives of positivism and its extremely critical stance toward metaphysics. On the other hand, it begins to make sense when we keep in mind that their insistence on verifiability, logical elaboration, and linguistics give a certain prominent role to the reality of mind, and to the world of ideas, over against the material world.

4. It is interesting to compare this comment by Bohr on the objectives of physics with the positivist Ayer's comment on the objectives of philosophy.

5. Those who are familiar with complex numbers will observe that $|a|^2 = 1$ is not equivalent to $a = 1$, but the theory implies that any factor of the modulus 1 *multiplying* the state has no physical implication, and can therefore be legitimately ignored.

6. We should recall that the argument now being made shows how in the framework of the new theory, it is formally correct and logically consistent to assume (as we did in chapter 3) that a photon passing through a polarized filter becomes polarized in the plane of that filter, the filter therefore playing no merely *passive* role in regard to the state of the photon.

Chapter Seven

1. To underline this point it is enough to recall that in 1931, when he proposed the names of Schrödinger and Heisenberg for the Nobel prize, he wrote: "In my judgment, such a theory undoubtedly contains a fragment of the ultimate truth."

2. Ernest Solvay was the inventor of the process of making the soda that bears his name and procured him so much fame and wealth. He cultivated a lively interest in the sciences and founded the Solvay Institute for Chemistry, Physics, and Sociology. He also financed important scientific conferences that provided a uniquely energizing meeting of the most brilliant scientific minds.

3. In fact, this is an inevitable implication of the theory of relativity. For just as that theory "mixes" space and time, so this theory requires an analogous "mixing" of the variables of momentum and energy. The uncertainty relation we are beginning to discuss here can be considered as simply another facet of the relations we have already discussed.

4. In fact, the principle of equivalence implies that the measurement of time by a clock depends on the gravitational field within which the clock is located.

NOTES TO CHAPTER EIGHT

CHAPTER EIGHT

1. The precise meaning of this assertion is relevant for the debate about microsystems that was going on at that time: "objective" must be understood to mean "independently of any observer," or "independently of whether or not a measurement is made."

2. In reality, as we will explain in chapter 14, quantum formalism implies the indistinguishibility of identical particles, so that it is somewhat inappropriate to designate the photon to the right as 1 and the photon to the left as 2. However, this fact really has no conceptual relevance for the analysis in which we are currently engaged. It will not affect the argument or its conclusions, and the treatment will be simpler without it. In other words, in the light of the requirements for "identical constituents" imposed by the formalism, the treatment will in fact be correct if every time photon 1 (or 2) is mentioned, one thinks of the photon in A (or B).

3. Note that this state is not to be confused with the quantum superposition of two states. For example, we know that the state $(1/\sqrt{2})\, [|1,V\rangle + |1,H\rangle]$ would represent a single photon, that is, photon 1, but furthermore characterized by a 45° polarization, and would not at all represent a photon "with horizontal and vertical polarization."

4. At the same time, it could be demonstrated that there does not exist any test of polarization—neither circular nor elliptical—that a photon could pass with certainty

5. Needless to say, the statement would be answered negatively.

6. As we will see, the reason why Einstein's argument is not conclusive derives from the fact that quantum correlations per se, that is to say, independently of any conceivable interpretation, imply the nonlocality of natural processes. But many years—and the fine contribution of Bell—will be needed for the scientific community to understand this revolutionary point which we are going to discuss in detail.

7. This is not meant as a criticism of Bohr, a profound thinker, and veritable giant of physics. As we will see, the general lack of comprehension of the EPR argument on the part of great physicists and epistemologists of the coming years (from Born to Pais to Popper) would be much worse than Bohr's.

8. In fact, as the reader will understand if he applies the quantum mechanical rules that he now has mastered, in both cases a measurement of 45° polarization upon the pair of photons would have a 1/4 probability of giving opposite outcomes (that is, one photon passes the test and the other does not), whereas, as we know, in the present case, the photons *both* either pass or fail any given test of polarization, so long as the tests are identical.

9. It may be of interest to the reader to know that Dr. Bertlmann was not Bell's fictional creation, but a real physicist. What Bell invented was the irresistible tendency to wear two different color socks. Bertlmann relates that after Bell's article appeared, anybody he met who worked in the field of quantum mechanics would try to steal a glance at his socks!

10. The usual meaning of this expression is as follows: if a measurement is later carried out to determine the position of the particles, there will be a certain prob-

ability of finding the projectile deflected into a certain direction and the target pushed in another direction, correlated to the other.

11. In order to understand this, we should refer to the first formula of section 8.5. In this way, as the state $[|1,V\rangle |2,V\rangle + |1,H\rangle |2,H\rangle]$ is identical to the state $[|1,n\rangle |2,n\rangle + |1,n_\perp\rangle |2,n_\perp\rangle]$, in a completely analogous way the state $|\Psi\rangle$ just mentioned can be written in terms of a state with definite positions. In this case it would assume the form $\Sigma_i\, d_i\, |1, r_i\rangle |2,R_i\rangle$, which expresses the fact that not only the impulses but also the positions of the constituents are perfectly correlated. It follows from this that in a measurement, if the projectile is found in position r_i it can be inferred from this that the target particle "is in position R_i." This is precisely the original EPR argument.

12. I think it proper to use the acronym *E.L.* in order to distinguish the demand for locality as formulated by EPR from the subtly different kind of locality that formed Bell's point of departure (*B.L.*). Technical texts do not (in my view) pay sufficient attention to the important shades of difference.

13. It is important to point out that, apart from a remark of Einstein made in a 1948 article (where he said explicitly, without reference to its incompleteness, that the Copenhagen interpretation is incompatible with the requirement of locality), no one—before Bell—had suspected that the quantum correlations by themselves implied the nonlocality of natural processes.

CHAPTER NINE

1. It should be pointed out that in the second phase of the debate, probabilistic hidden variable theories were considered, in such a way that assigning such variables determined only the probabilities that the observables would have definite values. But as we shall see in the course of the next chapter, there are no particular advantages at the conceptual level for using variables like that, especially since they offer no escape from the impasse of nonlocality, a fact first brought to light by Bell's profound analysis.

2. In his book *The Philosophy of Quantum Mechanics*, Max Jammer discusses Hermann's position, denies that von Neumann's argument is circular, and asserts that "what should have been criticized, instead, is the fact that the proof severely restricts the class of conceivable ensembles by accepting only those for which [the additivity assumption] is valid" (p. 275). This observation is clearly absurd. The demonstration given in Appendix 9A provides clear evidence to the contrary. For example, when the coefficients (of the combination of angular momentum components under examination) happen to be irrational numbers, *there is no sum at all* whose elements could satisfy the additivity assumption! Since one can always choose a direction n such that its components along the axes x,y,z are irrational, the component of the angular momentum along n would turn out to be measurable. By changing n and x,y,z, one would be led to conclude that no component of the angular momentum is measurable.

3. When there is a linear relation among compatible observables (i.e., observables that can be measured simultaneously), von Neumann's requirement is not only satisfied, but satisfied in all hidden variable theories presented in the literature. For example, if we take the observable A= bx + cy, with x and y as the po-

NOTES TO CHAPTER NINE

sition variables of a particle along two perpendicular directions, the relation would be $A(\lambda) = bx(\lambda) + cy(\lambda)$. This assertion, although true, would require a more detailed discussion due to the inevitable contextuality of every theory of the kind in question. This important matter will be considered later in the book. Here it is enough to say that in general the relationships in question can be satisfied only in a contextual sense.

4. It is important to point out here that this is a general problem that arises in the classical situation as well and is not at all dependent on the quantum nature of the process in question. To understand this point, the reader should refer to the method for preparing a particle with a well defined position as discussed above in section 3.7, and illustrated in Figure 3.12. Granting that the slit used to "localize" the particle can be made as small as we want (already an idealization in so far as no slit can be smaller than atomic dimensions), still it is not legitimate to think of reducing it to an *ideal line*, with no width at all.

5. The most obvious case is when a measurement is carried out by reading a position on a graduated scale. Furthermore, to cite a simple example, even when we want to determine the average velocity of a particle in a certain interval of time (t_1, t_2), we really carry out a measurement of the *position* of the particle at the two times under consideration, and then divide the distance given by the quantity $(t_2 - t_1)$.

6. We should recall here that quantum mechanics, being fundamentally probabilistic in nature, requires for any verification the statistical analysis of the results of experiments carried out on a lengthy series of identical systems.

7. The analysis just now undertaken shows how the modification of the situation in B that can be produced by the free choice of an observer at any moment (by removing the screen that covers the slit) implies an instantaneous change of quantum mechanical potential in all space. In itself, this does not lead to situations that are physically unacceptable or in violation of the requirements of relativity, since the quantum mechanical potential, unlike all the classical potentials, does not transport energy.

8. In fact, the demonstration that is usually referred to as the Kochen-Specker theorem, which will lead to the conclusions we are about to explain, was already contained in an elaborated general theorem worked out by A. M. Gleason ten years earlier. But the relevance of the theorem for the problem at hand was precisely identified by Bell, Kochen, and Specker.

9. Apart from such genuinely noncontextual observables, all the properties as well that at a given instant have definite values according to the standard quantum scheme (that is, they have a probability equal to 1 of being obtained) "objectively" have the same values in the hidden variable theory. This is interesting (and certainly ought to be) but, clearly, the type of properties possessed in this way depends on the state vector, and will vary from one instant to the next as a consequence of the state's evolution.

10. Of course, it would be possible to make an alternative choice and construct a theory equivalent to quantum mechanics in which the *velocity* of the particles is noncontextual and thus objective. The price of doing so would be to make the positions contextual, which is too high a price, as the reader will readily agree.

11. The contextuality of observable spin in the case of a particle of spin 1/2 is

NOTES TO CHAPTER NINE

not an inevitable element (see Appendix 9A), as was the contextuality discussed in the preceding section. This is because the space of spin has a dimensionality equal to two (there are only two possible outcomes of a measurement for spin). But this fact is not of much relevance. Indeed, it would be enough to have a system of two particles, each with spin 1/2, in order for the contextuality of the observable spin to become inevitable, since the spin space for a system of two particles is four dimensional.

12. It must be noted that to obtain the desired effect we do not simply invert the magnet (as has been naively thought, and as David Albert [1992] appears to think in his book), because doing so would invert both the direction of the field and the direction along which the field increases. The operation we are performing can be realized in practice by using another magnet, with an inverse shaping of its terminal expansions. If, in the magnet of the earlier illustration (Figure 3.6) the south pole had a V shape while the north pole was flat, in the new magnet it is the north pole that will be formed into a V, and the geometrical configuration will stay the same. Figure 9.6 shows the various possible orientations of the magnet in geometrical terms and the direction and intensity of the field; in white and gray, respectively, are shown a magnet with (first) the south pole, and then the north pole, formed into a V.

13. Since the observables are incompatible, this means having recourse to another group of identically prepared systems, subjecting them to the appropriate measuring processes, and averaging the results.

14. Of course, the sole instance where this equation would not be absurd is when both q and r + s + t are equal to zero. But this would be an extremely peculiar case, because, if ℓ is nonzero, the relation cannot be verified for all the values of the hidden variables λ of the ensemble group, since, according to quantum formalism, no state can always provide outcomes that make both sides of the equation equal to zero. But if we want to avoid even this irrelevant objection it should suffice to repeat the reasoning just made, with reference to the states of spin of a particle of spin 1/2, in which case q, r, s, and t can assume only one of the two values +1/2 and −1/2. The left side of the equation then assumes one of the two values +$\sqrt{3}$/2 or −$\sqrt{3}$/2, and the right side one of the values 3/2, −1/2, +1/2, or +3/2. The inescapable conclusion is that the assumption of additivity by itself eliminates the possibility of a deterministic completion of the theory.

15. It is worth mentioning here that the impossibility just demonstrated comes about because we are thinking of three-dimensional space. In two dimensions it would be perfectly possible to color a circle such that every pair of perpendicular vectors intersect it at one blue point and one red point. We would simply divide the circle into four quadrants and color them alternately red and blue. But now for a simple technical observation. The dimensionality relevant for the theorem is not the kind of space where the entities we speak of are "located" (such as the angular momentum vector) but instead is the kind typical of a system in the context of quantum formalism. This is connected to the possible outcomes of a measurement of an appropriate collection of compatible observables. For example, if we had considered a particle of spin 2, for which the component of the spin along any one direction can assume five values (as we have seen in Figure 3.4), the space ought to be considered pentadimensional.

NOTES TO CHAPTER TEN

Any physical system is associated in quantum formalism with multidimensional spaces—indeed, with infinite dimensionality. Thus the Kochen-Specker theorem is relevant for any real system.

CHAPTER TEN

1. In order to avoid any misunderstanding, it is important at this time to stress the fact, as Bell did, that the term *exact* was used, with regard to a theory, to indicate that the theory in question *neither needs nor is embarrassed by an observer.*

2. I must emphasize the essential role played, conceptually speaking, by the requirement that the variables λ represent the most complete specification possible for the state of the system. To illustrate this point we can make reference to the example of the two different colored balls in the two boxes from chapter 8. It is clear that if the state of the system is not specified as accurately as possible, the probability that the ball to the right would be white is equal to 1/2. Likewise equal to 1/2 is the probability that the ball in the left-hand box is black. On the other hand, the joint probability that "the particle on the right is white and the one on the left is black" is also equal to 1/2, since only the two alternatives are possible, namely, the situation mentioned above, and the one in which the colors are switched. Since 1/2 × 1/2 is 1/4, which is different from 1/2, it would appear that the condition B.L. is violated in this "classical" and certainly "local" example. The error in such a conclusion comes from the fact that the complete specification of the system has not in fact been considered. For the "classical" system in question, this requires specifying the color of the particle on the right-hand side (and consequently the one on the left as well). From this it follows that, if it is specified that the particle to the right is white and the one to the left is black, each of the two events has the probability of 1, and 1 × 1 = 1 turns out to be the probability that "the particle to the right is white and the one to the left is black." Analogously, if the state is completely specified as indicated above, the probability that the particle to the right is black is equal to 0, and the same will hold for the probability that the particle to the left is white as for the joint probability that "the particle to the right is black and the one to the left is white." Therefore, contrary to the hasty conclusion indicated at the beginning, the example perfectly satisfies the B.L. requirement.

3. For future reference it is worth noting that this expression, when it is convenient to associate the number +1 with YES and −1 with NO, coincides with the mean value of the product of the outcomes. This observation will be useful in what follows.

4. Of course, since we are placing ourselves within the viewpoint of the fact that the formalism is complete, that is to say, we are granting that the assignment of the wave function represents the most accurate specification possible of the state of the system, we will have to substitute the state vector Ψ for the λ variables in all the formulas.

5. For a more detailed exposition, consult the works of d'Espagnat (1979) and Shimony (1988) in *Scientific American* cited in the bibliography.

6. It should be mentioned that today the experiments that demonstrate a vio-

NOTES TO CHAPTER TWELVE

lation of Bell's inequality particularly in the field of quantum optics are much more numerous than in Aspect's times.

7. In order not to weigh down the discussion, it is not necessary to discuss this kind of measurement in detail. It should suffice to say that a measurement like this is very simple to carry out and gives only two outcomes: either the photon is found in a state of circular polarization to the left or in a state of circular polarization to the right (See chapter two, Figure 2.8, for a comprehensive description of such states).

8. We should recall that a photon in the state $|45\rangle$ overcomes with certainty a test of plane polarization of the type in question, and that a photon in the state $|135\rangle$ just as certainly fails to pass it.

CHAPTER ELEVEN

1. Of course, in modern, postrelativistic science, this view of gravitational phenomena has undergone a profound revision. The gravitational effect of my action here (that is, moving a certain mass in space) upon the various parts of the cosmos will take place only after enough time has passed for the gravitational wave I have generated (going as fast as the speed of light) to reach the affected regions. Thus the moon, which is one light-second away, will feel the effect of my action one second later, while the nearest star, Alpha Centauri, will feel it four years later.

2. In fact, the example we are discussing now is precisely the one used by Bohm in his valuable book, where he reformulated the EPR argument in a more simplified way.

3. Incidentally, this illustrates in a simple and incisive fashion the nonlocal character of contextuality. If Alice had not carried out any measurement, Bob's measurement would depend contextually on his choice of orientation of the magnet, but, after Alice has carried out a measurement and obtained an outcome (upward spin), Bob *must,* however he chooses to carry out his measurement, obtain a response perfectly correlated to the preceding measurement outcome according to the standard theory (i.e., downward spin).

4. In other words, the way the witch controls the process is equivalent to saying that, of the four possible alternatives A, B, C, D in Figures 11.7 and 11.8, the witch is allowing only A and C to happen.

5. The proof that within quantum mechanics one can clone orthogonal states but not their superpositions is now well known. It is referred to as the "no-cloning theorem" and plays an important role in quantum computation. It is appropriate to mention that it was proved, for the first time, in a letter from the author of this book to Professor A. van der Merwe.

CHAPTER TWELVE

1. It is helpful to be aware that in modern cryptography, especially of the commercial variety designed for intensive use, the code itself is in the public domain and the key is the only element that can guarantee secrecy.

Chapter Thirteen

1. It will interest the reader to know that the interests and drive of these two great thinkers in this field is connected to a certain extent with their interests in the field of cryptography. In fact, Shannon's fundamental work *Communication Theory of Secret Systems* (which can be considered the origin of modern cryptology), even if it appeared a year after the treatise *A Mathematical Theory of Communication* (which was the beginning of information theory), was written during the war years and its publication was delayed until 1949 for military reasons. In a similar way, Turing worked for the cryptanalysis department of the English Ministry of the Exterior during World War II.

2. The reader may object that—as he or she has learned very well—all other states of polarization other than the states mentioned here are possible for such a system. But the fundamental point consists in the need to have a nonambiguous distinguishability of states. We know very well that a if a photon is polarized for example at 45°, it can overcome or fail a test of vertical polarization. Nevertheless, if we want to use the photon to transmit precise information, we will have to clarify in what (mutually exclusive) states it is going to be prepared, and the recipient will have to carry out precisely a measurement that distinguishes one of these specific states from the other one. Only in this way will the recipient be able to ascertain the choice (among the possible options) purposely taken by the sender. The considerations I am here presenting will allow me to show, in section 13.5, how in fact the greater abundance of the possible states for a quantum mechanical system (such as our photon) will really be what makes possible an important advance, but the procedure for obtaining this result will be much more complex and will require having recourse to another genuinely quantum aspect, namely, entanglement.

3. It is worth mentioning that the operation in question allows us to implement the logical operation of negation at the quantum level: if $|0\rangle$ corresponds to "a certain assertion is true" and $|1\rangle$ corresponds to "the same assertion is false," while one of the two bits is stored in a logical circuit, the illumination with the laser, by transforming $|0\rangle$ into $|1\rangle$ and vice versa, corresponds to the logical negation.

4. Recall that it is only Bob's photon, that is, the second one in each formula, that undergoes one of the four transformations in question.

5. In fact it can be proven that knowledge of the outputs determines N linear equations in the N unknowns (0 or 1) that characterize, in order, the positions of the register (clearly, 0 corresponds to a disconnected register, 1 to a connected register) connected to the summator. But a system of equations like this admits of a unique solution.

6. Of course the same would be valid for any other typical English text, and would fail only for very peculiar cases such as the novel *Gadsby* by Ernest Vincent Wright, which the author deliberately wrote without a single "e'" in all 267 pages!

Chapter Fourteen

1. For the moment we will be considering particles without spin, for which the complete description of their state is represented, as is well known, by assigning

a wave function with its characteristic dependence on the spatial coordinates of the particles.

2. In fact, for the case under consideration, it could be shown that at any instant following the initial one, each initial wave function evolves in a function that is everywhere nonzero. Of course, its absolute value (which, by the rule of the square of the modulus, gives us the probability of finding the particle in question in a certain region) will be, for brief intervals of time, smaller in the various regions, the more distant they are from region A (or B).

3. It can easily be shown that even the consideration of an arbitrary linear combination of the states mentioned does not permit us to overcome our difficulty, that is, it does not allow us to construct an antisymmetrical function for the exchange of any given pair of particles.

4. It is clear from expression (14.7) that if 1 is replaced by 2, the first term turns into the second, and vice versa, and thus the state changes its sign—i.e., it is antisymmetrical.

5. The sole exception to this situation is represented by the case of bosons described by a factorized state in which all the factors are identical (that is, all the constituents are in exactly the same quantum state), insofar as a function of this kind is automatically symmetrical by the exchange of any pair of particles whatsoever, and thus satisfies the identity requirements.

CHAPTER FIFTEEN

1. Please note that in state (15.1) the outcome of a measurement at 45°, contrary to the vertical case, is certain: the photon *certainly* overcomes such a test.

2. Of course, an equivalent means of verifying the theory would consist in repeating the same experiment many times and noting each outcome. In this case, the ensemble to which we refer would consist of the total of all experiments.

3. The qualification "around" is needed because the ensemble is supposed to contain a finite number of systems, even though that number may be a very large one. We know that if a coin is flipped an enormous number of times, we will have "around" as many heads as tails, and not exactly the same number of each, even though the probabilities of either (as in the physical process we are studying) are in theory exactly equal.

CHAPTER SIXTEEN

1. To be even more precise on this point, it would seem appropriate to introduce a distinction between the members of the scientific community who have never been concerned in the least with foundational problems and those who have dedicated serious reflection to them. The scientists of the former group, which certainly includes towering figures who have made fundamental contributions to physics, are finding themselves in a position analogous to the one that Euan Squires confesses to have been his own during a large portion of his career. We can understand this position better by reading Squires' own words in his splendid book, *The Mystery of the Quantum World* (1994):

"For over thirty years I have used quantum mechanics in the belief that the problems discussed in this book were of no great interest and could, in any case, be sorted out with a few hours' careful thought. I think this attitude is shared by most who learned the subject when I did, or later. . . . It was only when, in the course of writing a book on elementary particles, I found it necessary to do this sorting out, that I discovered how far from the truth such an attitude really is. The present book has arisen from my attempts to understand things that I mistakenly thought I had already understood . . . and to discover what progress has been made in the task of incorporating the strange phenomena of the quantum world into a rational and convincing picture of reality."

2. The line of thinking we are investigating in this section was elaborated by various writers in a series of works that stretch over a remarkably long span of time. The pioneering work was done by Adriana Daneri, Angelo Loinger, and Giovanni Maria Prosperi, and appeared in 1962. A lucid contribution by Joseph M. Jauch should be mentioned, and more recently, the important and illuminating studies of E. Joos and H. Dieter Zeh, and of Wojciech Hubert Zurek.

3. The English translation of Borges' "Garden of Forking Paths" is by Donald A. Yates, in Jorge Luis Borges (1964).

4. In order to avoid misunderstandings, I would like to use this occasion to express myself clearly on this problem and to make it clear that in my opinion the argument that goes by this name is inappropriate and irrelevant, at least in its more widely taken meaning. Of course it is of some interest to observe that, keeping everything else the same about the reality, a certain even minimal change of one of the constants of nature would bring us to universes in which, for example, life as we know it would not exist. But to go beyond this simple truth (which in many cases is illegitimate because nobody can reasonably calculate the effect of such changes) and give it eschatological connotations that suggest that someone (God, of course) has regulated everything to produce life (if not specifically to lead up to the birth of a speaker holding a conference on the anthropic principle itself) seems to me like classical circular reasoning.

5. And here is another beautiful example of how the emotional embrace of a thesis can carry someone into meaningless assertions. If you want to say that an event is improbable, you need to have a probability space. It is legitimate to calculate the probability that by tossing a coin 100 times you could have 100 heads, and then assert that this sequence is extremely improbable, but to speak of probability without having well-identified the set to which the probability refers (in the case in question, this is constituted by all the possible sequences of outcomes of the hundred coin tosses), is absolutely devoid of any logical meaning.

6. It is worthwhile recalling the statement of Jordan: "We ourselves produce the results of measurement."

7. In what follows I will point out how, in fact, until a few years ago, such a program was considered impossible even by leading scientists.

8. In the original model, the localization function is not exactly equal to what we are considering, but instead has a bell shape (to be more precise, a Gaussian shape). However, it presents essentially the same characteristics as the function in question, that is, being appreciably different from zero only in an interval of amplitude δ centered around x^*. The reason why it is correct to use a function that

varies gently (and not one that makes an abrupt leap to the edges of the interval) is that the introduction of discontinuity into the wave function after localization occurs has some unfortunate physical consequences. I decided to resort to this schematization in so far as it makes the effect of spontaneous localization more immediately intuitive and allows us to understand directly the essential characteristics of the new dynamic. The assertions that we make in the following pages about some of the physical implications of the theory are exact only while the function here indicated is replaced by the "gentle" function of the original model.

CHAPTER SEVENTEEN

1. To cite a few examples: a detailed discussion of the the pilot wave theory can be found in P.R. Holland, *The Quantum Theory of Motion* and in D. Bohm and B. J. Hiley, *The Undivided Universe*; R. Omnès in *The Interpretation of Quantum Mechanics* discusses the quantum histories interpretation; finally, D. Z. Albert in *Quantum Mechanics and Experience* argues for the inevitability of accepting the many-minds interpretation.

2. That is, possessed independently of any process of measurement being carried out and/or any intervention being made on the part of an observer.

3. It has been demonstrated, I would like to point out, that where a GRW strategy is being adopted, positions are in fact the only properties that we can try to make definite. In fact, even though it is extremely easy to consider spontaneous processes that would, for example, make the velocities of the elementary constituents definite, the physical implications of such a move would be disastrous. The reader will have no trouble understanding the reason for the difference: if we take an atom for example—or better, a nucleus—and we attempt to visualize the movement of the electron or the nucleons, we will quickly be persuaded that the system is characterized by extremely diverse velocities (thinking, e.g., of the velocities of opposite extremities of the orbit that assume very large and opposing values). This huge dispersion of velocity values is such that any attempt to reduce it would immediately bring about a rupture of all the atomic and nuclear bonds. To avoid this, we would have to make the process too ineffective to be of any use in obtaining definite properties for macrosystems. Further, it is even possible to show that attempts to better define nonposition variables lead to consequences in a sense opposite to those that constitute the real benefit of the theory: the effect on microsystems would tend to diminish instead of being amplified when moving to systems with a larger number of constituents. This observation has some relevance insofar as it demonstrates that, if we try to follow the principles that have guided the elaboration of dynamic reduction models, the arbitrariness of the choice about which properties to make objective ends up, in fact, being only apparent.

4. It should be mentioned that another model, almost equivalent in its physical implications to the original GRW model, has been worked out by myself in collaboration with A. Rimini and R. Grassi. In this model, one of the two constants is replaced by Newton's law of universal gravitation, and thereby reduces the parameters to just one. The model also offers the advantage of linking suppression of the superpositions to gravitational effects.

5. It is interesting and curious to observe that in the heroic winter of 1925, Erwin Schrödinger had derived an equation (known afterwards as the Klein-Gordon equation, which is appropriate for the relativistic description of particles without spin) that represents a relativistic generalization of his fundamental equation. Because of certain interpretative difficulties he moved on from this equation to the one that now bears his name.

Chapter Nineteen

1. In the dialogues reported by Jeremy Bernstein in his book *Quantum Profiles,* John Bell recalled the embarrassment he felt from his earliest years of study about the indeterminacy principle and the role of the measurement apparatus in quantum-mechanical thinking. He recalled his arguments with one of his professors, Dr. Sloane, and said, "I was getting very heated and was accusing him, more or less, of dishonesty. He was getting heated too and said, 'You're going too far.'" In another brilliant interview Bell spoke in his lively and penetrating manner of his profound aversion to the orthodox position: "It was the inauguration of CERN. I got on the hotel elevator with Niels Bohr. I did not have the courage to tell him "I think your Copenhagen interpretation is horrible." The elevator trip didn't take very long. If the elevator had got stuck between two floors for a while, that would have made my day! Even if I can't imagine how . . ."

2. Organized in Erice, Sicily by Enrico Beltrametti and Jean-Marc Lèvy-Leblond.

3. I would like to mention that this great scientist has repeatedly expressed his lively interest in the GRW model, and in his latest book, *Shadows of the Mind,* makes it clear in the introduction that his ideas are closely connected to ours.

Bibliography

It is important to alert the reader's attention to the fact that the bibliography does not contain some of the more important works which are repeatedly cited and analyzed in the text, especially the works of de Broglie, Bohr, Heisenberg, and Schrödinger. The reason for this omission is very simple: many of these works appear in Italian translation in the collection edited by Sigfrido Boffi (cited as de Broglie, Schrödinger, and Heisenberg]1991]. Other important works, such as the one in which Schrödinger's famous cat appears, can be found in the collection edited by Wheeler and Zurek (1983).

Accardi, L., ed. 1994. *The interpretation of quantum theory: Where do we stand?*, Istituto della Enciclopedia Italiana, Rome.

Aicardi, F., Borsellino, A., Ghirardi, G. C. and Grassi, R. 1991. "Dynamical models for state vector reduction: do they ensure that measurements have outcomes?" *Foundations of Physics Letters* 4:109–128.

Albert, D. 1992. *Quantum mechanics and experience*, Harvard University Press, Cambridge, Mass.

———. (2000). *Time and Chance*. Harvard University Press, Cambridge, Mass.

Aspect, A, Dalibard, J., and Roger, G. 1982. "Experimental tests of Bells's inequality, using time-varying analyzers." *Physical Review Letters* 49:1804–1807.

Aspect, A., and Grangier, P. 1995. "Experimental tests of Bell's inequalities." *NATO Advanced Study Institute*, Series B, Physics, 347, 201–214.

Ayer, A. 1959. *Logical positivism*, The Free Press, Glencoe, Ill.

Baggott, J. 1992. *The meaning of quantum theory*, Oxford University Press, Oxford.

Bassi, A., and Ghirardi, G. C. 2000. "A general argument against the universal validity of the superposition principle." *Physics Letters A* 275:373–381.

Bell, J. S. 1987a. "Are there quantum jumps?" In *Schrödinger: Centenary celebration of a polymath*, C. W. Kilmister, ed., Cambridge University Press, Cambridge, England.

———. 1987b. *Speakable and unspeakable in quantum mechanics*, Cambridge University Press, Cambridge, England.

———. 1989. "Towards an exact quantum mechanics." In *Themes in contemporary physics, II: Essays in honor of Julian Schwinger's 70th birthday*, World Scientific, Singapore.

———. 1990. "Against measurement." In *Sixty-two years of uncertainty*, Plenum, New York.

Bernstein, J. 1991. *Quantum profiles*, Princeton University Press, Princeton, N.J.

BIBLIOGRAPHY

Bohm, D., and Hiley, B. J. 1993. *The undivided universe*, Routledge, New York.

Bohr, N. 1935. "Can quantum-mechanical description of reality be considered complete?" *Physical Review* 38:696–702.

Borges, J. L. 1964. *Labyrinths: Selected stories and other writings*, Donald A. Yates, trans., New Directions, New York.

Born, M. 1949. *Natural philosophy of cause and chance*, Clarendon Press, Oxford.

———. 1971. *The Born-Einstein letters*, Walker and Company, New York.

Clauser, J. F., Horne, M. A., Shimony, A., and Holt, R. A. 1969. "Proposed experiment to test local hidden-variable theories." *Physical Review Letters* 23:880–884.

Crease, R. P., and Mann, C. C. 1987. "Physics for mystics." *The Sciences*, July.

Cushing, J. T. 1994. *Quantum mechanics*, University of Chicago Press, Chicago.

Cushing, J. T., and McMullin, E. eds. 1989. *Philosophical consequences of quantum theory*, Notre Dame University Press, Notre Dame, Ind.

de Broglie, L., Schrödinger, E., and Heisenberg, W. 1991. *Onde e particelle in armonia—alle sorgenti della meccanica quantistica* Sigfrido Boffi, ed., Jaca, Milan.

d'Espagnat, B. 1979. "The quantum theory and reality." *Scientific American* 241(5):158–181.

———. 1989. *Conceptual foundations of quantum mechanics*, 2nd Ed., Addison-Wesley, Reading, Mass.

———. 1995. *Veiled reality*, Addison-Wesley, Reading, Mass.

Einstein, A. 1905. "On a heuristic point of view concerning the production and transformation of light" [translation of "Über einen die Erzeugung und Verwandlung des Lichtes betreffenden heuristischen Gesichtspunkt." *Annalen der Physik*, ser. 4, 17:132–148]. Pages 86–103 in Anna Beck, trans., and Peter Havas, consultant. 1989. *The collected papers of Albert Einstein*. Vol. 2, *The Swiss years, writings 1900–1909*, Princeton University Press, Princeton, N.J.

Einstein, A., Podolski, B., and Rosen, N. 1935. "Can quantum-mechanical description of physical reality be considered complete?" *Physical Review* 47:777–780.

Flew, Anthony, ed. 1984. *Dictionary of philosophy*, St. Martin's Press, New York.

French, A. P., and Kennedy, P. J., eds. 1985. *Niels Bohr: A centenary volume*, Harvard University Press, Cambridge, Mass.

Ghirardi, G. C. 1997. "Macroscopic reality and the dynamical reduction program." In *Structures and norms in science*, M. L. Della Chiara, K. Doets, D. Mundici, and J. van Benthem, eds., Kluwer Academic, Dordrecht.

———. 1998. "Dynamical reduction theories as a natural basis for a realistic world view." In *Interpreting bodies*, Elena Castellani, ed., Princeton University Press, Princeton, N.J.

———. 2000. "Beyond conventional quantum mechanics." In *Quantum reflections*, J. Ellis and D. Amati, eds., Cambridge University Press, Cambridge, England.
———. 2002. "John Stewart Bell and the dynamical reduction program." In *Quantum [un]speakables*, R. A. Bertlmann and A. Zeilinger, eds., Springer, Berlin.
———. 2002. "Making quantum theory compatible with reading." *Foundations of Science* 7:11–47.
Ghirardi, G. C., and de Stefano, F. 1996. "Il mondo quantistico: una realtà ambigua." In *Ambiguità*, C. Magris and G. O. Longo, eds., Moretti and Vitali, Bergamo.
Ghirardi, G. C., Rimini, A., and Weber, T. 1980. "A general argument against superluminal transmission through the quantum mechanical measurement process." *Lettere al Nuovo Cimento* 27:293–298.
———. 1986. "Unified dynamics for microscopic and macroscopic systems." *Physical Review D* 34:470–491.
Gribbin, J. 1990, "The man who proved Einstein was wrong." *New Scientist* 24 November:43–45.
Holland, P. R. 1993. *The quantum theory of motion*, Cambridge University Press, Cambridge, England.
Jammer, M. 1974. *The philosophy of quantum mechanics*, Wiley, New York.
Kahn, D. 1967. *The codebreakers: The story of secret writing*, Macmillan, New York.
Laplace, P. (1814). *Essai philosophique sur les probabilités* [Frederick Wilson Truscott and Frederick Lincoln Emory, trans., *A philosophical essay on probabilities*, Dover, New York, 1995].
Maudlin, T. 1994. *Quantum nonlocality and relativity*, Blackwell, Oxford.
Mermin, N. D. 1981. "Quantum mysteries for anyone." *Journal of Philosophy* 78:397–408.
———. 1990. *Boojums all the way through: Communicating science in a prosaic age*, Cambridge University Press, Cambridge, England.
Miller, A. I. 1987. *Imagery and scientific thought*, MIT Press, Cambridge, Mass.
Moore, W. 1992. *Schrödinger, life and thought*, Cambridge University Press, Cambridge, England.
Omnès, Roland. 1994. *The Interpretation of quantum mechanics*, Princeton University Press, Princeton, N.J.
Pais, A. 1982. *"Subtle is the Lord . . .": The science and life of Albert Einstein*, Oxford University Press, Oxford.
Penrose, R. 1989. *The emperor's new mind*, Oxford University Press, Oxford.
———. 1994. *Shadows of the mind*, Oxford University Press, Oxford.
Poincaré, H. 1903. *Science and method*, Francis Maitland, trans., Dover, New York.

BIBLIOGRAPHY

Rauch, H., Triemer, W., and Bonse, U. 1974. "Test of a single crystal neutron interferometer." *Physics Letters A* 47:369–371.

Redhead, M. 1987. *Incompleteness, nonlocality, and realism,* Oxford University Press, Oxford.

Schilpp, P. A. 1949. *Albert Einstein: Philosopher-scientist,* Tudor, New York, 1949.

Sgarro, A. 1993. *Crittografia,* Franco Muzzio, Padova, Italy.

Shannon, C. 1949. *The mathematical theory of communication,* University of Illinois Press, Urbana, Ill.

Shimony, A. 1988. "The reality of the quantum world." *Scientific American* 258(1):46–53.

———. 1989. "Conceptual foundations of quantum mechanics." In *The new physics,* Paul Davies, ed., Cambridge University Press, Cambridge, England.

———. 1993. *Search for a naturalistic world view,* Vols. I and II, Cambridge University Press, Cambridge, England.

Squires, E. 1994. *The mystery of the quantum world,* 2nd Ed., Institute of Physics Publishing, Bristol.

Tonomura, A., Endo, J., Matsuda, T., and Kawasaky, T. 1989. "Demonstration of single-electron buildup of an interference pattern." *American Journal of Physics* 57:117–120.

Wheeler, J. A., and Zurek, W. H. 1983. *Quantum theory and measurement,* Princeton University Press, Princeton, N.J.

Index

action at a distance, 264
Adleman, Leonard, 325–26
"Advances in Quantum Phenomena," 449
Aharanov, Y., 434
Aicardi, Franca, 431
Albert, David Z., 80; localization and, 427–29, 433–34; physical processes and, 394–95
Albert Einstein, Philosopher-Scientist (Schilpp), 420
American Journal of Physics, 163
angular momentum, 52–55, 457n4; time/energy relation, 159–64. See also spin
ASCII code, 302–4
Aspect, Alain, 176, 245–47, 449
atomic theory: Bohr and, 17–19, 112–14; charge and, 5–6; chemical bonds and, 331, 336–39; Coulomb attraction and, 338; electron orbits and, 17–19; fermions and, 336; hidden variable assignment and, 195–97; identity and, 331–43; indiscernibles identification and, 332–34; oscillators and, 114; planetary model of, 4–6; quantization and, 17–19; spin and, 25, 51–55; visualization and, 112–14; wave/particle dualism and, 19–24
Ayer, Alfred Jules, 129

Barenco, Adrian, 326
Bassi, A., 355
Bell, John Stewart, xvii–xviii, 165, 449; complementarity and, 77–78; death of, 226; diffraction and, 366; Dr. Bertlmann's socks and, 184–85; Einstein and, 226, 236–37; EPR paradox and, 180–81, 192, 194, 227–28, 231, 241, 243, 250–51; factor reduction and, 123; on identity, 377–78; localization and, 416, 419–20, 435–36; macroscopic systems and, 344, 355, 366; Schrödinger's cat and, 371–72; slit experiments and, 366–67; von Neumann and, 198

Bell's inequality, 194, 449; Bohm and, 226–27; determinism and, 238–39; different concept for, 251–58; experimental metaphysics and, 243–46; illustration of, 228–33; locality and, 237–39, 242; macrorealism and, 440–41; preliminaries for, 258–59; quantum telepathy and, 228–37, 243–46; realism and, 246–51; simplicity of, 228; theorem formality for, 239–43, 259–60
Benatti, Fabio, 405
Bernstein, Jeremy, 472n1
Bertlmann analogy, 184–85, 462n9
binary system, 300–301; ASCII code, 302–4; data compression and, 327; linear feedback shift registers, 320–23; perfect encoding and, 304–12; quantum bits and, 315–19
birefringent crystals. See crystals
black holes, 453
Bohm, David, 176, 449; Bell's inequality and, 226–27; Einstein and, 201; EPR paradox and, 188, 192, 194, 201; hidden variables and, 200–10; localization and, 435; McCarthy and, 201; pilot wave theory and, 202–18; Princeton and, 200–201; superluminal signals and, 281–84; von Neumann and, 198; witch of, 281–84
Bohr, Niels, xiv, 448; atomic model of, 17–19; boundary issues and, 365–66; complementarity and, 74–78; Copenhagen interpretation and, 121; EPR paradox and, 152–53, 180–82; equivalence principle and, 163; nonepistemic imprecision and, 151–52; nonlocality and, 153; objectification and, 399–400; on language, 130–31; particle/wave dualism and, 154–59; positivism and, 128, 131; realism and, 153–54; superposition principle and, 109; time/energy relation, 159–64; uncertainty principle and, 67,

477

INDEX

Bohr, Niels (*cont.*)
  151; visualization and, 112–14; wave theory and, 17–19
Bohr-Sommerfeld rules, 21
Boltzmann, Ludwig, 2
Borges, Jorge Luis, 300, 392–94
Born, Max, 132, 449; Bohr's atomic theory and, 113; Copenhagen interpretation and, 121; Einstein's rejection and, 120; EPR paradox and, 153, 192
boundary conditions: conscious observer and, 400–403; macroscopic systems and, 363–68

Caesar cipher, 299
calcite crystals, 38–39, 47
Caldirola, Piero, xvii
Carnap, Rudolf, 128
causality, 112
chemical bonds, 331, 336–39
Cheshire, Alice, 131–32
Chiara, Maria Luisa Dalla, 331
classical mechanics, 1; atomic theory and, 4–6, 17–19; Bohr's atom and, 17–19; de Broglie's hypothesis and, 19–22; defined, 2; diffraction and, 12–15; dualism and, 23–24; field concept and, 3; indiscernibles identification and, 332–34; macrorealism and, 439–40; measurement and, 152; noninvasive observation and, 445–47; nonlocality and, 263; objectification and, 378–81, 399–400; physical processes and, 387 (*see also* physical processes); Planck's hypothesis and, 15–17; polarization and, 6–42 (*see also* polarization); quantization and, 17–19; quantum computers and, 315–16; randomness and, 64–65; spectra and, 3–4; superposition principle and, 80; wave phenomena and, 6–17
closing the circle, 344
Codebreakers, The: The Story of Secret Writing (Kahn), 293
coding. *See* cryptography
cognitive science, 111
Communication Theory of Secret Systems (Shannon), 468n1

complementarity, 74–78
computers, xi, 451; quantum, 313–30
Comte, Auguste, 127
contextuality: hidden variables and, 210–18; pilot wave theory and, 213–18; reasons for, 222–25; spin variables and, 213–18
Copenhagen interpretation, 121, 133, 449; Bohr-Einstein dialogue and, 149–64; EPR paradox and, 187; superluminal signals and, 275–76
Copernican system, 112–13
Coulomb attraction, 338
Crease, R. P., xiii
cryptanalysis, 296–97
cryptogram, 296
cryptography: ASCII code and, 302–4; basic procedures of, 296–97; binary system and, 300–302; Caesar cipher, 299; computational complexity and, 323–27; Enigma and, 293; examples of, 297–300; interception, 297–98; letter substitution, 297, 299–300; linear feedback shift registers, 320–23; Magic and, 293–95; perfect system of, 304–5; permutations, 299–300; photons and, 306–12; plaintext, 296; Poe and, 292–93; public keys and, 326; Purple and, 293–95; quantum computers and, 319–23; RSA, 325–26; Shannon and, 304, 468n1; terminology of, 296–97; World War II and, 293–95
cryptology, 297
crystals, 37–42; distortion and, 100; macroscopic systems and, 350–55; measurement and, 350–55 (*see also* measurement); neutron interferometry and, 97–102; quanta and, 46–47; reversed, 84–85; Stern-Gerlach apparatus and, 55–60, 80, 84, 213–15, 278–84, 409, 429–30; superposition principle and, 80–90, 458; thought experiments and, 84–85; virtually perfect, 97
Cushing, James T., 226

Dalibard, J., 245
data compression, 327

478

de Broglie, Louis Victor, 134, 435, 449
de Broglie waves: interference and, 48–49; pilot wave theory and, 19–22, 202–10, 213–18; visualization and, 115–16
d'Espagnat, Bernard, 355
Deutsch, David, 314
diffraction, 12–15, 48–49, 455n6; Bell and, 366; complementarity and, 74–78; pilot wave theory and, 206–10; uncertainty principle and, 66–74
Dirac, Paul Adrien Maurice, 23; localization and, 433; superposition principle and, 79–80, 349; uncertainty principle and, 70, 79
Duerr, D., 452
dynamics: objectification and, 399–400; reduction program and 404–15 (*see also* localization)

Eckert, Arthur, 326
Einstein, Albert, vii, xiv, 43, 51, 448–49; accomplishments of, 456n7; Bell and, 226, 236–37; Bohm and, 201; boundary issues and, 365–66; changing position of, 135–37; Copenhagen interpretation and, 133; de Broglie's hypothesis and, 21; energy quanta and, 15–17; EPR paradox and, 152–53; equivalence principle and, 163; localization and, 420–23; nonepistemic imprecision and, 151–52; nonlocality and, 153, 176–78; objectification and, 399–400; particle/wave dualism and, 154–59; positivism and, 128; realism and, 153–54; superluminal communication and, 270; superposition principle and, 109–10; thought experiments and, 151–52; time/energy relation, 159–64; uncertainty principle and, 67, 151; visualization and, 116; wave theory and, 15–17
Einstein-Podolski-Rosen (EPR) paradox, 152–53, 179, 462n7, 463n11, 467n2; Bell and, 227–28, 231, 237–38, 241, 243, 250–51; Bohm and, 201; Bohr's reaction to, 180–82; Dr. Bertlmann's socks and, 184–85; entangled states and, 343; misunderstandings of, 182–88; nonlocality and, 175–78, 181, 188–89, 193–94, 227–28, 263; quantum nonseparability and, 189–91; superluminal signals and, 269–76; system properties and, 166–75; theory completion and, 191–93; uncertainty principle and, 188–89

electromagnetism: charge and, 5–6; field concept and, 3; frequency and, 9; light and, 2–4, 24–30 (*see also* light); locality and, 181, 188–89, 193–94; polarization and, 6–24 (*see also* polarization); rectilinear propagation and, 9, 12–15; spectra and, 3–4; spin and, 51–55 (*see also* spin); Stern-Gerlach apparatus and, 55–60, 80, 84, 213–15, 278–84, 409, 429–30; transmitted luminous intensity, 35; wave/particle dualism and, 19–24. *See also* wave mechanics

electrons, 17–19; angular momentum and, 52–55; chemical bonds and, 331, 336–39; contextuality and, 210–18, 222–25; entangled states and, 339–42; identity and, 331–43; indiscernibles identification and, 332–34; orbits of, 112–15, 124; spin and, 51–55, 210–18 (*see also* spin); Stern-Gerlach apparatus and, 55–60, 80, 84, 213–15, 278–84, 409, 429–30; tunnel effect and, 95–97; wave/particle dualism and, 19–24. *See also* atomic theory

Elitzur, A. C., 447
Elliot Cresson Medal, 201
Ellis, John, 453
energy: density of, 11; electron orbits and, 17–19; equivalence principle and, 163; of photon, 16; quanta of, 15–17; rectilinear propagation and, 12–15; time uncertainty and, 159–64

Enigma, 293

equations: amplitude, 29, 31–32, 41; angular momenta, 52; Bell's inequality, 240–41, 243, 256–60; binary, 300–302; Bohr's complementarity, 74–78; cryptographic, 297, 300–302, 306; de Broglie, 19; diffraction, 48–49; electron orbit energy, 18; entangled states, 231, 339–42; field state, 10; identity, 333, 335–36,

479

INDEX

equations (*cont.*)
339–41; law of Malus, 35; nonepistemic imprecision, 69; particle/wave dualism, 23–24, 154–59; photon energy, 16; polarization states, 29, 31–32, 41, 61, 139, 141–42, 146–47; propagation velocity, 9; quantum states, 169–70, 172–75, 177, 190, 234, 316–19, 329–30, 333, 335–36, 352–53, 385, 389, 407, 409–10; spin state, 57–58; superluminal signals, 272–73, 275, 277–78, 285, 290; superposition, 88, 103–4, 346–47, 349, 361, 373–76; time/energy relation, 159–61; transmitted luminous intensity, 35; trigonometric functions, 42; tunnel effect, 95–97; uncertainty principle, 69; variable contextuality, 223; von Neumann, 199, 219–22, 358; wave frequency, 9; wave packet reduction, 137–48
equivalence principle, 163
ethics, 129
Everett, H. III, 389–90
exponential complexity, 323

Faraday, Michael, 2
feedback registers, 320–23
Fermi, Enrico, 49
Fermi-Dirac statistics, 336
fermions, 336
Feynman, Richard Phillips, xi, 110, 118–19, 313–14
fields, 3, 457n3; energy density and, 11; identity issues and, 331–43; polarization states and, 26–33; quanta and, 43–47; state equation for, 10; superposition principle and, 92–95 (*see also* superposition principle); wave theory and, 6–24. *See also* polarization
filters: crystals and, 37–42; photon cloning and, 284–91; polarization and, 33–37 (*see also* polarization); superluminal speeds and, 264–67
Fleming, G. N., 427
Flew, Anthony, 126
formalism, 466n4; Bell's inequality and, 239–43, 258–60; Bohr-Einstein dialogue and, 149–64; conscious observer and, 400–403; entangled states and, 172–75, 339–42; EPR paradox and, 152–53, 165–94; incompleteness and, 381–83; macroscopic systems and, 345–50; many-worlds interpretation and, 388–95; microsystem identity and, 331–43; nonseparability and, 189–91; objectification of, 378–81, 399–400
Fourier transforms, 70
Francia, Giuliano Toraldo di, 331
Franklin Institute, 201
frequency, 9
Friedman, William, 294

Galilei, Galileo, vii, 2, 454
Gallis, M. R., 427
Garg, Anupam, 443
Gell-Mann, Murray, 450
Gerlach, Walter, 55
Gibbs, Willard, 2
God, 136, 196, 237, 452
Gödel, Kurt, 128
Gold Bug, The (Poe), 292
Goldstein, S., 452
Goudsmit, Samuel, 52
Grangier, Roger, 449
Grassi, Renata, xix, 405, 431, 434–35
gravity, 264, 452–54, 455n2;
Greenberger, Daniel M., 251–52
GRW theory, 405–6, 453–54, 471n3, 471n4; Bell and, 419–20, 422; dynamical reduction and, 406–15, 419–23; Einstein and, 422–23; historical parallel to, 423–25. *See also* localization

Hamilton, William Rowan, 2
Hawking, Stephen, 450
Heisenberg, Werner, xiv, 1, 135, 449; Copenhagen interpretation and, 121; matrix mechanics and, 22, 115; uncertainty principle and, 66–74; visualization and, 114–15, 131–32; wave theory and, 22–23
Hermann, Grete, 198
Hertz, Heinrich, 6
hidden variables, 379; Bell's inequality and,

259–60 (*see also* Bell's inequality); Bohm and, 200–10; contextuality and, 210–18, 222–25; description of, 195; incompleteness and, 381–83; localization and, 434–35; pilot wave theory and, 202–10; superluminal communication and, 276–84; uncertainty principle and, 195–97; von Neumann and, 197–200, 218–22

Hiley, Basil J., 195

Horne, Michael, 251–52

humanities, xii

I Ching, 300

idealism, 126

identity, 331; entanglement and, 339–43; implications of, 335–36; indiscernibles and, 332–34; objectification and, 378–81, 399–400; Pauli's principle and, 336–37; physical consequences of, 336–39; symmetry and, 336

information: computation complexity, 323–27; conscious observer and, 380, 384–88, 400–403, 428–33; cryptography and, 292–312; data compression, 327; linear feedback shift registers, 320–23; quantum bits and, 315–19; Schrödinger's cat and, 368–72, 379–83, 385–95, 413–14, 421, 445; superluminal signals and, 264–91; system evolution and, 102–3, 166–75, 352–53. *See also* probability

interference, 12–15, 48–49

interferometers, 93–94, 97–102, 446

International Symposium of the Foundations of Quantum Mechanics (ISQM), 371

Jammer, Max, 192, 463n2

Japan, 293–95

Jordan, Pascal, xiv, 132, 449

Josephson, Brian, 194

Kahn, David, 293, 295

Klein-Gordon equation, 433, 472n5

Kramer, Alwin D., 295

Kuhn, Thomas, 448

Lagrange, Joseph Louis de, 2

Laplace, Pierre-Simon de, 63, 65

Laub, J., xi

law of Malus, 35, 44, 82

Leggett, Antony J.: hypothesis of, 440–43; macrocoherence and, 443–45; macrorealism and, 437–47

Leibniz, Gottfried Wilhelm, 331

light: amplitude, 29–32, 41; crystals and, 37–42; de Broglie and, 19–22; diffraction and, 12–15, 48–49, 66–78, 206–10, 366, 455n6; electromagnetism and, 2–4; filters and, 33–37; intensity, 37–42; interference and, 12–15, 48–49; law of Malus and, 35, 44, 82; particle/wave dualism and, 19–24, 49–51, 154–59; photons and, 16–17, 159–61 (*see also* photons); polarization and, 6–24 (*see also* polarization); quantum systems and, 43–47, 166–75; rectilinear propagation and, 12–15; slit experiments and, 49–51 (*see also* slit experiments); spectra and, 3–4, 9–10; Stern-Gerlach apparatus and, 55–60, 80, 84, 213–15, 278–84, 409, 429–30; superluminal communication and, 264–91; superposition principle and, 79–110 (*see also* superposition principle); synchrotrons and, 455n3; temperature and, 3–4; transmitted luminous intensity, 35; uncertainty principle and, 66–74; wave packet reduction and, 137–48; wave theory and, 6–24

linear complexity, 323

linear feedback shift registers, 320–23

linearity, 103

localization, 453–54; amplitude interval of, 406; Bell and, 413–14, 416, 419–20, 435–36; Bohm and, 435; de Broglie and, 435; Dirac and, 433; dynamical reduction and, 406–15, 419–23; Einstein and, 420–23; Gaussian shape and, 470n8; graphical representation of, 407–8, 411–14; hidden variables and, 434–35; historical parallel to, 423–25; ideal line and, 464n4; location and, 407, 409–10; objectification and, 417–18; observer role and, 428–33; Ranvier nodes and,

INDEX

localization (*cont.*)
431–32; relativistic requirements and, 433–36; as rival to standard theory, 425–28; Schrödinger and, 413–14, 419–21, 433; Stern-Gerlach apparatus and, 409, 429–30; timing of, 407; van Fraassen and, 423–25; Wigner and, 431
Loewer, Barry, 394
logic: macrorealism and, 441–43; Occam's razor, 394; positivism and, 127–30; smoky dragon model, 105–10; superposition principle and, 86–88; visualization and, 111; von Neumann and, 197
Lorentz, Hendrik Antoon, 21, 111

Mach, Ernst, 128
Mach-Zehnder experiments, 93–94, 105, 108, 446
macrocoherence, 440, 443–45
macrorealism: Bell's inequality and, 440–41; coherence testing and, 443–45; hypothesis of, 438; Leggett and, 437–38, 440–43; noninteracting observance and, 445–47; noninvasive measurement and, 438–40; Schrödinger's cat and, 445
macroscopic systems, 344, 377; boundary ambiguity and, 363–68; conscious observer and, 380, 400–403; formalism and, 345–50; histories and, 395–99; incompleteness and, 381–83; limiting observables of, 384–88; localization and, 406–15 (*see also* localization); many-minds interpretation and, 394–95; many-worlds interpretation and, 388–95; measurement and, 350–55; objectification and, 378–81, 399–400, 417–18; realism and, 437–47; relativity and, 345–46; Schrödinger and, 352–53, 357, 368–72; slit experiments and, 366–67; superposition principle and, 345–50, 360–63, 373–76, 384–95; von Neumann chain and, 357–60; wave packet reduction and, 355–57; Wigner and, 400–403
Magic, 293–95
magnetism, 54, 465n12. *See also* electromagnetism; Stern-Gerlach apparatus
Mann, C. C., xiii

many-minds interpretation, 394–95
many-worlds interpretation, 388–94, 450
Marconi, Guglielmo, 6
Marshall, Leona, 49
Massachusetts Institute of Technology, 325
materialism, 127
Mathematical Foundations of Quantum Mechanics, The (von Neumann), 197
matrix mechanics, 22–23, 111, 115–16, 120
Maudlin, Tim, xii
Maxwell, James Clerk, 2–3, 6, 152
measurement, 464n5; angular momenta, 52; Bell's inequality and, 226–37 (*see also* Bell's inequality); Bohr-Einstein dialogues and, 149–64; Bohr's complementarity and, 74–78; conscious observer and, 380, 384–88, 400–403, 428–33; contextuality and, 210–18, 222–25; cryptography and, 306–12; Dr. Bertlmann's socks and, 184–85; dynamic reduction and, 404–15; effects of, 175; electron position and, 124; entangled states and, 368–72; EPR paradox and, 165–94; formalism and, 137–48; hidden variables and, 200–202 (*see also* hidden variables); histories and, 395–99; identity issues and, 331–43; incompleteness and, 381–83; indiscernibles identification and, 332–34; Laplace and, 63; limiting observables and, 384–88; localization function and, 406–15 (*see also* localization); macrorealism and, 437–47; macroscopic systems and, 350–55 (*see also* macroscopic systems); neutron interferometry, 97–102; nonepistemic, 62, 69, 151–52; noninvasive, 437–47; nonlocality and, 153; objectification and, 399–400, 417–18; particle/wave duality and, 23–24, 154–59; physical processes and, 378–81; pilot wave theory and, 202–10, 213–18; quantum systems and, 166–75; quantum telepathy analogy and, 228–37; randomness and, 46–47, 60–66; Ranvier nodes and, 431–32; reality and, 345–50; Schrödinger's cat and, 368–72; smoky dragon model, 105–10; Stern-Gerlach

482

apparatus and, 55–60, 80, 84, 213–15, 278–84, 409, 429–30; superluminal signals and, 267–76; superposition principle and, 79–110 (*see also* superposition principle); uncertainty principle and, 66–74, 123–25, 151 (*see also* uncertainty principle); von Neumann and, 197–200, 218–22; wave packet reduction and, 137–48, 197, 355–57
Mermin, David, 25, 236–37, 243, 247, 250
mesoscopic realm, 454
metaphysics, 121, 129; Bell's inequality and, 243–46; quantum telepathy and, 228–37, 243–46
microsystems. *See* measurement; quantum mechanics
Miller, Arthur, 112, 114, 117
Moore, Walter, 22–23
Mystery of the Quantum World (Squires), 228, 469n1

National Institute of Nuclear Physics, 445
neutron interferometry, 97–102
Newton, Isaac, 2, 453
Newtonian nonlocality, 263–64
nonepistemic measurement, 62, 69
nonlocality, 153, 261; action at a distance, 264; Bell's inequality and, 24, 237–39 (*see also* Bell's inequality); characteristics of, 262–64; classical mechanics and, 263; EPR paradox and, 175–78, 181, 188–89, 193–94, 227–28, 263; gravity and, 264; Newtonian, 263–64; pilot wave theory and, 210, 213–18; superluminal communication and, 264–91
normalization, 144–45, 340
nuclear energy, xi

observation. *See* measurement
Occam's razor, 394
Opat, G. I., 163
Oppenheimer, Robert, 1, 200
oscillators, 114
Otras Inquisitiones (Borges), 300

Pais, Abraham, 187, 194, 236
particles, 23–24, 455n1; angular momentum and, 52–55; Bohr's complementarity and, 74–78; field concept and, 3; ideal line and, 464n4; identical systems of, 331–43; light quanta and, 43–47, 49–51; macrorealism and, 437–47; pilot wave theory and, 202–10, 213–18; quanta and, 49–51; randomness and, 46–47, 60–66; spin and, 51–58 (*see also* spin); Stern-Gerlach apparatus and, 55–60, 80, 84, 213–15, 278–84, 409, 429–30; superposition principle and, 79–110, 345–50, 360–63, 373–76 (*see also* superposition principle); tunnel effect and, 95–97; uncertainty principle and, 66–74; wave dualism and, 23–24, 154–59. *See also* wave mechanics
Pauli, Wolfgang, xiv, 449–50; Bell and, 236; Copenhagen interpretation and, 121; EPR paradox and, 153; theory interpretation and, 133–34
Pearle, Philip, 405
Pearl Harbor, 293
Penrose, Roger, 169, 448, 450, 454
Peres, Asher, 163
philosophy, 121, 125, 449; Bell on, 377–78; Bell's inequality and, 246–51; Bohr-Einstein dialogue and, 149–64; conscious observer and, 380, 384–88, 400–403, 428–33; EPR paradox and, 152–53, 165–94; equivalence principle and, 163; idealism, 126; macroscopic systems and, 345–50; many-minds interpretation and, 394–95; materialism, 127; mind's plane and, 450–51; noninvasive observation and, 445–47; positivism, 127–30; realism, 126–27, 153–54, 437–47; Schrödinger's cat and, 368–72, 379–83, 385–95, 413–14, 421, 445; time/energy relation, 159–64; Vienna Circle and, 128–29
Philosophy of Quantum Mechanics, The (Jammer), 192, 463n2
photons, 16–17; absorbing screens and, 82–83; avalanches of, 46; Bell's inequality and, 226–27, 231–37 (*see also* Bell's

INDEX

photons (*cont.*)
inequality); Bohr's complementarity and, 74–78; cloning of, 284–91; crystals and, 80–88; diffraction and, 48–49; entangled states and, 172–75, 339–42; EPR paradox and, 188–89; factorized states and, 167–72; interference and, 48–49; law of Malus and, 35, 44, 82; macroscopic systems and, 346–55, 366–67 (*see also* macroscopic systems); measurement and, 123–25; nonlocality and, 263–64 (*see also* nonlocality); pilot wave theory and, 206–10; quanta and, 43–47, 166–75, 318–19; randomness and, 46–47, 60–66; Schrödinger's cat and, 368–72; smoky dragon model, 105–10; spin and, 51–55; superluminal speeds and, 264–91; superposition, 159–61 (*see also* superposition principle); thought experiments and, 84–85; time/energy relation, 159–61; tunnel effect and, 95–97; uncertainty principle and, 66–74; von Neumann chain and, 357–60; wave packet reduction and, 137–48, 355–57. *See also* light
physical processes, 377; conscious observer and, 380, 384–88, 400–403; dynamic reduction and, 404–15, 419–23; hidden variable theories and, 381–83; limiting observables and, 384–88; localization function and, 406–15; many-minds interpretation and, 394–95; many-worlds interpretation and, 388–95; measurement strategies for, 378–81; objectification and, 378–81, 399–400, 417–18; quantum histories and, 395–99; Wigner and, 400–403
physics: atomic theory and, 4–6; classical, 1–24; excitement of, xi; spectra and, 3–4; visualization and, 111–19; wave theory and, 6–24
pilot wave theory, 202, 204; Bell's inequality and, 226–27, 259–60 (*see also* Bell's inequality); contextuality and, 213–18; diffraction and, 206–10; examples of, 206–10; nonlocality and, 210, 213–18; quantum mechanical force and, 206; Schrödinger's equation and, 205, 207; Stern-Gerlach apparatus and, 213–14; wave function modulus, 203
plaintext, 296
Planck, Max, xiv, 4; constant of, 52, 69; energy quanta and, 15–17; uncertainty principle and, 69, wave theory and, 15–17
Podolsky, Boris, 135, 152–53; EPR paradox and, 165–94
Poe, Edgar Allan, 292–93
Poincaré, Jules-Henri, 64–65, 387
polarization, 456n2, 456n5; absorbing screens and, 82–83; amplitude, 29–32, 41; Bohr's atom and, 17–19; crystals and, 37–42; de Broglie and, 19–22; description of, 6–12; diffraction and, 12–15; Einstein and, 15–17; electromagnetic field and, 26–33; entangled states and, 172–75, 339–42; EPR paradox and, 165–94; filters and, 33–37; histories and, 395–99; horizontal, 82, 85; law of Malus and, 35, 44, 82; macroscopic systems and, 357–60; photon cloning and, 284–91; Planck's hypothesis and, 15–17; quanta phenomenology and, 17–19, 25, 43–47; quantum signals and, 318–19; quantum systems and, 166–75; randomness and, 46–47, 60–66; rectilinear propagation and, 12–15; spin and, 51–55; states of, 26–33, 237–39 (*see also* Bell's inequality; hidden variables); Stern-Gerlach apparatus and, 55–60, 80, 84, 213–15, 278–84, 409, 429–30; superposition principle and, 80–88; thought experiments and, 84–85; transmitted luminous intensity, 35; trigonometric functions for, 32, 35, 42; vertical, 81–82, 84, 468n2; von Neumann chain and, 357–60; wave packet reduction and, 137–48; wave/particle dualism and, 19–24, wave theory and, 6–24 (*see also* wave mechanics)
Popper, Karl, 185–86; superluminal speeds and, 267–76
positivism, 127–32
probability, 116, 470n5; Bell's inequality and, 233–43 (*see also* Bell's inequality);

INDEX

Bohr-Einstein dialogue and, 151–52; EPR paradox and, 165–94; Fermi-Dirac statistics and, 336; pilot wave theory and, 202–10; quantum systems and, 166–75; quantum telepathy and, 228–37; superluminal speeds and, 264–67; superposition principle and, 103–5 (*see also* superposition principle); wave packet reduction and, 137–48. *See also* randomness

propagation velocity, 9

Purple, 293–95

quadratic complexity, 323

quanta: Bohr's complimentarity and, 74–78; diffraction and, 48–49; interference and, 48–49; particles and, 49–51; randomness and, 46–47; spin and, 51–55; Stern-Gerlach apparatus and, 55–60, 80, 84, 213–15, 278–84, 409, 429–30; two-slit experiment and, 49–51; uncertainty principle and, 66–74

quantum bits, 315–19

quantum computers: computational complexity and, 323–27; data compression and, 327; Feynman and, 314–15; idea of, 313–14; information theory and, 314–15; linear feedback shift registers, 320–23; possibilities of, 328; quantum bits and, 315–19; random numbers and, 319–23; storage and, 317–18; superposition principle and, 316; teleportation and, 328–30; Turing and, 314

quantum mechanics: Bell's inequality and, 226–60; Bohm and, 200–202; Bohr-Einstein dialogue and, 149–64; Bohr's atom and, 17–19; Bohr's complementarity and, 74–78; Bohr's position on, 130–31; Born and, 132; classical physics and, 1–24; computers and, 313–30; conceptual changes of, 122–25; controversy of, 1–3; Copenhagen interpretation and, 121, 133, 187, 275–76, 449; cryptography and, 295–12; de Broglie and, 134; difficulty in understanding, 109–10; Einstein on, 135–37; energy density and, 11; entangled states and, 172–75, 339–42, 368–72; EPR paradox and, 152–53, 165–94; evolution and, 102–3, 352–53, 387–88; excitement of, xi–xv; formalism and, 120–21, 137–48 (*see also* formalism); Heisenberg's position on, 131–32; hidden variables and, 195–97, 200–202 (*see also* hidden variables); idealism and, 126; identity and, 331–43; incompleteness of, 165–66, 195, 381–83; indiscernibles identification and, 332–34; initial conditions and, 458n11; Jordan and, 132; linearity and, 103; materialism and, 127; measurement and, 137–48 (*see also* measurement); metaphysics and, 121; nonlocality, 153, 175–78, 181, 188–89, 193–94, 262–64 (*see also* nonlocality); nonseparability and, 189–91; particle/wave dualism and, 19–24, 154–59; Pauli and, 133–34; philosophical implications of, 125–30; positivism and, 127–30; predictions, 103–5; public assimilation of, xi–xiii; quantization and, 17–19; randomness and, 60–66; realism and, 126–27, 153–54; Schrödinger and, 134–35; scientific dignity for, 120–21; spin and, 25, 51–55; superluminal communication and, 264–91; superposition principle and, 79–110; system properties and, 166–75; teleportation and, 328–30; time/energy relation, 159–64; uncertainty principle and, 66–74; von Neumann and, 197–200; wave packet reduction and, 137–48; wave theory and, 6–24

Quantum Mechanics and Experience (Albert), 429

Quantum Non-locality and Relativity-Metaphysical Intimations of Modern Physics (Maudlin), xii

Quantum Profiles (Bernstein), 472n1

quantum states: conscious observer and, 380, 384–88, 400–403; entangled 172–75, 339–42, 368–72; Everett and, 389–90; histories and, 395–99; limiting observables and, 384–88; localization and, 406–15, 429–30 (*see also* localiza-

485

quantum states (*cont.*)
tion); macrorealism and, 437–47; many-minds interpretation and, 394–95; many-worlds interpretation and, 388–95; reduction and, 404–15; Schrödinger's cat and, 368–72, 379–83, 385–95, 413–14, 421, 445. *See also* superposition principle
quantum systems, 466n2; entangled states and, 172–75; factorized states and, 167–72; measurement effects and, 172–75; nonseparability and, 189–91; property problems of, 166–67
quantum telepathy, 228–37; experiments in, 243–46
Quantum Theory and the Schism in Physics (Popper), 185–86, 267
Quantum Theory (Bohm), 176, 200–201
quasars, 107

Rabi, Isaac, xi
radiation. *See* light
Rae, A.I.M., 427
randomness, 458n7; atomic theory and, 111–19; classical mechanics and, 64–65; Laplace and, 63; linear feedback shift registers, 320–23; nonepistemic, 62, 69; physical processes and, 60–66; probability density function and, 70–71 (*see also* probability); quanta and, 46–47; quantum computers and, 319–23; superposition principle and, 103–5 (*see also* superposition principle); uncertainty principle and, 66–74; vector analysis of, 61
Ranvier nodes, 431–32
realism, 126–27; Bell's inequality and, 246–51; EPR paradox and, 165–94; macroscopic systems and, 345–50, 437–47; realism and, 153–54
relativity, xii, 1, 261, 453; EPR paradox and, 152–53; localization and, 433–36; macroscopic systems and, 345–46
Rimini, Alberto, 405, 427
Rivest, Ron, 325–26
Roger, G., 245
Rosen, Nathan, 135; EPR paradox and, 152–53, 165–94

Rosenfeld, Leon, 162–63, 180
RSA (Rivest, Shamir, Adleman) code, 325–26
Russell, Bertrand, 128
Rutherford, Ernest, 4, 455n1

Salam, Abdus, 186–87
Schilpp, Paul Arthur, 181, 420
Schlick, Moritz, 128
School of Copenhagen, xiv–xv
Schrödinger, Erwin, xiv, 102, 449; complementarity and, 77; de Broglie waves and, 21–22; localization and, 419–21, 433; macroscopic systems and, 352–53, 357, 368–72; matrix mechanics and, 120; pilot wave theory and, 205, 207; quantum entanglement and, 165; system evolution and, 352–53; theory interpretation and, 132, 134–35; visualization and, 115–16; wave theory and, 21–23
Schrödinger's cat, 165–66; conscious observer and, 380, 385–86; entanglement and, 368–72; homogeneous states and, 379–81; incompleteness and, 381–83; localization and, 413–14, 421; macrorealism and, 445; many-worlds interpretation and, 388–95
Schumacher, Benjamin, 327
Schwinger, Julian, 366
Shamir, Adi, 325–26
Shannon, Claude, 304, 314, 327, 468n1
shift registers, 320–23
Shimony, Abner, 133, 228, 261, 264, 344
Shor, Peter, 326–27
signal compression, 318–19
Signal Intelligence Service (SIS), 294–95
Simon, H. A., 111
slit experiments, 49–51; Aspect-Dalibard-Roger, 245–46; Bell and, 366–67; ideal line and, 464n4; particle/wave dualism and, 23–24, 154–59; pilot wave theory and, 206–10; quantum telepathy and, 243–46; superluminal signals and, 267–76; superposition and, 154–59
Snow, Charles Percy, 149
Solvay, Ernest, 461n2
Solvay Conferences, 150, 154, 161, 399

Sommerfeld, Arnold, 114
spectra, 3–4, 9–10; atomic theory and, 112–14; oscillators and, 114
spin, 25; angular momentum and, 51–55, 457n4; contextuality and, 210–18, 222–25, 464n11; entangled states and, 339–43; identity and, 332–43; invariable nature of, 52–53; polarization and, 51–55; Stern-Gerlach apparatus and, 55–60, 80, 84, 213–15, 278–84, 409, 429–30; superluminal signals and, 276–84; superposition principle and, 89, 276–84, 345–50, 360–63, 373–76 (*see also* superposition principle); time/energy relation, 159–64
Squires, Evan, 127, 228, 469n1
Stern, Otto, 55, 236
Stern-Gerlach apparatus, 80, 84; description of, 55–60; localization function and, 409, 429–30; pilot wave theory and, 213–14; superluminal signals and, 278–84
"*Subtle is the Lord*" (Pais), 187
superconductors, xi
superluminal signals, 261; Bohm and, 281–84; Copenhagen interpretation and, 275–76; Einstein and, 270; EPR paradox and, 269–76; impossibility of, 264–67; Maxwell's demon and, 281; photon cloning and, 284–91; Popper proposal and, 267–76; spin and, 276–84; Stern-Gerlach apparatus and, 55–60, 80, 84, 213–15, 278–84, 409, 429–30; uncertainty principle and, 267–76; variable control and, 276–84
superposition principle, 79, 155, 462n3; crystal experiments and, 80–90; definite position states and, 88–91; delayed choice and, 92, 105–10; Dirac and, 349; law of Malus and, 35, 44, 82; Leggett and, 437–38; limiting observables and, 384–88; linearity and, 103; logic and, 86–88; macrorealism and, 437–47; macroscopic systems and, 345–50, 360–63, 373–76, 384–95; many-minds interpretation and, 394–95; many-worlds interpretation and, 388–95; neutron interferometry and, 97–102; phase and, 92–95; probability and, 103–5; quantum bits and, 316; reality and, 345–50; Schrödinger's equation and, 102; smoky dragon model, 105–10; spatially separate states and, 92–95; testing difficulties of, 360–63; theory formalism, 102–10; time/energy relation and, 159–61; thought experiments and, 84–85; tunnel effect and, 95–97
synchrotrons, 455n3

technology, xi; ACSII code, 302–4; artificial intelligence and, 451; binary numeration, 300–302; cryptography and, 292–312; data compression, 327; linear feedback shift registers, 320–23; quantum computers, 313–30; signal compression, 318–19; teleportation and, 328–30
teleporatation, 328–30
Tesche, C. D., 444
thermodynamics, 3–4, 15–17, 387
Thirring, Hans, 22
Thompson, George Paget, 24, 49
Thompson, Joseph John, 24
Thomsen, Hans, 293–94
thought experiments (*Gedankenexperimente*), 84–85; Bohr-Einstein dialogue and, 151–52; EPR paradox and, 152–53; objectification and, 399–400; Schrödinger's cat and, 368–72, 379–83, 385–95, 413–14, 421, 445; superluminal signals and, 264–91 (*see also* superluminal signals); time/energy relation, 161–64
time: energy uncertainty and, 159–64; equivalence principle and, 163
Time and Chance (Albert), 427
Tonomura, A., 49
traveling salesman problem, 323
trigonometry, 42
tunnel effect, 95–97
Turing, Alan Mathison, 314, 468n1

Uhlenbeck, George, 52
uncertainty principle, 123–25; Bohr and, 151; diffraction and, 66–74; Einstein

487

INDEX

uncertainty principle (*cont.*) and, 151; Fourier transforms and, 70; hidden variable assignment and, 195–97; probability density function and, 70–71; superluminal signals and, 267–76; time/energy relation, 159–64

United Sates, 293–95, 305

Unruh, W. G., 163

Vaidman, Lev, 428–29, 433, 447

van Fraassen, Bas C., 377, 423–24

vector analysis, 10; quantum systems and, 166–75; randomness and, 61; von Neumann theorem, 218–22; wave theory and, 7–12

Vienna Circle, 128–29

visualization: approximation and, 117–19; Bohr and, 112–14; cognitive science and, 111; Feynman diagrams and, 118–19; Heisenberg and, 114–15, 131–32; Schrödinger and, 115–16

von Neumann, John, 449–50; chain of, 337–60, 366; hypothesis issues and, 198–200; macroscopic systems and, 357–60; objectification and, 399–400; theorem requirement of, 218–22; wave packet reduction and, 197

wave function, 132, 459n13, 460n2; pilot wave theory and, 202–10; square of modulus, 203

wave mechanics, 455n5; amplitude, 29–32, 41; Bohr and, 17–19, 74–78; cryptography and, 306–12; de Broglie and, 19–22; diffraction and, 12–15, 48–49, 66–78, 206–10, 366, 455n6; Einstein and, 15–17; energy density and, 11; field concept and, 3; frequency, 9; general description of, 6–12; Hertz and, 6; interference and, 12–15, 48–49; localization function and, 406–15; macrorealism and, 437–47; macroscopic systems and, 355–57; matrix mechanics and, 120; Maxwell and, 6; particle dualism and, 19–24, 154–59; photons and, 16–17 (*see also* photons); pilot wave theory and, 202–10, 213–18; Planck's hypothesis and, 15–17; polarization and, 6–24 (*see also* polarization); propagation velocity, 9; quanta and, 17–19, 43–51; rectilinear propagation and, 12–15; spin and, 51–55; Stern-Gerlach apparatus and, 55–60, 80, 84, 213–15, 278–84, 409, 429–30; superluminal signals and, 264–91; superposition principle and, 79–110, 155, 159–61, 345–50, 360–63, 373–76 (*see also* superposition principle); time/energy relation, 159–61; uncertainty principle and, 66–74; vector analysis of, 7–12; visualization and, 111–19; von Neumann and, 197–200, 218–22; wave function, 132, 202–10, 459n13, 460n2; wave packet reduction and, 137–48, 197, 355–57

Weber, Tullio, 405

Weyl, Hermann, 23, 433

Wheeler, John Archibald, 92, 105–10

Wigner, Eugene, 197, 400–403, 431, 450

Wittgenstein, Ludwig, 128

World War II, 293–95

Zanghì, N., 452

Zeilinger, Anton, 251–52